T0289121

Harbour Ecology

Harbour Ecology

Environment and Development in Poole Harbour

edited by
John Humphreys and Alice E. Hall

Pelagic Publishing | www.pelagicpublishing.com

First published in 2022 by
Pelagic Publishing
20–22 Wenlock Road
London N1 7GU
UK

www.pelagicpublishing.com

Harbour Ecology: Environment and Development in Poole Harbour

British Library Cataloguing in Publication Data
A catalogue record for this book is available from the British Library

ISBN 978-1-78427-403-0 Hbk
ISBN 978-1-78427-335-4 ePub
ISBN 978-1-78427-336-1 PDF

https://doi.org/10.53061/LGUG2858

2023
Poole Harbour Study Group
www.pooleharbourstudygroup.org.uk

Every effort has been made to trace copyright holders and obtain their permission of reproduction of material. The editors and the publisher apologise for any omissions and would be grateful if notified of any acknowledgements, which will be duly incorporated in future editions.

Cover image: Aerial photograph of Poole Harbour looking west from Poole Bay, with the harbour entrance in the foreground.

Contents

Contributors

Ruth Barden has over 20 years' experience in the water industry, particularly focused on environmental and wastewater aspects. Ruth leads Wessex Water's catchment and environmental teams, covering environmental investigation programmes, catchment management, partnerships and environmental planning and policy. This involves elements of research, innovation and a high degree of collaboration with a wide range of stakeholders to ensure that the right investment solutions are delivered.

Sarah Birchenough is an Inshore Fisheries and Conservation Officer with the Southern IFCA and is part of the Fisheries Management and Policy Team. Her role includes development of fisheries management and policy for the Southern IFCA District with a particular focus on the Poole Harbour clam and cockle fishery. Sarah has a PhD from the University of Southampton, with her thesis focusing on sustainable fisheries management in Poole Harbour as an estuarine environment and a Marine Protected Area.

Jessica Bone is a marine biologist and PhD researcher at Bournemouth University and has lived in Poole for most of her life. As both a Master's and undergraduate student she studied the ecology of the Poole Harbour lagoons and worked briefly in public engagement with Birds of Poole Harbour. Jess currently works in the field of eco-engineering and oversees the monitoring of artificial rockpools in Poole Harbour.

Katharine Bowgen is a research ecologist at the British Trust for Ornithology working for BTO Cymru and the Wetland and Marine Team. Following a PhD on predicting the effects of environmental change on wader populations through modelling, she uses her analytical skills alongside fieldwork to study bird populations and behaviours.

Fiona Bowles is Chair of the Poole Harbour Catchment Initiative and sits on the national support group for the catchment-based approach. As an ecologist with 35 years in Wessex Water, Fiona's interests cover water quality, fisheries and aquatic biodiversity, with a particular interest now in river restoration through natural processes.

Niall Burton is Head of the Wetland and Marine Research Team at the British Trust for Ornithology. Working closely with partners of the Wetland Bird Survey and Seabird Monitoring Programme, the team's research explores the factors that affect waterbird and seabird population dynamics, including climate change and the impacts of coastal and offshore developments.

Greg Campbell has been a professional archaeologist for over 30 years and a freelance archaeo-malacologist for over 20 years. Current interests are standards for retrieving and archiving, identifying and quantifying fragmentary shells, and cataloguing the British medieval scallop pilgrim-badges from Santiago de Compostela.

Elena Cantarello is Principal Academic in Sustainability Science in the Department of Life and Environmental Sciences at Bournemouth University. Her research focuses on examining how ecosystems' resilience has changed and is likely to change in the future, accounting for factors such as climate change and human impacts.

Robert W.E. Clark the Chief Officer of the Association of Inshore Fisheries and Conservation Authorities (IFCAs). He lives in Poole and has worked in fisheries and conservation management for over 20 years. He was previously the CEO of Southern IFCA, establishing the award-winning regulation of the Poole clam and cockle fishery and the Poole Harbour Fishery Several Order. Robert has published extensively on inshore fisheries and marine protected area management.

Leo Clarke is a postdoctoral marine and fisheries scientist at Bangor University, Wales, working on sustainable marine and fisheries management. Leo's PhD studied the impacts of novel shellfishing methods on the benthic infaunal communities of Poole Harbour and interactions with protected shorebird communities on the site through a range of practical fieldwork and quantitative modelling and remote-sensing approaches.

William Davies has a background in sustainable development, environmental policy and natural resource management, especially in a coastal and marine context. His work involves measuring and evaluating social impact and developing innovative ways to better integrate social and environmental aspects into economic modelling.

Daniel J. Franklin is Associate Professor at Bournemouth University. A marine biologist, he specialises in the ecology and physiology of the photosynthetic microbes that make up the phytoplankton.

Iain D. Green is Senior Lecturer in Biological Science at Bournemouth University. He has been researching the environmental effects of pollutants for 25 years, and this work has encompassed terrestrial, marine and freshwater ecosystems, covering all levels of biology from the cell to communities.

Alice E. Hall is a marine biologist at the University of Plymouth, who specialises in marine ecology and ecological engineering. Alice has conducted research within the harbour for the last six years. She is the environmental commissioner for Poole Harbour Commissioners.

Andrew Harrison is a Principal Environmental Consultant at Bournemouth University Global Environmental Solutions (BUG). His areas of expertise include aquatic and fish ecology, environmental impact assessment and drone aerial surveys. He is the chief drone pilot at Bournemouth University and is responsible for maintaining BU's drone-operating permissions.

Julie Hatcher is a marine biologist based at the Fine Foundation Wild Seas Centre, Kimmeridge. She works for Dorset Wildlife Trust to raise awareness of marine wildlife in Dorset. Julie initiated the Dorset Seal Project, with Sarah Hodgson, in 2014 and is co-author of three books about marine life.

Roger J.H. Herbert is a marine biologist at Bournemouth University and has worked on the ecology of Poole Harbour for the past 18 years. During this period he has published work on the mudflats and lagoons of the harbour, Manila Clam fisheries and bird feeding behaviour. He has also led student field courses around the harbour and particularly on Brownsea Island.

Sarah Hodgson works for Dorset Wildlife Trust at the Coastal Visitor Centres to promote Dorset's marine wildlife and raise awareness of conservation issues. Sarah has played an integral role in all aspects of the Dorset Seal Project since its inception.

David A. Humphreys is an archaeologist with a specialism in geoarchaeology at Museum of London Archaeology (MOLA). Raised in London and Dorset, he has worked as a commercial archaeologist for the past five years after graduating from the University of Bristol.

John Humphreys is Professor of Marine Biology and formerly pro vice-chancellor of the University of Greenwich and Chair of the Southern Inshore Fisheries and Conservation Authority. His edited books include *The Ecology of Poole Harbour* (2005) with Vincent May and *Marine Protected Areas: Science, Policy and Management* (2020) with Rob Clark, both published by Elsevier.

Annesia L. Lamb is a postdoctoral researcher at Bournemouth University working on the RaNTrans project. She has expertise in macroalgae mat nitrogen dynamics and genetic identification of the green opportunistic genus *Ulva*.

Sinead E. Morris is a PhD student at Bournemouth University, specialising in coastal ecology. She is currently researching nuisance algal mats and their impacts on estuarine invertebrate and bird communities.

Philip Pickering was born in Poole and has lived there all his life. Being observed by his parents fishing in a puddle at the end of their garden in Hamworthy with a piece of string on a stick resulted in the gift of a Woolworths fishing kit for the author's fifth birthday. And so began a lifelong fascination with the creatures that dwell in the world beneath the surface, none more so than those that inhabit the waters of Poole Harbour. He currently works as a boatman and countryside ranger for the National Trust, Brownsea Island, and takes a special interest in the marine ecology surrounding the island.

Andrew Powell teaches A-Level Biology at Canford School and writes the Singapore A-Level Biology paper for Cambridge Assessment. He has been monitoring the zooplankton in Poole Harbour since studying for an MSc in Quantitative Ecology at Bournemouth University in 2011.

Emma Rance is a consultant at Noctiluca Marine and has over 20 years' experience in marine biology and conservation. Leading the Great Dorset Seafood campaign and fishery liaison work, on behalf of the Poole and District Fishermen's Association, were highlights during her time with Dorset Wildlife Trust (2004–20).

Sophie Richier graduated from the University of Nice-Sophia Antipolis and obtained her PhD in 2004 in marine biology and molecular and cellular aspects of biology. After more than ten years in academic research, she joined the CEVA in September 2017 to lead the Ecology and Environment department.

Eleanor Rivers is a research assistant and a PhD student at Bournemouth University. She has a focus on bird behavioural ecology and has previously studied Black Brant Geese in Alaska and Mexico, and their relationship with intertidal seagrass. She is currently working on Eurasian Curlew in the New Forest.

Kathryn E. Ross is an academic staff member at Toi Ohomai Institute of Technology in Rotorua, New Zealand. She completed her PhD at Bournemouth University in 2014 on modelling the effects of environmental change on Poole Harbour's Avocet population and subsequently worked as a research ecologist at the British Trust for Ornithology.

Richard Stillman is Professor of Conservation Ecology in the Department of Life and Environmental Sciences, Bournemouth University. His research aims to predict how environmental change affects coastal and wetland birds, to advise policy makers, conservationists and industry on the best ways of reconciling the interests of the birds with those of humans.

Ann Thornton is an ecologist with a particular interest in estuarine and coastal ecosystems. She completed her PhD at Bournemouth University, studying the impact of macroalgal mats on benthic invertebrates and wading birds in Poole Harbour. Ann currently works at the University of Cambridge in the Conservation Science Group and is the managing editor of the *Conservation Evidence* journal.

Gordon J. Watson is a marine ecologist whose expertise lies in investigating the effects of humans on aquatic organisms. His research focuses on using benthic species and habitats to assess the role of management in protection from fisheries and aquaculture exploitation, as well as the impacts of key pollutants. He is a lead partner on the Interreg-funded project RaNTrans (https://rantransproject.com), which is testing interventions that can reduce nutrient concentrations in coastal habitats.

Chris Williams is a biologist and social scientist working on fisheries, aquaculture, marine ecosystem services, Marine Protected Areas and the local economic contribution of fisheries, angling and aquaculture to coastal communities. He has published in *Marine Policy, Environmental Scientist* and *Human Geography* and has been working on fisheries resources management problems and coastal community economic development.

Jessica Winder was a professional research biologist for most of her working life and investigated marine shells (particularly oysters) from archaeological excavations for over 45 years before retiring in 2020. She remains an avid nature photographer and blogger.

Suzy Witt trained as a marine biologist and has been working for the Environment Agency as an Environmental Monitoring Officer since 2001. She has been involved in most ecological monitoring programmes carried out in Poole Harbour. Suzy grew up in Poole so has a familiarity with and interest in the harbour.

Preface

Poole Harbour is protected and recognised, both nationally and internationally, for its ecological importance. However, along with its designations as a Marine Protected Area (MPA) it has also been classified as polluted and 'eutrophic'. These twin designations – protected yet polluted – exemplify the condition of many English estuaries, making Poole Harbour a useful case study for elucidating the circumstances behind this apparent paradox.

This book is the outcome of a conference organised by the Poole Harbour Study Group as part of the Poole Maritime Festival and hosted by Poole Harbour Commissioners on 6 June 2018. The conference, entitled 'Spotlight on Poole Harbour: Environment & Economics', included presentations across the themes of water quality, ecology, industrial activities and fisheries. This is the second book produced by the Poole Harbour Study Group; the first, *The Ecology of Poole Harbour*, was published in 2005.

This volume builds on the earlier book in terms of aspects of both the ecology and industries of the harbour, with a particular focus on ecological service industries. This phrase is used to distinguish those industries that profoundly depend on the ecological functions of the harbour, from those for which the harbour provides a location for recreational or industrial vessel use or manufacture. Our focus on ecological service industries allows an examination of the question of sustainable development in a coastal MPA and, in the concluding chapter, consideration of some of the factors that determine whether it is achieved.

The book consists of four main parts. After a short introduction, Part I, 'Background', provides a broad introduction to the harbour in terms of its prehistoric and historic significance for human communities and a conceptual overview of its modern character and uses. Part II, 'Ecology', adds to the earlier book by extending or updating that account of the harbour, with chapters ranging from plankton to mammals. In so far as they also deal with anthropogenic aspects of the ecology, these chapters anticipate the remaining sections of the book which deal specifically with aspects of the ecological service industries supported by the harbour. Part III, 'Fisheries', covers recreational and commercial fishing and aquaculture, examining economic value and key shellfish species. Part IV, 'Water Quality', addresses those industries for which the harbour's chemical and biological processes remediate various effluents, as well as some of the environmental consequences thereof and noteworthy efforts to reduce such impacts. Part V, 'Conclusion', by the editors then seeks to elucidate some general shortcomings of environmental legislation and regulation in the case of Poole Harbour.

Acknowledgements

The Poole Harbour Study Group (PHSG) aims to encourage, coordinate and report research on Poole Harbour. PHSG's membership includes representatives from statutory, public sector and non-governmental organisations involved directly with the harbour. The core members include Natural England, the National Trust, BCP Council, the Environment Agency, Southern Inshore Fisheries and Conservation Authority, Dorset Wildlife Trust, Bournemouth University, along with several individuals with a particular interest in the objective study of the harbour.

PHSG is grateful to all these organisations for their support over many years, along with the port authority, Poole Harbour Commissioners, who have kindly now hosted three PHSG conferences, all of which have resulted in published collections such as this

The conference on which this book is based was organised by the editors, who wish to express their gratitude to the presenters and particularly to our authors who have delivered excellent chapters even in a difficult time of pandemic. We are also grateful to our publisher, Nigel Massen and production editor David Hawkins, both of Pelagic Publishing.

Disclaimer

The views expressed in this book should not be taken as those of the Poole Harbour Study Group or any of its member organisations or individuals.

About Poole Harbour Study Group

The objectives of the Poole Harbour Study Group (PHSG) are to further the study of the geology, hydrology, ecology, physiography and biological communities, and the monitoring of environmental change, in Poole Harbour by:

- maintaining a database and archive of studies of Poole Harbour.
- undertaking and encouraging further harbour studies.
- acting as a centre for advice and information on the harbour.
- publishing studies and holding seminars and conferences.
- maintaining links with the Dorset Environmental Records Centre and other bodies with an interest in the harbour.

Membership

Natural England, Bournemouth University, Environment Agency, National Trust, Poole Harbour Commissioners, Dorset Wildlife Trust, BCP Council, Birds of Poole Harbour, Southern Inshore Fisheries Conservation Authority, University of Southampton.

Website: www.pooleharbourstudygroup.org.uk

Aerial photograph of Poole Harbour looking north from the harbour entrance, with Brownsea Island on the left and Sandbanks on the right.

CHAPTER 1

Introduction: Ecology and Economics in an Industrialised Harbour

JOHN HUMPHREYS and ALICE E. HALL

Sustainability is the ubiquitous concept brought to bear in all plans for development. From global through national to local policy, 'sustainability' is a keyword. But what does it mean? And is it really achieved in practice, especially in relation to the marine environment? The reality or otherwise of sustainability can best be assessed at the local level, most of all where nature and humanity brush up close together. So, in the aquatic coastal sphere, Poole Harbour is a perfect place to consider these questions.

In an earlier volume from the Poole Harbour Study Group (PHSG), it was observed that this harbour represents in microcosm the worldwide tensions between environment and development. Perhaps 'microcosm' is the wrong word. With a shoreline exceeding 100 km, an area of about 3,600 ha and five main islands, the harbour is one of Europe's largest lowland estuaries (Humphreys and May 2005). It serves a river catchment area of 800 km², mainly from its two largest rivers: the Frome and Piddle. Yet its connection with the English Channel is only about 350 m across. This combined with a microtidal regime (1.8 m at spring tides) and a prolonged high-water stand can give the impression of a vast lagoon (Humphreys 2005).

The harbour itself is a Marine Protected Area with a number of overlapping statutory designations, most significantly as a Site of Special Scientific Interest since 1991 and a Special Area of Protection (SPA) since 1999. Through these, a variety of habitats and features are protected including internationally significant populations of overwintering estuarine wading birds. Such protections require that the harbour is maintained in, or where necessary restored to, 'favourable' condition.

Yet the harbour also has a history of water quality issues, which include high dissolved nitrate levels originating from agriculture in the catchment and sewage effluent released directly or from upstream sources. In 2016 the rate of nitrogen load received into the harbour was reported as 2,100 tonnes N y^{-1} (MMO 2016). As a consequence, in addition to its protected status the harbour has also been classified as 'Polluted Waters (Eutrophic)' and a 'Sensitive Area (Eutrophic)' under European directives.

These two contrasting styles of designation – protected yet polluted – are reflected in the character of the harbour's adjacent landscapes. On the south side of the harbour, terrestrial protections cover habitats including Dorset heathland and on Studland peninsula, a precious and largely intact dune and slack ecosystem. In contrast, the northern shores are largely residential and industrial with a maritime economy including a commercial port, inshore fishing and boat building, whose supply chain tentacles reach deep into the town of Poole and beyond. The harbour environment itself, completely sheltered from the storms and Atlantic driven swells of the English Channel, also offers a benign place for water-based recreation, from sailing to angling and windsurfing, and all the commercial opportunities this creates.

John Humphreys and Alice E. Hall, 'Ecology and economics in an industrialised harbour' in: *Harbour Ecology*. Pelagic Publishing (2022). © John Humphreys and Alice E. Hall. DOI: 10.53061/XXOI6415

Table 1.1 Major Poole Harbour-dependent commercial activities

Industry	Activity	Organisations
In the water		
Water treatment	Wastewater disposal	Wessex Water
Commercial fishing	Netting, trawling, dredging	100+ inshore boats
Aquaculture	Oysters and mussel cultivation	Othniel oysters and so on
Agriculture	Run-off of fertilisers and effluent	Harbour catchment farms
Charter boat angling	Recreational fishing	Various
On the water		
Commercial shipping port	Freight, Channel ferries, aggregates, cruises	Port authority (Poole Harbour Commissioners)
Boat-/ship building	Motor cruisers, lifeboats	Sunseeker, RNLI
Hydrocarbon extraction	Oil and gas extraction/refining	Perenco
Commercial marinas	Berths, boatyards, training and so forth	Various
Marine engineering	Vessel maintenance and repair	Numerous
Defence	Commando base and training	UK military
Recreational sailing clubs	Moorings, leisure activities	Various
Tourism and ecotourism	Boat trips	Pleasure cruise companies

Poole Harbour industries

In terms of ecological significance such commercial interests can be usefully divided into two distinct categories: 'on the water' and 'in the water' (Table 1.1). The former includes activities and associated industries, whose dependence on the harbour is essentially confined to the surface: to using the water as a means of transport or recreation. The second category engages with the substance of the water itself and as such is dependent more profoundly on the aquatic ecosystem: in particular its chemistry and biology. This use of the harbour relies on it delivering 'ecological services' from food supply to waste treatment. It is these more profound uses of the harbour that provides the focus of the industrial aspects of this book.

The question of sustainability

In a sense it is imperative that some version of sustainable development is achievable. A functioning economy is what people depend on for livelihoods and wellbeing. It also supplies their food and deals with their waste. One way or another most established statutory and non-governmental environmentally focused organisations seek to balance environmental and socio-economic needs to find sustainable solutions.

The idea of sustainability can be traced back to the concept of sustainable development introduced into international policy by the World Commission on Environment and Development in 1987. The concept has great political appeal in its assumption that economic development can be achieved in such a way as to be compatible with conserving the natural environment. It therefore allows for economic growth as a legitimate component of conservation thinking (Humphreys and Herbert 2018).

However, despite widespread adoption of sustainable development as a policy goal, there is evidence to suggest that the concept may be flawed: development, in the currently configured economies of the world, seems invariably to come at the expense of the natural environment. For some, perhaps not least national governments, the suggestion of such a basic contradiction may be contentious. But a continuing documented and general decline in the marine environment must raise questions about the validity of the concept, at least in practice: that is to say in its implementation, if not also in principle.

In addition to expanding and updating ecological information in the earlier PHSG volume, *The Ecology of Poole Harbour*, this book can be seen as a case study examining the question of sustainable development in the coastal estuarine/marine context. As such it assembles some of the available information on harbour-dependent industries, with particular reference to 'in the water' economy. Then in our final chapter, we venture into some initial conclusions on the reality of sustainability in this case.

References

Humphreys, J. 2005. Salinity and tides in Poole Harbour. In: Humphreys, J. and May, V. (eds) *The Ecology of Poole Harbour: Proceedings in Marine Science 7*. Elsevier, Amsterdam, pp. 35–48. https://doi.org/10.1016/S1568-2692(05)80008-7

Humphreys, J. and Herbert, R.J.H. 2018. Marine protected areas: Science, policy and management. *Estuarine, Coastal and Shelf Science* 215: 215–18. https://doi.org/10.1016/j.ecss.2018.10.014

Humphreys, J. and May, V. (eds). 2005. *The Ecology of Poole Harbour: Proceedings in Marine Science 7*, Elsevier, Amsterdam.

MMO. 2016. Evidence supporting the use of environmental remediation to improve water quality in the south marine plan areas. A report produced for the Marine Management Organisation. Marine Management Organisation Project No: 1105. ISBN: 978-1-909452-44-2.

PART I

Background

CHAPTER 2

An Industrial History of Poole Harbour

DAVID A. HUMPHREYS

Abstract

The Poole Harbour environment, with its sheltered waters, natural resources, south coast location and connected river system, has proved attractive and economically fruitful for human habitation since the Stone Age. In the Neolithic period, around 2000 BC, permanent agricultural settlements began to emerge around the harbour. By the Iron Age, the area was settled by Celts who established a significant clay manufacturing centre and extensive European trade networks for which Poole Harbour provided an important cross-Channel port. This strategic aspect was exploited in AD 43 when the harbour hosted a major military invasion port for the Roman conquest of the south. By the late eighth century AD, Wareham, which was now under Saxon control, had become a major centre of commerce and administration, whose affluence was sufficient to attract Viking attention. Despite extensive damage, both during and after the Norman Conquest, Poole Harbour's economy would remain important, with the town of Poole becoming a prominent settlement and overshadowing Wareham by the thirteenth century. Throughout the period of human settlement, Poole Harbour provided valuable fisheries, evidenced most spectacularly by the medieval oyster midden under Poole Town Quay. The industrialisation of fishing effort after the Newfoundland fishing grounds became available following the discovery of the New World resulted in a major economic boom. In summary, much of Poole's modern economy can trace its origins to the port, fishing, boat building, military and clay manufacturing industries that have characterised the harbour for millennia.

Keywords: Poole Harbour, maritime history, industrial archaeology

Correspondence: davidhumphreys218@yahoo.co.uk

The area around Poole Harbour has been a site of modern human settlement since at least the Neolithic period, with various flint tools including arrowheads being uncovered at more than a dozen archaeological sites in the area (Calkin 1951). While these Neolithic artefacts show a human presence, the period of permanent human settlement in this region would not truly begin until approximately 2000 BC, when the surrounding landscape had fully transitioned into a region of fixed agricultural production. This coincided with the clearance of the woodlands that occupied the region around the harbour until that time (Darvill 2010).

Evidence of third-century trading activities around the harbour includes the remains of two jetty-type structures ('moles') located on either side of South Deep, one projecting from Cleavel Point, the other from Green Island. Comprised of both stone and wood, and carbon dated to approximately 300 BC, these structures are thought to have functioned

David A. Humphreys, 'An Industrial History of Poole Harbour' in: *Harbour Ecology*. Pelagic Publishing (2022).
© David A. Humphreys. DOI: 10.53061/MYBO1590

as harbour wharfs and were likely progenitors of a Green Island trading settlement (Westwood 2018; Wilkes 2019). The significance of these structures as trading quays for an extensive network both along the coast and cross-Channel is indicated by their size: the southern mole being 170 metres long and 80 metres wide (Wilkes 2019). That Poole Harbour was an established centre of trade in the third century BC is also supported by the discovery of the famous Poole Harbour log boat, which has been carbon dated to approximately 295 BC (see Berry, Parham and Appleby 2019). Cullingford (2003) suggests that as well as people, such boats would have been used to ferry corn, peat and various construction materials such as local clay and stone throughout the harbour.

Archaeological evidence for cross-Channel trade in the Poole Harbour region during the pre-Roman period includes a specific style of cauldron vessel discovered in various sites in the north of France as well as Hamworthy and Green Island (Cullingford 2003). The existence of these cross-Channel trade routes is supported by Roman sources such as Julius Caesar's *Commentaries on the Gallic War*, in which he wrote about the Gaelic Veneti people of Brittany sailing to Britain. Caesar's book also provides a detailed description of the flat-bottomed boats the Veneti employed for the journey and the uniquely maritime focused nature of their culture (Handford 1951). Additionally, the Roman geographer Strabo (63 BC–AD 19) wrote of extensive exports from Britain to the continent, mentioning such goods as gold, cattle, slaves and corn (Jones 1989).

Other material evidence suggesting trade between the Veneti and the Celtic population of the Poole Harbour region consists of a series of iron artefacts, including a chain and anchor of possible Veneti origin excavated at Bulbury Camp. The camp's proximity to Poole Harbour, located just to the north-west, makes it likely that any goods from Europe found in the hill fort will first have passed through Poole Harbour. However, Cullingford (2003) has articulated an alternative theory, that rather than a result of trade, these findings are the belongings of Veneti refugees fleeing Brittany following Julius Caesar's invasion. This is argued on the basis of the uniquely nautical nature of both the Bulbury Camp artefacts and the Veneti culture as described by Caesar (Handford 1951).

It has been argued that Poole Harbour along with Hengistbury Head located to the east were central to a trade bloc set up by the Durotriges, the Celtic nation that inhabited the lands around Poole. This is supported by the widespread circulation of Durotrigean coins (Westwood 2018). According to Field (1992), however, Poole Harbour had fully taken over from Hengistbury Head as the central part of the Durotrigean trade network with the continent by the middle of the first century AD. Examples of continental trade flowing into southern Britain through Poole Harbour include the pottery discovered in Ower, which has been identified as European in origin and dating to before the Roman invasion (Field 1992).

Rivers such as the Piddle and the Frome served as valuable trade routes and will have made Poole Harbour an ideal location, allowing merchants from elsewhere along the southern coast or the continent, such as the Veneti, to access inland communities (Westwood 2018). Similarly, the Frome in particular could have provided access to Poole Harbour and the wider maritime networks for merchants and craftsmen from West Dorset (Woodward 1987). Nevertheless, the Poole Harbour locality is itself well known for pottery manufacture based on the extensive local clay deposits, not least the Black Burnished Ware from the late Iron Age period which can be found throughout southern Britain (Jones 2017).

In AD 43 Emperor Claudius launched the Roman conquest of Britain. This campaign lasted a total of 39 years, ending in AD 84, and resulted in the near-complete subjugation of all British territories south of Caledonia (modern-day Scotland). The Durotrigean lands in Dorset were among the earliest territories taken, and following the initial landing the various settlements around Poole Harbour served as resupply ports for the legions.

In particular, there is strong evidence that Hamworthy served such a function with an early Roman road built by the army from the harbour to a major military site at Lake Farm near modern-day Wimborne.

Such was the importance of the Lake Farm military site that numerous other roads were in turn built connecting it to various other strategically significant settlements, including Durotrigean hill forts such as Badbury Rings and Hod Hill. Additional archaeological evidence of the military utility of Poole Harbour during the Roman invasion includes various Holton Heath earthworks, the location of which would have been ideal for use by an auxiliary force to survey the Hamworthy Peninsula during the landing of soldiers and supplies (Field 1992). The fact that Poole Harbour will have served as a key component of this early Roman invasion network helps further illustrate its strategically valuable location for movement between Britain and the continent, whether this be the movement of legionaries, supplies for an invasion or simply trade and other forms of commerce.

Following the completion of the Roman conquest of southern Britain, the various industries and commercial ventures present in the Poole Harbour region did not end. Rather it appears that continued production was encouraged by the various Roman garrisons and administrators, so as to maintain the economic stability of the newly conquered territories and the cross-Channel trade that resulted from it. Additionally, there is evidence of Hamworthy serving as an early Roman trading port, with pottery dating from the reign of Claudius being uncovered there (Field 1992). While it is possible that the presence of this pottery was the result of cross-Channel trade dating just prior to the Claudian invasion, this is unlikely due to the invasion of southern Britain (AD 43–47) commencing in the first few years of Claudius's reign (AD 41–54).

Further evidence of the continuation of cross-Channel movement of goods has been uncovered at Ower, consisting of pottery dating to both before and after the invasion. However, it is important not to overemphasise such archaeological evidence, as while it does show Poole Harbour continuing to function as a destination for imports from the continent, little evidence has appeared of these imports reaching native, non-Roman settlements. Additionally, the luxury Samian ware pottery discovered at Ower dating to after the invasion seems only to have been imported while the region was under direct military control (Field 1992).

Pottery was not the only commodity passing through Poole Harbour during the invasion period, however. By AD 49 the Romans had reached the Mendips in Somerset and according to Field had begun exploiting the silver and lead deposits located there. Field (1992) argues that the most appropriate port from which to export these resources to the rest of the empire will have been in Poole Harbour.

As well as being a significant element of cross-Channel movement, there is also compelling evidence of more local industries around Poole Harbour. Hamworthy in particular appears to have been the location of a major salt-boiling industry, supported by pieces of black Iron-Age C pottery and briquetage uncovered in the 1920s. These were used for salt production during both the pre-invasion and post-invasion periods (Cullingford 2003). This pottery production is also discussed by Field, who references the large number of potsherds excavated in the Durotrigean territories around Poole Harbour. The unique and uniform nature of these sherds is consistent with the growth of a specialised pottery industry that developed along the southern coast of Greater Dorset, taking advantage of the region's plentiful clay deposits: an industry that continued throughout the Roman period (Field 1992) right up to the modern day.

With its strategic geographical location and attributes, trade routes connecting it with the rest of southern Britain and an abundance of post-invasion archaeological evidence such as the pottery at Ower, it has been argued that the period of economic importance for Poole Harbour will likely have continued after the conclusion of the Roman conquest, with Hamworthy in particular becoming a major civilian port. However, it is noteworthy that while significant communities do appear to have developed on the harbour's coastline, there is no written or archaeological evidence of the modern settlement of Poole at this time. For example, there does not seem to have been any Roman settlement there, possibly discouraged by acidic heathlands (Cullingford 2003). This lack of long-term civilian settlement would continue through to AD 410 when in the face of increasingly frequent raids by various Germanic peoples, such as the Jutes and the Saxons, Emperor Honorius withdrew the legions from the province and the age of Roman rule in Britain came to an end.

Few written records survive from the period between the Roman withdrawal and the conclusion of the Saxon settlement, but entries in the *Anglo-Saxon Chronicle*, written approximately 200 years after the event, suggest that by AD 682 the Germanic invaders had reached as far as the south coast under the leadership of King Centwin of Wessex. While this should not be taken as proof of seventh-century Anglo-Saxon control of the region, a later AD 784 entry offers stronger evidence with the earliest available reference to 'Wareham' (Giles 1914), which, with the use of the Saxon term 'ham' meaning village, confirms a long-term Saxon presence around Poole Harbour. In addition to this, Anglo-Saxon terminology can also help in understanding the economic activity that existed around Poole Harbour in this period: specifically, the use of the term 'weir' (from which the name 'Wareham' is derived) meaning 'fish trap' (Fagersten 1978). That it was prolific enough to influence the naming of the settlement strongly suggests that fishing was a prominent industry during the Anglo-Saxon settlement of this region (Hill and Rumble 1996).

Wareham would continue to be a prominent settlement during the next few centuries and was utilised in a military capacity during the Viking invasions. The *Anglo-Saxon Chronicle* refers to the arrival of a Danish Viking fleet in AD 876 led by Guthrum, in response to which the population fled to Wareham, which is referred to as 'a fortress of the West-Saxons' (Giles 1914). However, the settlement was taken and Poole Harbour's surrounding regions were devastated (Bettey 1974). This military significance is further exemplified by Wareham's function as a burg (fortified settlement) by King Alfred the Great of Wessex during his post-invasion reconstruction and defensive build up, best illustrated by the large earthen ramparts that still stand today. This is further evidenced in the *Burghal Hidage*, which describes Wareham as consisting of 1,600 hides of land, which if true, would have made it the third-largest settlement in the kingdom, exemplifying its importance (Butler 2014) and the more general significance of Poole Harbour at that time.

Wareham's status as a burg does not just reinforce its military importance to the West Saxons; it also reinforces its status as a key commercial and potentially administrative centre, which burgs often were. In the case of Wareham with its location on the point where the River Frome meets Poole Harbour, this seems likely. Additionally, Wareham's importance, particularly as an administrative centre, is further exemplified by its status as the location of multiple royal Saxon burials, examples of which can again be found in the *Anglo-Saxon Chronicle* which lists both King Bentric (802) and King Edward (979) as having been entombed there (Giles 1914). It was also the location of two royal mints, from which the currency of Wessex and later the larger English kingdom were produced, again exemplifying its role as a centre of state administration (Ebsworth 1965). Wareham continued in this capacity, at least until 1015 when it was taken by King Cnut of Denmark who utilised it as one of the earliest staging points for his larger invasion of England (Bettey 1974).

It is also during this period that evidence of oyster harvesting around Poole Harbour first appears, an industry that would persevere into the twenty-first century (Humphreys *et al.* 2014). Excavations carried out in the 1970s and 1980s on the medieval warehouse called the Town Cellars in Poole revealed a midden comprised of oyster shells beneath a layer of pottery sherds dated to the 1300s. Radiocarbon dating of these shells, as well as on additional deposits found beneath Hamworthy, has confirmed that they date from the tenth, eleventh and twelfth centuries. This, combined with the enormous quantity of the shells, suggests the existence of an oyster harvesting industry that predated both the Norman Conquest and potentially the founding of Poole itself (Horsey and Winder 1992).

Following the Norman invasion of 1066, Wareham and the areas around Poole Harbour experienced considerable damage from William the Conqueror's army, with, according to Bettey (1974), 150 of its 285 houses being destroyed. In the centuries that followed partially as a result of damages caused by the Civil War between Stephen and Matilda, and partially as a result of the silting of the River Frome, trade and industry began to shift away from Wareham towards Poole, which before this period will have been little more than a minor fishing village (Internet Archive 2021). As a result, Poole overshadowed Wareham and rose to prominence.

By 1239 Poole received royal consent to hold a weekly market and an annual religious fair, and in 1248, the increased presence of merchants and shipowners gained it a town charter through the efforts of Earl William Longespee of Salisbury (Bettey 1974). This rapid growth into prosperity was aided significantly, as with Wareham and earlier settlements, by its strategic location on the shores of an ideal natural harbour and the trade that resulted from this, but also from the increasing demand for limestone mined on the Isle of Purbeck. While Purbeck Marble had been mined and utilised since Roman times, in the Middle Ages its popularity surged as it became a key building material in virtually all the cathedrals of Southern England (Leach 1975). Poole benefited greatly from this increased demand.

An additional aspect of Poole's development and its status as a port was the role it played in the naval forces of the medieval English kings. Multiple examples of this can be found throughout the thirteenth and fourteenth centuries as requests for ships to join various invasions or military supply fleets, in particular for campaigns in France, were made by the English government. An early example of this was in 1295 when, at the request of Edward I, Poole provided three ships for an invasion of Guyenne. In 1324 Poole was again called upon to provide ships, this time for an expedition in Aquitaine at the request of Edward II. Poole's military responsibilities further increased at the advent of the 100 Years' War, starting in 1337 on the initiative of King Edward III. At various times during this conflict Poole was again required to provide ships, often for expeditions in northern France. These included the 1338 landing in Flanders for which the town contributed six ships and the 1347 capture of Calais to which it contributed four ships (Cullingford 2003).

This again emphasizes just how important Poole Harbour's highly strategic geographical character and location on the south coast has been to the development of the various surrounding settlements and suggests a prominent shipbuilding industry. This role was not without risk, however, as it attracted counter-attacks from the French with a particularly devastating attack occurring in 1377 in which part of Poole was burnt to the ground. The increasingly prominent position of Poole during the fourteenth century is further emphasised by the 1341 enquiry into a petition organised by the Earl of Surrey and William Montacute. This resulted in Poole being confirmed as a free borough, allowing it to be run by its own local government and to tax merchants' customs duties for anchorage and dockage. Additionally, the town's port was greatly extended, giving it increased jurisdiction over the waters of the harbour (Cullingford 2003).

However, Poole's period of growth would come to a traumatic end with the advent of the Black Death in 1348. Arriving with Black Rats *Rattus rattus* on trade ships and spread by flea bites, the bubonic plague hit Melcombe Regis first and before long had spread to the rest of England. Due to its status as a major centre of population and trade, and its location only 30 miles east of Melcombe Regis, Poole was hit particularly hard and fast with an estimated third of the population dying and severe disruption to both trade and the surrounding region's agricultural industry (Bettey 1974). Poole was not the only settlement around the harbour to be affected by the Black Death, however, with Wareham reporting the deaths of a prior and three priests. Again, however, this period provides keen insights into the prominent position Poole had earned by the mid-fourteenth century as while the plague was ravaging the kingdom, many wealthier citizens attempted to flee to Europe taking their money and other valuables with them. Poole was one of the primary ports from which these people attempted to emigrate as exemplified by King Edward III's proclamation to the port authorities that with the exception of merchants, notaries and accredited messengers, no one was to be allowed to leave the country. That Poole was a recipient of this proclamation illustrates the continuingly important role it played in the movement of goods and people in and out of the kingdom (Cullingford 2003).

Despite the setbacks resulting from the Black Death, Poole recovered with the rest of England and in 1433 it was granted the status of Port of the Staple by King Henry VI. With this status Poole became one of a small group of ports which had the right to export various staple goods, in particular wool, to the continent. Taking into account the large numbers of sheep-farms around Poole and the fact that approximately four-fifths of England's total customs revenue during that period was a result of the wool trade, the reputation and prosperity of the town were greatly enhanced in the region. Further developments came when in 1453 King Henry again recognized the growing prominence of Poole by granting its citizens the right to hold a market every Thursday and two additional annual fairs each one lasting eight days (Cullingford 2003).

Because of its harbour and port, Poole continued to grow in both wealth and status during the next two centuries. As well as wool exports this growth was largely sustained by the various transatlantic industries that grew up following the discovery of the Americas. For Poole, this came primarily from the fishing fleets which, following reports from John Cabot's voyage in 1479 of large quantities of cod off the coast of Newfoundland, established a triangular network of trade linking Poole, Newfoundland and the nations of the Iberian and Italian peninsulas (Granger 2010). With this trade network established, a new variety of valuable goods were imported and then exported to and from Poole. Poole's merchant fleet would ship goods such as oil, cod and various animal skins to the Mediterranean, and then return home with wine, salt, fruit and olive oil (Beamish, Hillier and Johnstone 1976).

As these new trade links developed, Poole continued to grow in wealth and prosperity and by the second half of the sixteenth century the town had gained official status as a county in its own right (Bettey 1974). Over the next few centuries, the Newfoundland trade routes continued to be at the centre of Poole's economic and industrial activities and by the late eighteenth-century merchant fleets numbering as many as 78 vessels were setting sail for North America. With this new prosperity came redevelopment as the medieval portions of the town were replaced by structures more fitting for Georgian tastes and styles (Beamish, Hillier and Johnstone 1976). However, as the Napoleonic Wars drew to a close in 1815, Poole's vital North Atlantic fishing grounds were opened up to increasing competition from French and American fishing fleets, resulting in the merchants of the area relocating their operations over the next few decades (Granger 2010). With this, Poole's Newfoundland ventures entered a period of rapid decline – to such an extent that by 1844 the port's merchant fleet numbered a fifth of what it had been just 30 years earlier. The impact of this collapse in trade rippled out into Poole's hinterland, resulting in mass unemployment, and many farms faced bankruptcy due to the influx of European corn (Cullingford 2003).

Poole would emerge from this period of decline with the arrival of the railways. In 1847 a line was opened, linking Dorchester to Southampton and passing through numerous Dorset towns along its route. This included Poole Harbour settlements like Poole and Wareham (Bettey 1974). Expansions to this line would take place multiple times throughout the second half of the nineteenth century. Of particular note were the 1872 and 1874 extensions, which resulted in both Poole and Bournemouth being increasingly linked up with the west and north of England. While the development of the railways initially increased the decline in Poole's maritime trade, the increased interconnectedness of the county and the country in general led to greater prosperity for the town after its period of stagnation as the railway replaced the port as the town's primary economic driver (Lucking 1982). Nevertheless, the economically strategic position of Poole Harbour in the towns' economy was diminished.

The economic revival experienced by Poole, however, was greatly overshadowed by extensive developments at Bournemouth, to such an extent that by the 1880s Poole had ceased to be Dorset's largest town. Despite this, Bournemouth's increased prosperity and the development boom that resulted from it would greatly benefit Poole in the following decades as various building materials required for Bournemouth's numerous construction projects were imported via the harbour. These imports included timber from the Baltic and the Americas and slate tiles from North Wales. Additionally, the lands around the harbour would provide raw materials for construction. Specifically, they would provide clay, which in factories such as those built on Brownsea Island by Colonel William Waugh (Humphreys 2016) would be used to make bricks, tiles and piping.

As Poole entered the twentieth century it was hard hit by the First World War due to the significant manpower contributions towards the British forces in France. To the west, on the outskirts of Wareham on Holton Heath, on the orders of First Lord of the Admiralty Winston Churchill, a factory was built to supply the Royal Navy with cordite.

In the years following the Great War, Poole's population continued to grow and throughout the 1920s and 1930s among the town's largest employers were the shipbuilding yards, which had been experiencing a boom during the opening decades of the new century due to increasing interest in yachting and a general growth in maritime trade.

During the Second World War the harbour area received financial support from the government for shipbuilding and became the location of various government agencies such as the southern HQ of the British Overseas Airways Corporation (PFBC 2010). As the Battle of Britain began, Poole became a potential target due to its increasing role in Britain's war effort. However, unique among major settlements on the south coast, Poole was spared from most of the damages from the Luftwaffe bombing campaigns. A key reason for this was the Major Strategic Night Decoy located on Brownsea Island, which – through the clever use of flares and fireworks – successfully misled the German bombers into dropping their explosives onto the relatively deserted island, saving both Poole and Bournemouth from destruction (Humphreys 2016). The harbour itself would also be used as a major staging area for the D-Day invasion fleet (Beamish, Bennett and Hillier 1980).

In the decades following the end of the war, as the harbour attracted substantial numbers of recreational yachtsmen, Poole Harbour's boat-building sector would experience further revival, not least through the establishment of Sunseeker in 1969. Additionally, tourism and cross-Channel trade in various cargoes such as timber, steel and yachts have benefited the harbour's commercial port through its roll-on-roll-off ferry services (Poole Harbour Commissioners 2021).

In summary, the Poole Harbour environment, with its sheltered waters, natural resources, south coast location and connected river system, has proved both attractive and economically fruitful for human habitation since the Stone Age, and much of Poole's modern economy can trace its origins to the port, fishing, boat building, military and clay manufacturing industries that have characterised the harbour for millennia.

References

Beamish, D., Bennett, H., and Hillier, J. 1980. *Poole and World War II*. Poole Historical Trust, Dorset.

Beamish, D., Hillier, J., and Johnstone, H.F.V. 1976. *Mansions and Merchants of Poole and Dorset*. Poole Historical Trust, Poole.

Berry, J., Parham, D., and Appleby, C. (eds). 2019. *The Poole Iron Age Logboat*. Archaeopress, Oxford. https://doi.org/10.2307/j.ctvndv5rj

Bettey, J.H. 1974. *Dorset*. David & Charles, England.

Butler, M. 2014. *The Burghal Hidage: The Text of the A-Version Burghal Hidage*. CreateSpace Independent Publishing Platform, England.

Calkin, J.B. 1951. The Bournemouth area in Neolithic and early Bronze Age time. *Proceedings of the Dorset Natural History and Archaeological Society* 73: 32–37.

Cullingford, C.N. 2003. *A History of Poole*. Phillimore, England.

Darvill, B., Dyer, B., and Darvill, T. 2010. *The Book of Poole Harbour*. The Dovecote Press Ltd, England.

Ebsworth, N.J. 1965. The Anglo-Saxon and Norman mint of Warwick. *British Numismatic Journal* 34: 53–70.

Fagersten, A. 1978. *The Place Names of Dorset* (2nd edn). Amberley Publishing, England.

Field, N. 1992. *Dorset and the Second Legion*. Dorset Books, England.

Giles, J.A. 1914. *The Anglo-Saxon Chronicle*. G. Bell and Sons Ltd, London.

Granger, D., Dyer, B., and Darvill, T. 2010. *The Book of Poole Harbour*. The Dovecote Press Ltd, England.

Handford, S.A. 1951. *Caesar: The Conquest of Gaul*. Penguin Classics, London.

Hill, D. and Rumble, A.R. 1996. *The Defence of Wessex: The Burghal Hidage and Anglo-Saxon Fortifications*. Manchester University Press, England.

Horsey, I.P. and Winder, J.M. 1992. The late-Saxon and conquest-period oyster middens. In: Jarvis, K. S. (ed.) *Excavations in Poole 1973–1983, Monograph Series: Number 10*. Dorset Natural History and Archaeological Society, Dorset, pp. 60–61.

Humphreys, D. 2016. *Maryland Conservation Management Plan*. Bristol University, Bristol.

Humphreys, J., Herbert, R.J.H., Robert, C., and Fletcher, S. 2014. *A Reappraisal of the History and Economics of the Pacific Oyster in Britain Aquaculture*. Elsevier, Amsterdam, pp. 117–24. https://doi.org/10.1016/j.aquaculture.2014.02.034

Internet Archive. 2021. Available from: https://web .archive.org/web/20110107095832/http://www .wareham-tc.gov.uk/WTC_pages/wtc_history .htm (accessed January 2021).

Jones, G.P. 2017. Sourcing the clay: Iron Age pottery production around Poole Harbour and the Isle of Purbeck, Dorset, UK. Bournemouth University (in collaboration with the Poole Harbour Heritage Project).

Jones, H.L. 1989. *Strabo. Geography*. Loeb Classical Library, England.

Leach, R. 1975. *An Investigation into the Use of Purbeck Marble in Medieval England*. E.W. Harrison and Sons Ltd, England.

Lucking, J.H. 1982. *Dorset Railways*. Railways Correspondence & Travel Society, England.

PFBC (Poole Flying Boats Celebration), Dyer, B., and Darvill, T. 2010. *The Book of Poole Harbour*. The Dovecote Press Ltd, England.

Poole Harbour Commissioners. 2021. Available from: https://www.phc.co.uk/commercial/con-ventional-cargo/ (accessed January 2021).

Westwood, R. 2018. *Ancient Dorset*. Inspiring Places Publishing, England.

Wilkes, E. 2019. Environmental and archaeological background to prehistoric Poole Harbour. In: Berry J., Parham D., and Appleby C. (eds) *The Poole Iron Age Logboat*. Archaeopress, Oxford. https://doi.org/10.2307/j.ctvndv5rj.9

Williams, A. and Marton, G.H. 2003. *Domesday Book*. Penguin Classics, England.

Woodward, P.J. 1987. Romano-British industries in Purbeck. In: Sunter, N. and Woodward P. J. (eds) *Excavations at Norden, Ower and Rope Lake Hole, Monograph Series: Number 6*. Dorset Natural History and Archaeological Society, Dorset, pp. 180–84.

Disclaimer

The views expressed in this chapter do not represent those of any party other than the author.

CHAPTER 3

The Archaeology of the Marine Shells of Poole Harbour

JESSICA WINDER and GREG CAMPBELL

Abstract

Poole is built on oysters, on the shells from over a century's worth of commercial-scale oystering that began before the Norman Conquest and ceased long before the town was founded. This was far from the first time people made use of shells in the harbour; archaeology shows how people began eating shellfish in substantial numbers at least a thousand years earlier. This chapter summarises how archaeologists make use of shells (especially oysters) and outlines the important groups of archaeological shells found around the harbour. A plea is made to those using and researching the present-day harbour and its shells to help archaeologists to refine their techniques.

Keywords: Poole Harbour, shellfish, history, archaeology, archaeomalacology

Correspondence: g.v.campbell@btinternet.com

Introduction

Archaeology has shown that humans began eating marine molluscs at least 164,000 years ago (Jerardino and Marean 2010). Europeans have been consuming them for so long that it is not clear whether this practice actually began with us or our Neanderthal cousins (Fa *et al*. 2016). Occupants of Britain began eating them at least 9,500 years ago (Ashmore and Wickham-Jones 2009), and by 7,000 years ago these Mesolithic hunter-gatherers were discarding shells in heaps all around the coasts of Britain and Ireland (Gutiérrez-Zugasti *et al*. 2011). Archaeological shellfish are sparse from the Neolithic (4000–2500 BC) (Sargeantson 2011) and the Bronze Age (2500–750 BC), and almost absent from the Iron Age (750 BC–AD 50), except at a few coastal sites; only a tiny proportion of the human bones studied from these periods have the chemical signatures of a marine diet (Jay and Richards 2007). It would seem that once prehistoric societies had shifted to farming in Britain, few people ever ate shellfish.

This changed shortly after the conquest of what would become England and Wales by the forces of the Roman Empire. Romans had a very long history of mollusc consumption and management (Günther 1897) because they were part of the ancient Classical Mediterranean tradition of eating shellfish (Voultsiadou *et al*. 2010). Almost every archaeological site of the Romano-British period (AD 50–400) has at least a smattering of shells, however far from the coast, so at the time there was an extensive transport network capable of delivering bulky perishable goods (Wilkinson 2011: 49). Shellfish are found regularly and sometimes copiously in sites far from the sea as well as on the coast in the Anglo-Saxon (AD 400–1066) and the medieval period (AD 1066–1536) (Wilkinson 2011: 52).

Of course the use of molluscs was neither uniform across Britain nor constant over time in any of these periods but was the product of personal preference, family tradition and regional cuisine, all being altered by the movement of people or by their exchanging ideas and practices. How a particular household, settlement or region at a given time made use of marine molluscs as food has much to say about the sophistication of their cuisine, their cultural affinities and their familiarity with the seascape. Some of the key questions archaeologists wish to answer are as follows:

> Were they harvesting throughout the year from a wide range of shore types (e.g. rocky, sandy, muddy; estuaries, bays, offshore beds; upper shore, throughout the tidal range or subtidal), or were they restricted?
> Were they familiar with the preparation of a wide range of shellfish species or only a selected few?
> Were they harvesting from across the species' range or limiting themselves to the microhabitats that were most accessible or most productive?
> Were they harvesting throughout the year or limiting themselves to the most accessible or most productive seasons?
> What methods were they using for harvesting each species (scrabbling, digging, diving or dredging)?
> Was their impact on the shellfish stock insignificant, habitat-altering, sustainable or locally extinctive?

Archaeologists' methods

Fortunately, the shells preserve much of the information needed to answer these archaeological questions (Claassen 1998; Thomas 2015a and 2015b; Somerville *et al.* 2017). Each species being consumed tends to flourish in particular conditions (limpets on intertidal rocks and cockles on intertidal and shallow subtidal soft seabeds), so past peoples must have harvested from those conditions. Changes in the balance between the consumed shellfish species can be used to infer changes in the shores being exploited (Jerardino 1997; Shackleton and van Andel 1986). However, most of the consumed species are numerous enough to harvest because they flourish in quite a wide range of conditions (they occupy a broad range of microhabitats), so it is helpful to have a more refined picture to know whether past peoples were harvesting across the species' range or in particular parts thereof.

Mollusc species have shells with shapes that suit their usual conditions, and the reason there are so many species is that they have evolved shell shapes suitable for various habitats (Stanley 1970). Most mollusc species adopt different shapes in different microhabitats (Seed 1968; Eschweiler *et al.* 2008): they exhibit eco-phenotypic plasticity (Rosenberg 2000). This variation in shape has been used by archaeologists to reconstruct the microhabitats exploited (Andrews *et al.* 1985).

In fact, it is the manner in which the shape of the shell changes as the mollusc grows bigger (its allometry) that differs between microhabitats (Baxter 1983; Seed 1973). The microhabitat of archaeological shells can be inferred by comparing their allometry (the growth of body parts at different rates) to shells of the same species from known conditions (Cabral and da Silva 2003).

In most molluscs growth slows with age, so the average size tends to reach an upper limit over time; in a particular species, the growth rate and the maximum average size differ between habitats (Seed 1980). In many molluscs annual growth slows or stops as the season becomes too hot (in the Tropics) or too cold (in the seas off Britain) forming a distinct annual ring, although these rings can be hard to measure in older shells because they are packed close together at the margin, and trauma also induces similar rings (Richardson 2001). The age in years of each individual shell can be found by counting these rings, and the habitat of the archaeological shells can be inferred by comparing their age-profiles to the age-profiles of modern shells from known modern conditions (Laurie 2008).

The annual rings can also be used to measure an individual shellfish's size at each successive year. The habitat of the archaeological shells can be inferred by comparing their growth curve of length at successive ages to growth curves of the same species from known modern conditions (Campbell 2009: 9).

Marine molluscs grow their shells by adding calcium carbonate crystals along the shell edge at a rate that waxes and wanes with the tides, even in species that live well below low tide, forming microscopic growth-bands (Richardson 2001). Counting the number of tidal growth-bands since the last annual ring can be used to estimate the season of harvest (Laurie 2008). Since the minerals to make these crystals are primarily drawn from the surrounding seawater, the crystals' composition reflects the individual mollusc's local seawater temperature, salinity, acidity, nutrient content and concentration of heavy metals at the time the crystals were laid down (Butler and Schöne 2017). Techniques developed by palaeontologists to reconstruct sea conditions in the geological past with fossil shells (Ivany and Huber 2012) can be used by archaeologists for the more recent past (see Andrus 2011 for a review), for instance to study harvest season (Mannino *et al.* 2003) and the level and sources of past heavy-metal pollution (Labonne *et al.* 1998).

Archaeological deposits often include the shells or traces of smaller creatures that live on, in or among the main shellfish being harvested (commensal species), sometimes still attached to the shells. These often are restricted to a narrower habitat than the main species harvested, and this can be used to infer the microhabitat exploited (Langejeans *et al.* 2017) and the use of marine resources that otherwise decay, such as seaweeds and seagrass (Ainis *et al.* 2014).

Oysters and archaeology

Oyster shells, the valves of the shell of the Common or Flat Oyster *Ostrea edulis*, the native oyster of Europe, are regular finds in European archaeological excavations of all periods, and there is a growing body of research on how they were used in the past (reviewed by Winder 2017). The shells of the oysters preserve many indications of how their beds were used and managed. When very young the oyster cements its left valve to some solid stable surface and grows there for the rest of its life, unless it is moved. The base of the left (lower) valve near the hinge grows over the adjacent surface, cementing it to the shell or preserving an impression of it, such as stones, gravel, other oysters, twigs or cultch (dead shells purposely laid onto a soft seabed to give young oysters places to cement).

Oysters are famous for their growing a shell that is shaped to suit their immediate conditions, and this eco-phenotypic plasticity is useful for reconstructing the microhabitat. Despite their wide range of shapes, the great majority of *O. edulis* can be put into four broad categories (Winder 2011): rounded, with the length (anterior-posterior distance) as long or longer than height (dorso-ventral distance); elongate or boat-shaped (higher than long); irregular (the lower valve is partly flattened or incurved); and facetted (the outer surface of the lower valve has two or three flat faces which meet at a distinct corner; Campbell 2009), the French *huitres carenées* ('keeled oysters') (Gruet 1998).

Rounded and elongated oysters tend to grow well separated on the seabed, with rounded oysters from soft seabeds in slow currents and elongated oysters from harder beds in faster currents; there is a gradation in oyster shape from rounded, thin-shelled, lobed, small-hinged shells on muddy, sheltered seabeds to boat-shaped, heavy thick-shelled broad-hinged shells in high tidal flows offshore (Campbell 2010).

Irregular and facetted oysters tend to grow in oyster reefs, with irregular oysters in broader extensive reefs or parts of reefs and the facetted oysters in deep close-packed reefs or their close-packed portions. This can be demonstrated because they are found in archaeological deposits cemented to each other in multi-oyster masses, fragments of the reef in which they lived, or can be refitted together to form these masses, what French archaeologists call *bouquets d'huitres* (Gruet 1998).

Only archaeologists see native oyster reefs. Natural reefs of native oysters are unknown and probably extinct in Europe, and any found would almost certainly get statutory protection (Holt *et al.* 1998: 17). Competition from rapidly growing reefs of imported oyster species may prevent the return of native oyster reefs (Reise 2008), although reefs made by those imported species may provide settlement sites for native oysters, producing reefs of mixed occupancy (Christeanen *et al.* 2018). Reefs of oysters were once a substantial part of the world's shallow-water ecosystems but are nearing extinction globally through over-exploitation (Beck *et al.* 2011). This has implications for near-shore hydrodynamics: many present-day seabeds that are easily eroded would formerly have been stabilised by a pavement of oyster shells.

Most facetted oysters and many irregular oysters are also 'socketed' (Campbell 2009): the umbo in the left (lower) valve has an umbonal cavity, a distinct hollow under the hinge (Stenzel 1971: 994). This makes them look like the Portuguese Oyster (once *Gryphaea angulata*, then *Crassostrea angulata*, now *Magallana angulata*), in which the umbonal cavity is almost universal, while the socket is weak and usually absent in the native *Ostrea* (Stenzel 1963; Ranson 1940). The two species can be distinguished by the edges of the hinge in *Ostrea* bearing chomata (small regularly spaced ridgelets along the edge of the right valve which fit into matching slots in the left valve), which are absent in *M. angulata* (Harry 1985: 154). Distinguishing the two genera is important. The Portuguese Oyster was long thought to be native to Europe, but its genes are almost identical to the Pacific, or Miyagi, Oyster (formerly *C. gigas*, now *M. gigas*) of temperate Pacific Asia (Ó Foighil *et al.* 1998). There are just enough differences between the two species in behaviour (Batista *et al.* 2017) and genetics (Lapègue *et al.* 2020) to make it likely they are distinct species. It remains to be seen with certainty whether the Portuguese Oyster is an ancient survival of a genus which made its way from Asia around the Arctic Sea in a previous inter-glacial stage of warm climate or the first accidental colonial introduction of a foreign animal into Europe (via the hull of a Portuguese merchantman).

Sometimes oysters are moved by waves and currents or by humans, accidentally during dredging or intentionally by 're-laying' (transferring oysters between beds to improve their growth rate or market quality). This transfer between beds is clear in most oysters because it causes an abrupt change in the shape of the shell (to the shape suitable for the new habitat) or in the hinge (Campbell 2010).

Oysters do lay down annual rings which can be used to estimate age-profiles and yearly growth rates, although they are not as obvious as other bivalves, and the right (upper, flatter) valve is clearer (Winder 2011: 14). Oysters also lay down tidal growth-bands, clearest in the hinge (Richardson *et al.* 1993), and these have been used to estimate the season of harvest (Milner 2002). Oyster tidal growth-band shell chemistry does alter to reflect the surrounding sea (Kirby *et al.* 1998), and this is being increasingly used in native oysters to distinguish the number of sources of supply for a site (Mouchi *et al.* 2018) and season of harvest (Hausmann *et al.* 2019).

The use of commensal organisms surviving on archaeological oysters ('infestation' or 'encrusting organisms') to reconstruct the habitats exploited is fairly advanced, thanks to the work of Winder (1992b). Recording the presence on individual valves of eight types of 'infestation' or 'encrustation' (Winder 2011: 14–17) can distinguish sources of oyster supply for individual archaeological deposits, group those deposits together by source and infer the likely region that supplied those oysters (Winder and Gerber-Parfitt 2003).

A small proportion of archaeological oysters have a V- or W-shaped notch on the shell margin, or slash-marks on the inner surface, the result of knives or other tools being used to open the oyster ('shucking-marks', usually called 'opening marks' by archaeologists) (Winder 2017). Having observed hundreds of opening marks it is clear to these authors that in archaeological oysters most marks are on the ventral edge, so in the past oysters were opened with a long blade inserted opposite the hinge, rather than the modern practice of inserting a short broad blade from the back edge.

Poole Harbour's excavated shells

There are few reports of archaeological marine shells from around the Poole Harbour. This in part may be due to ancient shells not surviving in the acidic sandy soils in the heathlands around Poole. For example, there are no marine shells or animal bones from the excavations of Mesolithic, Iron Age and Roman occupation of Hengistbury Head (Barton 1992; Cunliffe 1987), and no shells reported from the extensive excavations at Bestwall Quarry, Wareham (Ladle and Woodward 2009). It may also be in part due to some reports of shells not being accessible, but only existing in paper records in archaeological archives or practitioners' files – what academics call 'grey literature' (Seymour 2010).

The earliest definite evidence of people harvesting shells around the harbour is from the middle Iron Age (300–100 BC), a small midden on the north shore of Furzey Island (Winder 1991) composed exclusively of cockle shells (*Cerastoderma* sp.), which were not measured.

A settlement of the late Iron Age (100 BC–AD 50) on Cleavel Point on the Ower Peninsula (Winder 1991) had several deposits rich in shells, principally cockles, Winkles *Littorina littorea*, carpet shells (*Ruditapes decussata*, from one deposit) and fragments of Mussels *Mytilus edulis* and of Native Oysters *Ostrea edulis*, which could easily have been washed ashore rather than being food remains. There is no solid evidence from the area around Poole Harbour, or anywhere else in England, for the consumption of oysters in the Iron Age. There were some Roman oysters, but no pre-Roman shells of any species, from excavations in Dorchester (Winder 1993). Those thought to be Iron Age at the Roman villa at Owslebury, Hants (Winder 1988) were later recognised to be from deposits of Romano-British date. The 2,100 marine shells from the Iron Age middens at Mount Batten, Plymouth, included five identifiable oyster shells, but also 40 accidentally gathered sea-shells, some collected dead (Cunliffe and Hawkins 1988).

Poole is built on heaps of oysters much older than the town. There are two huge oyster middens either side of the Holes Bay gat, one under Poole Quay, the other in Hamworthy on the site of the former Shipwrights Arms public house (Winder 1992a). The Poole-side midden occupies almost the full length of the modern quay and contains somewhere between four and eight million oysters (Horsey and Winder 1992), putting it on the same scale as the prehistoric megamiddens of South Africa (Jerardino and Yates 1997), the sambaquis of South America (Wagner *et al.* 2011) or the shell-matrix sites of North America (Roksandic *et al.* 2014). Radiocarbon dates show the Hamworthy midden began about the time England became a single unified state, under King Athelstan (AD 935 ± 80), with the Poole Quay midden beginning about half a century later (AD 995 ± 80). The Hamworthy midden continued building up until about the time of the Norman Conquest (AD 1075 ± 90), and the build-up of the Poole Quay midden lasted about a generation longer (AD 1095 ± 110) (Horsey and Winder 1992). At this point Poole did not exist and Hamworthy was a tiny hamlet, so the oysters were not for local consumption but shucked and the preserved meat was shipped elsewhere.

The balance of sources being harvested for oysters was quite consistent for each midden and differed between the two middens. The Hamworthy midden tended to have very substantial oysters (averaging about 90 mm) and twice as many rounded as elongated oysters throughout, while the Poole Quay midden consistently tended to have smaller but still quite large oysters (averaging about 80 mm) with elongated and round oysters in roughly equal numbers. The one exception was a portion of the Poole Quay midden, dominated by very large oysters (averaging about 93 mm) and by rounded oysters (five times as many rounded oysters as elongated). This portion also had a different pattern of encrustations from the rest of the middens, the only portion with appreciable numbers of the tubes of sand-worms (Sabellidae) and boreholes of predatory gastropods, and with more oysters bored by *Polydora hoplura* (a bristle-worm of muddy seabeds in still waters) than *Polydora ciliata* (another more widely distributed bristle-worm of stiffer sandy and clayey beds). Modern oysters from Poole Harbour were massively encrusted, while modern oysters from offshore in Poole Bay were only slightly affected, and archaeological oysters even less so, so nutrient levels from run-off into the seawater are likely greater in the present day than in antiquity, especially in the harbour.

About a century later (twelfth century–early thirteenth century, about the time Magna Carta was signed), across the harbour at Ower Farm (Winder 1991), the regular processing of shellfish produced a midden much smaller than at Poole (about 5 m wide and over 12 m long), and therefore on a much more domestic scale – probably to supply the monks of Milton Abbey who were the landlords of Ower Farm. Cockles were most common (80%), winkles were relatively less common (20%) and oysters were rare (2%), but the midden was formed of hundreds of distinct small layers of shells heaped one upon the other, each layer almost entirely composed of an individual species, the result of targeting parts of the harbour where those species were known to be rich. Some of the distinct layers included Mussels and carpet shells *R. decussata* in enough quantity to suggest they were also targeted species. The few tellins, *Tellina* sp., Variegated Scallops *Mimachlamys varia*, Common Whelks *Buccinum undatum*, razor shells (Solenidae) and a piece of Lobster carapace *Homarus vulgaris* suggest these species were gathered opportunistically or collected accidentally as dead shells. The very few saddle oysters, *Anomia* sp., Sting Winkles *Ocenebra erinacea*, Netted Whelks *Nassarius reticulatus* and rice-shells (Rissoidae) were probably collected accidentally with the main targeted species.

Cockles were of a consistent size throughout the midden's accumulation, with the average size for all the cockles in a stratum ranging from 28 mm to 30 mm. Only one distinct small layer (from the northernmost end of the midden) had significantly smaller cockles, suggesting the beginning of over-exploitation. Winkles were also of a consistent size, with average height in a stratum ranging from 23.7 ± 2.9 mm to 25.9 ± 2.5 mm, and significantly larger only in one stratum.

The Ower midden oysters were quite different from those in the Poole megamiddens, being much sparser in the midden (only one stratum had enough for statistical comparison), smaller (these averaged 70 mm), and their infestation and shell features showed they grew on a rough seabed with some dead cockle shells (possibly cultch), not on muds or sands.

An excavation in the centre of medieval Poole, on the site of the former foundry across the street from St James's Church (Winder 1994), found significant numbers of shells on an early fifteenth-century beach used as a shipwright's timber-store. The oysters were of intermediate size (averaging 77 mm), smaller than Poole megamidden oysters but larger than Ower Farm's, and of a bimodal distribution, so they probably came from two sources. Infestation was similar to the Poole megamidden oysters, so the medieval harbour beds were probably one source. Winkles were 25 mm on average, about the same size as the typical winkles at Ower Farm, and 60% were badly wave-worn, probably by being rolled about on the shipwright's beach. A few of the oysters were also wave-rounded as though left on the beach, but about 6% had opening marks, so some were definitely food-waste.

Early in the sixteenth century, the beach was reclaimed for building by the dumping of a thick layer of the town's rubbish, which included considerable numbers of shells. These oysters were smaller than the earlier beach oysters (averaging 69 mm), like the Ower Farm midden oysters, and were more commonly flakey (due to rainwater percolation) and stained (due to nearby faecal matter). The rubbish's oysters had fewer shells with *P. hoplura* perforations than the earlier beach oysters, and more with barnacles and with sand-tubes (like the eccentric portion of the Poole megamidden), so the beds being harvested had likely changed. The winkles from the rubbish were only slightly smaller than the beach's (averaging 24 mm), like Ower Farm midden, but only 30% were wave-worn. The dumped rubbish was the only layer with cockles in any numbers, and these were the same size (averaging 28 mm) as the majority of cockles from Ower Farm midden. All the site's cockles were Common Cockles *Cerastoderma edule* with no convincing examples of the Lagoon Cockle *C. glaucum*, despite the site being so near the shore.

Next steps

It might seem that surprisingly few archaeological excavations have reported on their marine shells. The main reason for this is that the usefulness of marine shells is not widely appreciated. Marine biologists (assessing modern-day ecology or fishery sustainability) do not know that archaeologists find well-dated, well-understood 'spot samples' of marine shells, preferentially clustered at the times in the past when their exploitation was high, in which indicators of the ecology and sustainability at those times are fossilised in the shells. Archaeologists, having almost no conversations with marine biologists about shells, tend to give much greater priority during retrieval and analysis to materials with an established track record of being informative about the past. This makes it difficult to gather the detailed data required to refine the analysis of archaeologists' marine shells.

The indicators of past conditions retained by the sizes and shapes of the shells (average and distribution of ages and sizes, change of shape with size) and preserved in the shells (daily and annual growth rate, element concentration, isotope ratios) vary within a particular species between the individual shell's microhabitat (substrate, current speed, position in the tidal range, temperature and its variation through the year, salinity, water chemistry) in complex ways that are not fully understood (Eschweiler *et al.* 2008; Freitas *et al.* 2008). Identifying reliable ways of converting the stored information into inferences about past sea conditions requires a sizeable number of samples, each with large numbers of shells, for each of the main species harvested, across as much of each species' range as feasible. Because archaeologists and palaeontologists are not marine biologists, they have little means to gather enough samples, and big enough samples, of shells from known parts of the sea. Marine biologists do not know the shells they collect can have a wider use for archaeologists and palaeontologists, so they do not keep their shells in archives like archaeologists and palaeontologists do. The result is inadequate modern samples for archaeological and palaeo-ecological inferences (Claassen 1998: 144, 169). Each sample location would need the state of the sea (temperature, salinity, pH, chlorophyll level, etc.) and the state of the bed (ideally its particle size profile, at least rock/gravel/sand/mud) measured directly or estimated reliably (by interpolation from adjacent sampling points).

While this might seem a tall order, it is quite likely that researchers and commercial fishermen are already gathering these in the harbour. Poole Harbour is one of the most studied bodies of water in the world, with boats regularly putting to sea to record its water quality, currents, movement of sediment, benthic ecology and its stocks of commercial species (by fishermen and by biologists to see if the fisheries are sustainable). All that is required is some cooperation. If sample site information was stored centrally in an accessible database, if slightly larger samples of shells were taken (large enough to have adequate statistical power and for destructive sampling) and if those shells were permanently stored in an archive rather than discarded, we would gain the ability to work out how to use shells to understand their biology and the wider ecology (modern, archaeological, geological).

References

Ainis, A.F., Vellanoweth, R.L., Lapeña, Q.G., and Thornber, C.S. 2014. Using non-dietary gastropods in coastal shell middens to infer kelp and seagrass harvesting and paleoenvironmental conditions. *Journal of Archaeological Science* 49: 343–60. https://doi.org/10.1016/j.jas.2014.05.024

Andrews, M.V., Gilbertson, D.D., Kent, M., and Mellars, P.A. 1985. Biometric studies of morphological variation in the intertidal gastropod *Nucella lapillus* (L.): Environmental and palaeoeconomic significance. *Journal of Biogeography* 12: 71–87. https://doi.org/10.2307/2845030

Andrus, C.F.T. 2011. Shell midden sclerochronology. *Quaternary Science Reviews* 30(21–22): 2892–905. https://doi.org/10.1016/j.quascirev.2011.07.016

Ashmore, P. and Wickham-Jones, C. 2009. Section 4: Radiocarbon determinations in context. In: Hardy, K. and Wickham-Jones, C. (eds) *Mesolithic and Later Sites around the Inner Sound, Scotland: The Work of the Scotland's First Settlers Project 1998–2004.* Scottish Archaeological Internet Reports (31), Edinburgh. Available from: http://www.sair.org.uk/sair31 (accessed 23 September 2020).

Barton, R.N.E. 1992. *Hengistbury Head, Dorset (Vol. 2: The Late Upper Palaeolithic & Early Mesolithic Sites).* Oxford University Committee for Archaeology (Monograph 34), Oxford.

Batista, F.M., Fonseca, V.G., Ruano, F., and Boudry, P. 2017. Asynchrony in settlement time between the closely related oysters *Crassostrea angulata* and *C. gigas* in Ria Formosa lagoon (Portugal). *Marine Biology* 164: 110–18. https://doi.org/10.1007/s00227-017-3145-6

Baxter, J.M. 1983. Allometric relationships of *Patella vulgata* L. shell characters at three adjacent sites at Sandwick Bay in Orkney. *Journal of Natural History* 17(5): 743–55. https://doi.org/10.1080/00222938300770581

Beck, M.W., Brumbaugh, R.D., Airoldi, L., Carranza, A., Coen, L.D., Crawford, C., Defeo, O., Edgar, J.G., Hancock, B., Kay, M.C., Lenihan, H.S., Luckenbach, M.W., Toropova, C.L., Guofan, Z., and Ximing, G. 2011. Oyster reefs at risk, and recommendations for conservation, restoration, and management. *BioScience* 61: 107–16. https://doi.org/10.1525/bio.2011.61.2.5

Butler, P.G. and Schöne, B.R. 2017. New research in the methods and applications of sclerochronology. *Palaeogeography, Palaeoclimatology, Palaeoecology* 465(B): 295–9. https://doi.org/10.1016/j.palaeo.2016.11.013

Cabral, J.P. and da Silva, A.C.F. 2003. Morphometric analysis of limpets from an Iron-Age shell midden found in northwest Portugal. *Journal of Archaeological Science* 30(7): 817–29. https://doi.org/10.1016/S0305-4403(02)00254-6

Campbell, G.E. 2009. Southampton French quarter 1382 specialist report download E3: Marine shell. In: Brown, R. (ed.) *Southampton French Quarter 1382 Specialist Report Downloads.* Oxford Archaeology OA Library Eprints, Oxford. http://library.thehumanjourney.net/42/1/SOU_1382_Specialist_report_download_E3.pdf (accessed 25 September 2020).

Campbell, G.E. 2010. Oysters ancient and modern: Potential shape variation with habitat in flat oysters (*Ostrea edulis* L.), and its possible use in archaeology. In: Álvarez-Fernández, E. and Carvajal Contreras, D. (eds) *Not Only Food: Proceedings of the 2nd ICAZ Archaeomalacology Working Group Meeting, Santander, 2008,* Munibe 31, pp. 176–87.

Christianen, J.A., Lengkeek, W., Bergsma, J.H., Coolen, J.W.P., Didderen, K., Dorenbosch, M., Driessen, F.M.F., Kamermans, P., Reuchlin-Hugenholtz, E., Sas, H., Smaal, A., van den Wijngaard, K.A., and van der Have, T.M. 2018. Return of the native facilitated by the invasive? Population composition, substrate preferences and epibenthic species richness of a recently discovered shellfish reef with native European flat oysters (*Ostrea edulis*) in the North Sea. *Marine Biology Research* 14(6): 590–7. https://doi.org/10.1080/17451000.2018.1498520

Claassen, C. 1998. *Shells.* Cambridge University Press (Cambridge Manuals in Archaeology), Cambridge.

Cunliffe, B.W. 1987. *Hengistbury Head (Vol. 1: The Prehistoric and Roman Settlement, 3500 BC–AD 500)*. Oxford University Committee for Archaeology (Monograph 13), Oxford.

Cunliffe, B.W. and Hawkins, S. 1988. Section 2.5: The shell midden deposits. In: Cunliffe, B.W. (ed.) *Mount Batten, Plymouth: A Prehistoric and Roman Port*. Oxford University Committee for Archaeology (Monograph 26), Oxford, pp. 35–8.

Eschweiler, N., Molis, M., and Buschbaum, C. 2008. Habitat-specific size structure variations in periwinkle populations (*Littorina littorea*) caused by biotic factors. *Helgoland Marine Research* 63(2): 119–27. https://doi.org/10.1007/s10152-008-0131-x

Fa, D.A., Finlayson, J.C., Finlayson, G., Giles-Pacheco, F., Rodríguez-Vidal, J., and Gutiérrez-López, J.M. 2016. Marine mollusc exploitation as evidenced by the Gorham's Cave (Gibraltar) excavations 1998–2005: The Middle–Upper Palaeolithic transition. *Quaternary International* 407: 16–28. https://doi.org/10.1016/j.quaint.2015.11.148

Freitas, P.S., Clarke, L.J., Kennedy, H.A., and Richardson, C.A. 2008. Inter- and intra-specimen variability masks reliable temperature control on shell Mg/Ca ratios in laboratory and field cultured *Mytilus edulis* and *Pecten maximus* (Bivalvia). *Biogeosciences Discussions* 5(1): 531–72. https://doi.org/10.5194/bgd-5-531-2008-supplement

Gruet, Y. 1998. Morphologie des *O. edulis*. In: Laporte, L. (ed.) *L'Estuaire de la Charent de la Protohistoire au Moyen Âge: La Challonnière et Mortantambe (Documents d'Archéologie Française 72)*. Éditions de la Maison des Sciences de l'Homme, Paris, pp. 76–77.

Günther, R.T. 1897. The oyster culture of the ancient Romans. *Journal of the Marine Biological Association of the United Kingdom* 4: 360–5. https://doi.org/10.1017/S0025315400005488

Gutiérrez-Zugasti, I., Andersen, S.H., Araújo, A.C., Dupont, C., Milner, N., and Monge-Soares, A.M. 2011. Shell midden research in Atlantic Europe: State of the art, research problems and perspectives for the future. *Quaternary International* 239(1–2): 70–85. https://doi.org/10.1016/j.quaint.2011.02.031

Harry, H.W. 1985. Synopsis of the supraspecific classification of living oysters (Bivalvia: Gryphaeidae and Ostreidae). *The Veliger* 28(2): 121–58.

Hausmann, N., Robson, H.K., and Hunt. C. 2019. Annual growth patterns and interspecimen variability in Mg/Ca records of archaeological *Ostrea edulis* (European Oyster) from the Late Mesolithic site of Conors Island. *Open Quaternary* 5(9): 1–18. https://doi.org/10.5334/oq.59

Holt, T.J., Rees, E.I., Hawkins, S.J., and Seed, R. 1998. *Biogenic Reefs: An Overview of Dynamic and Sensitivity Characteristics for Conservation Management of Marine SACs (UK Marine SACs Project IX)*. Scottish Association for Marine Science, Oban.

Horsey, I.P. and Winder, J.M. 1992. The late Saxon and conquest-period oyster middens. In: Horsey, I.P. (ed.) *Excavations in Poole 1973–83*. Dorset Natural History and Archaeological Society (Monograph Series No. 10), Dorchester, pp. 60–1.

Ivany, L.C. and Huber, B. (eds). 2012. *Reconstructing Earth's Deep-Time Climate: The State of the Art in 2012*. Yale University Press for the Paleontological Society (Paper 18), New Haven.

Jay, M. and Richards, M.P. 2007. British Iron Age diet: Stable isotopes and other evidence. *Proceedings of the Prehistoric Society* 73: 169–90. https://doi.org/10.1017/S0079497X0002733X

Jerardino, A. 1997. Changes in shellfish species composition and mean shell size from a Late-Holocene record of the west coast of South Africa. *Journal of Archaeological Science* 24: 1031–44. https://doi.org/10.1006/jasc.1997.0182

Jerardino, A. and Marean, C.W. 2010. Shellfish gathering, marine paleoecology and modern human behavior: Perspectives from cave PP13B, Pinnacle Point, South Africa. *Journal of Human Evolution* 59: 412–24. https://doi.org/10.1016/j.jhevol.2010.07.003

Jerardino, A. and Yates, R. 1997. Excavations at Mike Taylor's midden: A summary report and implications for a re-characterisation of megamiddens. *The South African Archaeological Bulletin* 52: 43–51. https://doi.org/10.2307/3888975

Kirby, M.X., Soniat, T.M., and Spero, H.J. 1998. Stable isotope sclerochronology of Pleistocene and Recent oyster shells (*Crassostrea virginica*). *Palaios* 13(6): 560–9. https://doi.org/10.2307/3515347

Labonne, M., Ben Othman, D., and Luck, J.-M. 1998. Recent and past anthropogenic impact on a Mediterranean lagoon: Lead constraints from mussel shells. *Applied Geochemistry* 13(7): 885–92. https://doi.org/10.1016/S0883-2927(98)00016-X

Ladle, L. and Woodward, A. 2009. *Excavations at Bestwall Quarry, Wareham, 1992–2005*. Dorset Natural History and Archaeological Society (Monograph 19), Dorchester.

Langejans, G.H., Dusseldorp, G.L., and Thackeray, J.F. 2017. Pleistocene molluscs from Klasies River (South Africa): Reconstructing the local coastal environment. *Quaternary International* 427: 59–84. https://doi.org/10.1016/j.quaint.2016.01.013

Lapègue, S., Heurtebise, S., Cornette, F., Guichoux, E., and Gagnaire, P.A. 2020. Genetic characterization of cupped oyster resources in Europe using informative single nucleotide polymorphism (SNP) panels. *Genes* 11(4): 451–68. https://doi.org/10.3390/genes11040451

Laurie, E.M. 2008. *An Investigation of the Common Cockle (Cerastoderma edule (L.)): Collection Practices at the Kitchen Midden Sites of Norsminde and Krabbesholm, Denmark*. British Archaeological Reports (International Series 1834), Oxford. https://doi.org/10.30861/9781407303185

Mannino, M.A., Spiro, B.F., and Thomas, K.D. 2003. Sampling shells for seasonality: Oxygen isotope analysis on shell carbonates of the inter-tidal gastropod *Monodonta lineata* (da Costa) from populations across its modern range and from a Mesolithic site in Southern Britain. *Journal of Archaeological Science* 30(6): 667–79. https://doi.org/10.1016/S0305-4403(02)00238-8

Milner, N. 2002. *Incremental Growth of the European Oyster Ostrea edulis: Seasonality Information from Danish Kitchenmiddens*. British Archaeological Reports (International Series 1057), Oxford. https://doi.org/10.30861/9781841714370

Mouchi, V., Briard, J., Gaillot, S., Argant, T., Forest, V., and Emmanuel, L. 2018. Reconstructing environments of collection sites from archaeological bivalve shells: Case study from oysters (Lyon, France). *Journal of Archaeological Science: Reports* 21: 1225–35. https://doi.org/10.1016/j.jasrep.2017.10.025

Ó Foighil, D., Gaffney, P.M., Wilbur, A.E., and Hilbish, T.J. 1998. Mitochondrial cytochrome oxidase I gene sequences support an Asian origin for the Portuguese oyster *Crassostrea angulata*. *Marine Biology* 131(3): 497–503. https://doi.org/10.1007/s002270050341

Ranson, G. 1940. La charnière de la dissoconque de l'huitre. *Bulletin du Muséum National de l'Histoire Naturelle* (2nd Series) 12: 119–28.

Reise, K. 2008. Oysters: Natives gone and aliens coming. *Neobiota* 7: 258–62.

Richardson, C.A. 2001. Molluscs as archives of environmental change. *Oceanography and Marine Biology: An Annual Review* 39: 103–64.

Richardson, C.A., Collis, S.S., Ekaratne, K., Dare, P., and Key, D. 1993. The age determination and growth rate of the European flat oyster, *Ostrea edulis*, in British waters determined from acetate peels of umbo growth lines. *ICES Journal of Marine Science* 50: 493–500. https://doi.org/10.1006/jmsc.1993.1052

Roksandic, M., de Souza, S.M., Eggers, S., Burchell, M., and Klokler, D. (eds). 2014. *The Cultural Dynamics of Shell-Matrix Sites*. University of New Mexico Press, Albuquerque.

Rosenberg, G. 2000. Ecophenotypic variation in mollusks. *American Conchologist* 28: 22–3.

Sargeantson, D. 2011. Review of animal remains from the Neolithic and early Bronze Age of Southern Britain (4000–1500 BC). English Heritage (Research Report 29-2011), Portsmouth.

Seed, R. 1968. Factors influencing shell shape in the mussel *Mytilus edulis*. *Journal of the Marine Biological Association of the United Kingdom* 48(3): 561–84. https://doi.org/10.1017/S0025315400019159

Seed, R. 1973. Absolute and allometric growth in the mussel, *Mytilus edulis* L. (Mollusca, Bivalvia). *Proceedings of the Malacological Society of London* 40: 343–57.

Seed, R. 1980. Chapter 1: Shell growth and form in the Bivalvia. In: Rhoads, D.C. and Lutz,

R.A. (eds) *Skeletal Growth of Aquatic Organisms*. Plenum Press, New York, pp. 23–67. https://doi.org/10.1007/978-1-4899-4995-0_2

Seymour, D.J. 2010. In the trenches around the ivory tower: Introduction to black-and-white issues about the grey literature. *Archaeologies: Journal of the World Archaeological Congress* 6(2): 226–32. https://doi.org/10.1007/s11759-010-9130-z

Shackleton, J.C. and van Andel, T.H. 1986. Prehistoric shore environments, shellfish availability, and shellfish gathering at Franchthi, Greece. *Geoarchaeology* 1(2): 127–43. https://doi.org/10.1002/gea.3340010202

Somerville, E., Light, J., and Allen, M.J. 2017. Marine molluscs from archaeological contexts: How they can inform interpretations of former economies and environments. In Allen, M.J. (ed.) *Molluscs in Archaeology: Methods, Approaches and Applications*. Oxbow Books, Oxford, pp. 214–37. https://doi.org/10.2307/j.ctvh1dk5s.19

Stanley, S.M. 1970. *Relation of Shell Form to Life Habits of the Bivalvia (Mollusca)*. Geological Society of America (Memoir 125), Boulder.

Stenzel, H.B. 1963. A generic character: Can it be lacking in individuals of the species in a given genus? *Systematic Zoology* 12(3): 118–21. https://doi.org/10.2307/sysbio/12.3.118

Stenzel, H.B. 1971. Oysters. In: Moore, R.C. (ed.) *Treatise on Invertebrate Paleontology Part N, Vol. 3: Mollusca 6 (Bivalvia)*. Geological Society of America and University of Kansas Press, Lawrence, pp. 953–1224.

Thomas, K.D. 2015a. Molluscs emergent, Part I: Themes and trends in the scientific investigation of mollusc shells as resources for archaeological research. *Journal of Archaeological Science* 56: 133–40. https://doi.org/10.1016/j.jas.2015.01.024

Thomas, K.D. 2015b. Molluscs emergent, Part II: Themes and trends in the scientific investigation of molluscs and their shells as past human resources. *Journal of Archaeological Science* 56: 159–67. https://doi.org/10.1016/j.jas.2015.01.015

Voultsiadou, E., Koutsoubas, D., and Achparaki, M. 2010. Bivalve mollusc exploitation in Mediterranean coastal communities: An historical approach. *Journal of Biological Research* 13: 35–45.

Wagner, G., Hilbert, K., Bandeira, D., Tenório, M.C., and Okumura, M.M. 2011. Sambaquis (shell mounds) of the Brazilian coast. *Quaternary International* 239(1–2): 51–60. https://doi.org/10.1016/j.quaint.2011.03.009

Wilkinson, K. 2011. Review of environmental archaeology in the Southern Region: Molluscs. English Heritage (Research Report 52/2011), Portsmouth.

Winder, J.M. 1988. Oyster Shells from Owslebury, Hamphsire. Historic Buildings and Monuments Commission for England Ancient Monuments Laboratory (Research Report 53/88), London.

Winder, J.M. 1991. Marine Mollusca. In: Cox, P.W. and Hearne, C.M. (eds) *Redeemed from the Heath: The Archaeology of the Wytch Farm*

Oilfield (1987–1990). Dorset Natural History and Archaeological Society (Monograph Series No. 9), Dorchester, pp. 212–16.

Winder, J.M. 1992a. The oysters. In: Horsey, I.P. (ed.) *Excavations in Poole 1973–83*. Dorset Natural History and Archaeological Society (Monograph Series No. 10), Dorchester, pp. 194–200.

Winder, J.M. 1992b. A study of the variation in oyster shells from archaeological sites and a discussion of oyster exploitation. Unpublished PhD Thesis. Archaeology Department, University of Southampton.

Winder, J.M. 1993. Section 8.6: Oyster and other marine mollusc shells. In: Woodward, P.J., Davies, S.M., and Graham, A. (eds) *Excavations at the Old Methodist Chapel and Greyhound Yard, Dorchester, 1982–1984*. Dorset Natural History and Archaeological Society (Monograph Series No. 12), Dorchester, pp. 347–8.

Winder, J.M. 1994. The marine mollusc shells. In: Watkins, D.W. (ed.) *The Foundry: Excavations on Poole Waterfront 1986–7*. Dorset Natural History and Archaeological Society (Monograph Series No. 14), Dorchester, pp. 84–8.

Winder, J.M. 2011. Oyster shells from archaeological sites: A brief illustrated guide to basic processing. https://oystersetcetera.files.wordpress.com /2011/03/oystershellmethodsmanualversion11 .pdf

Winder, J.M. 2017. Chapter 14: Oysters in archaeology. In: Allen, M.A. (ed.) *Molluscs in Archaeology: Methods, Approaches and Applications*. Oxbow Books, Oxford, pp. 238–58. https://doi.org/10 .2307/j.ctvh1dk5s.20

Winder, J.M. and Gerber-Parfitt, S. 2003. Section 7.22: The oyster shells. In: Malcolm, G., Bowsher, B., and Cowie, R. (eds) *Middle Saxon London: Excavations at the Royal Opera House, Covent Garden*. Museum of London Archaeology Service (Monograph 15), London, pp. 325–32.

CHAPTER 4

Developing Conceptual Ecosystem Models for Poole Harbour

ELEANOR RIVERS and ROGER J.H. HERBERT

Abstract

Conceptual models, typically presented as flow charts or diagrams, are a means of organising and presenting complex bodies of information about an ecological system in a way that can graphically illustrate key features, processes, system drivers and linkages. Information from key sources of literature was used to create visualised models of key habitats in Poole Harbour: mudflats, saltmarsh and saline lagoons. Physical features coupled with relevant anthropogenic constructions, such as marinas or drainage inputs, are superimposed on notable wildlife assemblages, and linkages with species upon which they depend for survival are highlighted. Natural and anthropogenic stressors were mapped onto these elements, potentially allowing visualisation of risk areas and gaps in monitoring effort. Model development was an iterative process, with different perspectives offered by stakeholders within the Poole Harbour Study Group and staff and students from Bournemouth University. It is proposed that these types of conceptual models could provide valuable foundations for a discussion of new harbour management plans and other sites of nature conservation importance.

Keywords: conceptual models, ecosystem models, management plans, stakeholder engagement, environmental change

Correspondence: erivers@bournemouth.ac.uk

Introduction

Conceptual ecological models

Natural systems are highly complex, and in many cases there is a paucity of knowledge about how they function, especially where human-natural systems are coupled and inter-dependent (Liu *et al.* 2007). In these challenging times of high anthropogenic impact and a warming climate, there is a pressing need for a method of capturing and communicating large bodies of empirical evidence about a system and distilling them into an accessible format to prioritise practical actions such as management planning or monitoring (Watson *et al.* 2018). Conceptual models have the potential to fulfil this need and as a result this approach has been gaining popularity (Walters *et al.* 2000; Ogden *et al.* 2005; Briggs *et al.* 2013; Wingard and Lorenze 2014).

Eleanor Rivers and Roger J.H. Herbert, 'Developing Conceptual Ecosystem Models for Poole Harbour' in: *Harbour Ecology*. Pelagic Publishing (2022). © Eleanor Rivers and Roger J.H. Herbert. DOI: 10.53061/BRFB1578

In essence, conceptual models are a means of organising and presenting complex bodies of information about an ecological system in a way that can graphically (often through diagrams or flowcharts) illustrate key features, processes, drivers, linkages and system functioning (Suter 1999; Gentile *et al.* 2001; Healey *et al.* 2007). This may include connections between species or abiotic processes, or the sphere of impact of anthropogenic stressors. Conceptual models can represent systems at multiple spatial or temporal scales, for example ranging from detailed snapshots of micro processes such as spatio-temporal plankton dynamics (Medvinskii *et al.* 2002) through to harbour-wide models with coarse-scale information (Barnes 2005). Models may vary widely in the level of detail presented and the techniques used to present this information. If scientific knowledge is to be successfully integrated into management and policy development and increase consensus and understanding, then it needs to be presented in a way that can realistically be accessed by all stakeholder groups, not just those with specialised scientific understanding (McInerny *et al.* 2014).

In this series of models, physical features such as benthic sediment, geological features and shoreline structure were included, coupled with relevant anthropogenic constructions such as marinas or drainage inputs. Layered over these elements were notable wildlife assemblages and linkages with species upon which they depend for survival. Natural and anthropogenic stressors were mapped onto these elements, potentially allowing visualisation of risk areas and gaps in monitoring effort. We chose to develop models which aimed to be highly visually accessible and easy to read, in line with our objective of achieving inclusivity in dissemination.

Although these models may generally lack the quantitative power of agent/individual-based models, which can allow for scientific prediction, their strength lies in their potential to identify and communicate complex interactions, highlight critical gaps in the knowledge base and point to directions for future research or monitoring (Walters *et al.* 2000). Not only this but a tangible benefit of the development of conceptual models can be the development process itself: encouraging communication and cooperation between stakeholder groups who may otherwise not have engaged in a constructive process such as this. However, despite the many positives inferred by this method, it has been largely absent from UK planning processes, with the most numerous examples of its use occurring in the United States.

The current development planning approach

One of the core objectives of the Poole Adopted Core Strategy is to enhance protection of the natural environment (Bournemouth and Poole Council 2009). This is addressed by ensuring that planning applications are subjected to a process of assessment to identify the level of risk they pose to the natural environment. For example, if a project is likely to impact a European protected site such as Poole Harbour, a system known as Habitats Regulations Assessment (HRA) may be implemented (Natural England 2017). Part of the HRA screening includes a 'Judgement of Likely Significant Effects'. This preliminary assessment stage requires that the site features are characterised, alongside conservation objectives and 'possible impact pathways'. However, the process of site characterisation appears to lack a stage where drivers and linkages are captured in the way which is possible with conceptual models. The key elements currently included are qualifying interests, conservation objectives, condition status, key environmental conditions (factors that support site integrity) and vulnerabilities (pressures and trends affecting site integrity). However, this system is not very effective at capturing the potential importance of apparently less influential species, where some species may be scarce yet have a disproportionate influence on the provision of ecosystem services. Therefore, in the absence of capturing their effect on ecosystem functioning, it may be that they are not adequately protected by the current procedures (Dee *et al.* 2019). In North America, for example, the removal of the Purple Marsh Crab *Sesarma reticulatum* from an ecosystem indirectly triggered the die-off of Saltmarsh Cordgrass *Spartina alternifolia*, which then in turn reduced wave attenuation and increased the likelihood that shorelines could become unstable (Bertness *et al.* 2014). These relationships are easily identified and presented using a conceptual modelling framework.

Aims

Through this project we intended to create a set of conceptual ecosystem models relevant to Poole Harbour, which could be employed as a tool to introduce this approach to management planning. From the outset it was accepted that the models would always be subject to change and development as more evidence came to light, more perspectives were offered and situations changed through time. The models needed to be accessible and easily understood by stakeholders who may have little or no scientific or ecological knowledge, so they can facilitate open and inclusive discussion of issues surrounding the management of the harbour for multiple complex needs.

Methods

Study site

Poole Harbour covers approximately 3,600 ha (Poole Harbour Study Group 2021) and is a busy site with many varied habitats in close proximity to each other. There are differing and sometimes conflicting activities occurring within its boundaries at any one time. The harbour has high leisure value with boating, kite surfing and paddle boarding being very popular. Alongside these activities are many industrial businesses associated with a busy port and cross-Channel ferries. All these activities occur upon the backdrop of the importance of Poole Harbour's biodiversity, a number of statutory designations, including European Special Protection Area (SPA), and its inherent vulnerability to damage from human activity.

It would have been unwieldy to attempt to present the whole of the harbour in one diagram, which would have obscured the clarity of the models with densely packed information. Therefore, we selected three key habitats within the harbour, based on those which were considered to be of either high wildlife or human amenity value: the intertidal mudflats, the saltmarsh and the saline lagoon.

Laying the foundations

Since the purpose of these models was to take a large, disparate and potentially inaccessible body of evidence and distil it into three comprehensible diagrams, the first task was to source and obtain all relevant publications for review. Once we were satisfied that the most relevant literature had been identified, key themes were extracted based on features which were deemed important to statutory designations or themes which appeared frequently in the literature. Therefore, for every habitat we established the following:

- Protected species (SPA designation).
- Key supporting species.
- UK Biodiversity Action Plan species recorded in this habitat.
- Species notified in Special Site of Scientific Interest (SSSI) citation.
- Economic value of the area.

We then established the main impacts on these species, habitats or industries, using a table to describe the source of the impact, the effect of the impact and any linked effects (see Appendix 4.1). This was an effective way of capturing connections which could be illustrated in the model.

Based on these tables and discussions between the research team, we created a basic series of models. From this point onwards some subjectivity entered the process, which required group discussion to ensure a balanced perspective was maintained. For this reason we then took the initial models to a group discussion with people representing other areas of academia, Poole Harbour Study Group (PHSG) and students at Bournemouth University. These discussions were extremely fruitful, and it was apparent this stage could be considered an outcome in itself, with the potential to strengthen cooperation between stakeholder groups and participants from groups who might not normally engage in activities together. The discussions and feedback were recorded with a voice recorder and also captured through feedback sheets. Following this process, the models were subsequently revised and many edits were applied. These versions were then later presented at a meeting of PHSG for further discussion and review.

Key aspects described by the models

Intertidal mudflats

The mudflats represent one of the most significant habitats in Poole Harbour, supporting internationally important numbers of wintering birds every year. Between 3,500 and 4,000 ha is covered by this habitat.

Mudflat key aspects

- Site of a significant shellfishery with introduced stocks of Manila Clam *Ruditapes philippinarum*.
- The primary source of energy for thousands of birds annually.
- There are a number of discharges into the harbour – the rivers Piddle, Frome, Sherford and Corfe; Holes Bay sewage treatment plant and several combined sewage outfalls (CSO) which are licensed to discharge in times of heavy rain when the sewerage systems are at risk of becoming overwhelmed.
- Many recreational activities take place over the mudflats at high tide, such as kite surfing, boating and fishing.
- Bait digging is a popular activity on the mudflats.
- The surrounding catchment area is heavily urbanised.

Saltmarsh

Saltmarsh extent is approximately 400 ha and is a habitat with significant ecological value. Some areas of the harbour have seen dramatic losses of area through pressures such as land reclamation, erosion and sea level rise (Edwards 2005; Gardiner *et al.* 2011; Gardiner 2015).

Saltmarsh key aspects

- Common Cordgrass *Spartina anglica* dieback may cause cadmium release.
- Cadmium causes reductions in phytoplankton and zooplankton levels, and can be found in fish and molluscs.
- Run-off and sewage input from various CSOs around the harbour create nitrogen and phosphate inputs.

- Non-native Sika Deer *Cervus nippon* grazing can cause trampling of saltmarsh plants.
- Sea level rise could negatively affect saltmarsh extent and bird breeding/roosting habitat.
- Disturbance of birds by recreational activities.
- The surrounding catchment area is heavily urbanised.

Lagoon

Poole Park Lagoon is a brackish human-made waterbody, extending for 21 ha at an average water depth of 1 m (BCP 2016). The lagoon receives heavy recreational use and large numbers of wildfowl such as Greylag *Anser anser* and Canada Geese *Branta canadensis*.

Lagoon key aspects

- Nitrogen and phosphate input from various sources including wildfowl.
- E-coli levels from sewage inputs.
- Raised hydrogen sulphide levels in sediment.
- Chironomid midge swarms (through an indirect link: this is driven by reduced competition since other species cannot tolerate such low sedimentary oxygen levels).
- Significant population of the protected anemone *Nematostella vectensis*.
- Human recreation.
- Salinity levels.
- The surrounding catchment area is heavily urbanised.

Discussion

Through this process of development it has become clear how this kind of model can never be considered entirely complete. It is an iterative process which has to encompass many differing perspectives and much inherent subjectivity, as well as the varying strength of different evidence sources. However, the intention was never to produce a 'complete' model but more generally to establish a discussion point and to create the foundation of future model development. It is anticipated that scientific and ecological understanding of this area will increase through time, so these models should be considered reference points to where we currently are.

The sessions where models (Figures 4.1, 4.2 and 4.3) were discussed with stakeholders proved highly productive, and it was clear how they could form an effective bridge for communication between potentially disparate groups, even before any collaboration or planning based on the models has been undertaken. Discussion of this kind could lay foundations where stakeholders may feel encouraged that their views are important to those in ecological fields, and this could assist in future communications when the stakes may be higher or more controversial. Although the models might be used as stand-alone diagrams for personal study and research, they are much more valuable when presented in an open forum where they can be examined and discussed.

These models were primarily based on the published scientific literature, which was available at this point in time. However, it can be argued that there is a vast body of knowledge and understanding which is not formally and scientifically validated, yet likely has a place in a model such as this, where local knowledge and observations could be highly relevant. A site such as Poole Harbour, which attracts a lot of interest from a wide range of stakeholders including the public and special interest groups such as birdwatchers and

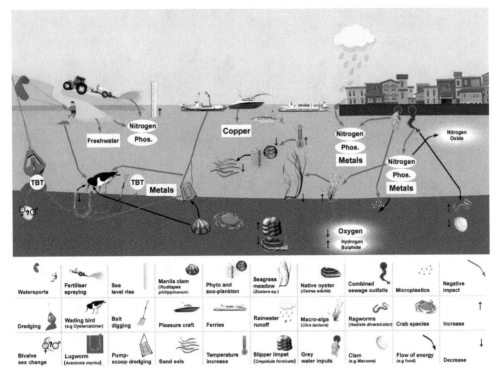

Figure 4.1 Conceptual model representing the mudflat habitats within Poole Harbour. E. Rivers.

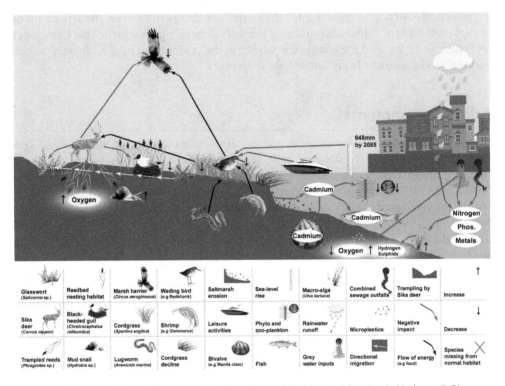

Figure 4.2 Conceptual model representing the saltmarsh habitats within Poole Harbour. E. Rivers.

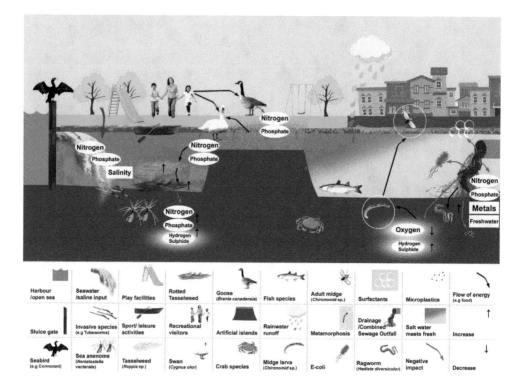

Figure 4.3 Conceptual model representing the lagoon habitats within Poole Harbour. E. Rivers.

citizen scientists, will be understood in a number of different ways, not all of which can be measured easily. One direction for the future development of these models might be to explore the use of different connecting symbols depending on whether the knowledge was scientific or lay, and the level of confidence in the evidence, or even to develop a comparative model using only lay knowledge of the area.

Acknowledgements

We are very grateful for all the assistance and discussions with many students and staff from Bournemouth University and with stakeholder representatives within the Poole Harbour Study Group. Particular thanks to Martin Whitchurch (Bournemouth and Poole Council), John Humphreys (Southern IFCA), Andrew Powell (Canford School), Jez Martin (Bournemouth and Poole Council), Luke Johns (Dorset Wildlife Trust), Adrian Newton (BU), Luciana Esteves (BU), Stephen Watson (BU), Marin Cvitanovic (BU), Frank Grandfield (BU), Andrew Powell (Canford School), Alice Hall (BU), Dan Franklin (BU), Jess Bone (BU).

Appendix 4.1 Process of Identifying Key Impacts and Linked Effects

Key: > higher; < lower

MUDFLAT

Impacts

Sewage outfall

Source	Effect	Impacted species	Linked effects
Sewage treatment works (STW with) outfall into Holes Bay + farm run-off via Frome and Piddle rivers into Wareham Channel	Increase polychaetes, decrease large bivalves (Alves *et al.* 2011)	Positive for large waders but less diversity near outfalls	
	Introduction of nutrients/ ammonia < oxygen > hydrogen sulphide > ammonia > algae > phosphorous	Positive for Ragworm *Hediste diversicolor.* Can be harmful to Eelgrass *Zostera marina*	Macroalgal mats (*Ulva* spp.) may smother invertebrates and limit birds feeding Fluctuations in dissolved oxygen from algae can cause sub-lethal effects (Thornton 2016)
	Introduction of metals (Underhill-Day 2006; Hübner 2009)	Negative for all species but *Hediste* tolerate contaminated areas	< clam *Scrobicularia*. Contaminated *Hediste* may transfer metals up food chain Dredging may remobilize (Underhill-Day 2006)
STW + storm drains + local urbanisation	Introduction of plastics	Negative for all species	

TBT from boat hulls

Effect	Impacted species	Linked effects
Retained by sediment – dredging may remobilize	Causes sex changes in molluscs (Rittschoff and McClellan-Green 2005; Langston *et al.* 2015) Toxic to *Arenicola* (Cefas 2002), *Corophium volutator* and Copepod *Tisbe battagliai* (Underhill-Day 2006) Hübner (2009). Pacific Oyster *Magallana* (*Crassostrea*) *gigas* shell thickening (Dyrynda 1992)	*Scrobicularia* declined (Langston *et al.* 2003) Limits aquaculture and fishery potential

Hydrocarbons

Effect	Impacted species
Polycyclic Aromatic Hydrocarbons (PAH) in sediment at threshold effect level (Woodhead *et al.* 1999) *Source*: Wytch Farm	Non-selective deposit feeders may be at risk, as opposed to those living in sediment but feeding selectively or on detritus May cause neoplasms in fish (Malins 1988)
Possible spill event	All

More research needed on compound effect of all substances.

Disturbance to sediments

Activity	Effect	Impacted species
Dredging	< Water clarity	Tern, seagrass beds (English Nature 2000)
	Removal of species	Invertebrates, with potential impact on fish and birds
	Release of contaminants	All (Underhill-Day 2006)
Bait digging/ dragging	Direct damage	Infaunal communities (Watson *et al.* 2017, Underhill-Day 2006)
	Re-sorting of sediment types	
	Release of contaminants	Potentially all species (Underhill-Day 2006)
	Change to intertidal topography + release of sediment from the harbour to Poole Bay	
	Change in community composition	Large older molluscs replaced by fast-growing deep-burrowing polychaetes (Underhill-Day 2006)
	Disturbance	Feeding waders, infaunal communities (English Nature 2000)
	Remove species	Infaunal communities (Durrell *et al.* 2006)

Shellfish aquaculture

Effect	Impacted species
> Water clarity	Tern, seagrass (Underhill-Day 2006)
Localised eutrophication and sedimentation	Invertebrates (Kaiser *et al.* 1996)
Disturbance of seabed	Seagrass (Ferriss *et al.* 2019)

Introduced species

Species	Impact
Manila Clam	Positive for waders, who may limit density of species (Caldow *et al.* 2007; Humphreys *et al.* 2015)
	Positive for local economy. May become issue with future warming but could replace species that move north
Slipper Limpets *Crepidula fornicata*	Outcompete Native Oysters *Ostrea edulis* and now occupy much of the seabed (Dyrynda 2005)

SALTMARSH

Impacts

Cadmium

Source	Impacted species	Mitigation	Future risk
Poole STW, Holes Bay	Zooplankton, phytoplankton, fish, molluscs (accumulation – factor for fisheries)	Retained by *Spartina*	Sea level rise, damage to saltmarsh or die-back of *Spartina* could cause release of cadmium (Hübner *et al.* 2010)

Sika Deer *Cervus nippon*

Source	Impact	Linked effects
Introduced species which rapidly reached high numbers	Grazed areas showed higher numbers of invertebrates – due to < plant biomass (Diaz 2005)	Better food supply for birds such as Shelduck and Dunlin.
	Intermediate grazed areas had > plant diversity	This can increase Redshank *Tringa totanus* densities (Diaz *et al.* 2005)
	Heavily grazed areas could be devoid of biomass	Bare areas can become hypersaline and cause the appearance of lower saltmarsh species (e.g. glasswort *Salicornia ramosissima*) in higher regions (Diaz *et al.* 2005). This has low density and may not support nesting waders such as Redshank
	Trampling can increase sediment density (Diaz *et al.* 2005)	This can negatively affect infaunal communities
	Grazing has eliminated two reedbeds (Diaz *et al.* 2005)	Loss of habitat for threatened species such as marsh harriers, cetti's warbler, reed warblers, bearded tits, water rail, water voles and wainscot moths

Poor management

Source	Impact	Linked effects
Poor water management Deer damage or lack of appropriate management	Saltmarsh may dry out > salinity in bunded freshwater beds	Loss of saltmarsh/reedbed plants, area returns to mudflat or bare ground, loss of infaunal communities and wildlife interest (Cook 2005)
Encroachment of scrub into reedbeds (sallow and bog myrtle) (South Middlebere beds) Over-grazing/trampling by cattle (Middlebere and Wych Lake)	Loss of reedbed, loss of valuable habitat for threatened species	

Rising sea levels

Prediction	Impact
Estimated 12–67 cm sea level rise in coming 80 years (Cook 2005)	Loss of reedbed and saltmarsh, erosion already evident on western beds, for example, Swineham Point (Cook 2001) Breach of freshwater bunds

Obstruction to site lines

Source	Impact	Linked effects
Construction of piers, jetties, telecoms towers + power lines	Visual range across area is lost for waders or Brent geese (Underhill-Day 2006)	Wildlife is more vulnerable to predation or abandons area completely

Disturbance

Source	Impact	Linked effects
Boats/recreational use of the adjoining waters	Birds may be prevented from feeding, roosting or nesting (West *et al.* 2002)	Birds may be forced to use less profitable or suitable sites, leading to negative energy balance or failed reproduction > competition on less disturbed sites

Erosion/siltation

Source	Impact	Linked effects
Aggregate/channel dredging	Siltation of saltmarsh creeks and gullies	Loss of nesting or feeding areas for birds such as Avocet *Recurvirostra avosetta* (on *Corophium volutator*) (Herbert *et al*. 2010) Increased tidal energy and pressure on sea defences when buffers are congested
Wash from passing boats	Erosion of saltmarsh substrate exacerbates effect of sea level rise	

LAGOON

Impacts

Commercial fisheries

Source	Impact	Linked effects
Overfishing of sand eel in harbour entrance and Poole Bay	< food for nesting Common Tern (English Nature 2000)	Poor reproductive success

Eutrophication – lagoons are particularly vulnerable due to low levels of flushing

Source	Impact	Linked effects
STW with outfall into Holes Bay + farm run-off via Frome and Piddle rivers into Wareham Channel	> levels chlorophyll *a* across harbour	Sudden changes in dissolved oxygen levels during die-offs can lead to fish and shellfish mortality (Underhill-Day 2006)
	> macroalgae *Ulva* spp. throughout harbour	Can smother mudflats and reduce diversity (Thornton *et al*. 2019)
	Japanese Wireweed *Sargassum muticum* (invasive species) – regarded as problem plant (blocks propellers, fouls nets) (Langston *et al*. 2003)	Can be biologically productive providing useful habitat for some species of invert such as gastropods/amphipods
	> Ammonia levels	Can be toxic to fish and shellfish (Langston *et al*. 2003)

Pollution

Source/ contaminant	Impact	Linked effects
Metals – historical trade discharges, Poole STW in Holes Bay, shipping (Underhill-Day 2006)	Fish and marine invertebrates (Hübner *et al*. 2010)	Poor recruitment
Pesticides/herbicides – discharges, sediment (Langston *et al*. 2003)	Crustaceans, fish	Possible endocrine disruption
Run-off from roads (Langston *et al*. 2003)	Nutrients, phosphates, detergents, sewage	

Disturbance

Source	Impact	Linked effects
Boats/recreational use of waters	Birds may be prevented from feeding, roosting or nesting in nearby areas	Birds may be forced to use less profitable or suitable sites, leading to negative energy balance or failed reproduction > competition on less disturbed sites (Ross 2013)

Abundant species

Chironomids (midges)	Irritation to visitors (Harrison *et al.* 2016)	
Greylag and Canada Geese	Nitrogen input	Algae growth
Tasselweed (*Ruppia* spp.)	Inconvenience to recreational users	

References

Barnes, T. 2005. Caloosahatchee Estuary conceptual ecological model. *Wetlands* 25: 884. https://doi.org/10.1007/BF03173126

Bertness, M.D., Brisson, C.P., Coverdale, T.C., Bevil, M.C., Crotty, S.M., and Suglia, E.R. 2014. Experimental predator removal causes rapid salt marsh die-off. *Ecology Letters* 17: 830–5. https://doi.org/10.1111/ele.12287

Bournemouth and Poole Council. 2009. *Poole Adopted Core Strategy.* Poole, Dorset.

Bournemouth and Poole Council. 2016–20. Poole park life: Lakes & Lagoon [online]. Poole: BCP Council. Available from: https://pooleprojects.net/pooleparklife/project-themes/lakes-drainage/ (accessed 10 March 2021).

Briggs, V.S., Mazzotti, F.J., Harvey, R.G., Barnes, T.K., Manzanero, R., Meerman, J.C., Walker, P., and Walker, Z. 2013. Conceptual ecological model of the Chiquibul/Maya Mountain Massif, Belize. *Human and Ecological Risk Assessment* 19: 317–40. https://doi.org/10.1080/10807039.2012.685809

Dee, L.E., Cowles, J., Isbell, F., Pau, S., Gaines, S.D., and Reich, P.B. 2019. When do ecosystem services depend on rare species? *Trends in Ecology & Evolution* 34(8): 746–58. https://doi.org/10.1016/j.tree.2019.03.010

Diaz, A., Pinn, E., and Hannaford, J. 2005. Ecological impacts of Sika Deer on Poole Harbour saltmarshes. In: Humphreys, J. and May, V. (eds) *The Ecology of Poole Harbour.* Elsevier, London, pp. 175–88. https://doi.org/10.1016/S1568-2692(05)80019-1

Edwards, B. 2005. The vegetation on Poole Harbour. In: Humphreys, J. and May, V. (eds) *The Ecology of Poole Harbour.* Elsevier, London, pp. 49–59. https://doi.org/10.1016/S1568-2692(05)80009-9

Gardiner, S., Nicholls, R., and Tanton, T. 2011. Management implications of Flood/Ebb tidal dominance: Its influence on saltmarsh and intertidal habitat stability in Poole Harbour. In: EDP Sciences (EDPS) (ed.) *Littoral 2010: Adapting to Global Change at the Coast: Leadership, Innovation and Investment,* London, 21–23 September 2010. EDP Sciences, Paris. https://doi.org/10.1051/litt/201106004

Gardiner, S.C. 2015. *Physical Drivers of Saltmarsh Change in Enclosed Microtidal Estuaries.* PhD Thesis. University of Southampton.

Gentile, J.H., Harwell, M.A., Cropper Jr., W., Harwell C.C., DeAngelis, D., Davis, S., Ogden, J.C., and Lirman, D. 2001. Ecological conceptual models: A framework and case study on ecosystem management for South Florida sustainability. *The Science of the Total Environment* 274: 231–53. https://doi.org/10.1016/S0048-9697(01)00746-X

Healey, M.C., Angermeier, P.L., Cummins, K.W., Dunne, T., Kimmerer, W.J., Kondolf, G.M., Moyle, P.B., Murphy, D.D., Patten, D.T., Reed, D.J., Spies, R.B., and Twiss, R.H. 2007. Conceptual models and adaptive management in ecological restoration: The CALFED Bay–Delta Environmental Restoration Program. 40 pp.

Liu, J., Dietz, T., Carpenter, S.R., Alberti, M., Folke, C., Moran, E., Pell, A.N., Deadman, P., Kratz, T., Lubchenco, J., Ostrom, E., Ouyang, Z., Provencher, W., Redman, C.L., Schneider, S.H., and Taylor, W.W. 2007. Complexity of coupled human and natural systems. *Science* 317: 1513. https://doi.org/10.1126/science.1144004

McInerney, G.J., Chen, M., Freeman, R., Gavaghan, D., Meyer, M., Rowland, F., Spiegelhalter, D.J., Stefaner, M., Tessarolo, G., and Hortal, J. 2014. Information visualization for science and policy: Engaging users and avoiding bias. *Trends in Ecology & Evolution* 29(3): 148–57. https://doi.org/10.1016/j.tree.2014.01.003

Medvinskii, A.B., Petrovskii, S.V., Tikhonova, I.A., Tikhonov, D.A., Venturino, E., Malchow, H., and Ivanitskii, G.R. 2002. Spatio-temporal pattern formation, fractals, and chaos in conceptual ecological models as applied to coupled plankton-fish dynamics. *Physics - Uspekhi* 45: 1. https://doi.org/10.1070/PU2002v045n01ABEH000980

Natural England. 2017. Natural England standard habitats regulations. Available from: https://www.legislation.gov.uk/uksi/2017/1012/contents/made (accessed 4 April 2022).

Ogden, J.C., Davis, S.M., Jacobs, K.J., Barnes, T., and Fling, H.E. 2005. The use of conceptual ecological models to guide ecosystem restoration in South Florida. *Wetlands* 24(4): 795–809. https://doi.org/10.1672/0277-5212(2005)025[0795:TUOCEM]2.0.CO;2

Suter, G.W. 1999. Developing conceptual models for complex ecological risk assessments. *Human and Ecological Risk Assessment* 5(2): 375–96. https://doi.org/10.1080/10807039991289491

Walters, C., Korman, J., Stevens, L.E., and Gold, B. 2000. Ecosystem modeling for evaluation of adaptive management policies in the Grand Canyon. *Conservation Ecology* 4: 2. https://doi.org/10.5751/ES-00222-040201

Watson, S., Herbert, R.J.H., Grandfield, F., and Newton, A. 2018. Detecting ecological thresholds and tipping points in the natural capital assets of a protected coastal ecosystem. *Estuarine and Coastal Shelf Science* 215: 112–23. https://doi.org/10.1016/j.ecss.2018.10.006

Wingard, G.L. and Lorenz, J.J. 2014. Integrated conceptual ecological model and habitat indices for the Southwest Florida coastal wetlands. *Ecological Indicators* 44: 92–107. https://doi.org/10.1016/j.ecolind.2014.01.007

Appendix References

Alves, J.A., Sutherland, W.J., and Gill, J.A. 2011. Will improving wastewater treatment impact shorebirds? Effects of sewage discharges on estuarine invertebrates and birds. *Animal Conservation* 2011: 1–9. https://doi.org/10.1111/j.1469-1795.2011.00485.x

Assessment (HRA) Standard. 2017. Available from: http://publications.naturalengland.org.uk/publication/8740045 (accessed 22 January 2019).

Caldow, R., Stillman, R.A., dit Durell, S.E., West, A.D., McGrorty, S., Goss-Custard, J.D., Wood, P.J., and Humphreys, J. 2007. Benefits to shorebirds from an invasion of non-native shellfish. *Proceedings of the Royal Society B-Biological Sciences* 274: 1449–55. https://doi.org/10.1098/rspb.2007.0072

Cefas. 2002. Development and application of a chronic sediment bioassay. Final Report to DEFRA. CW0832.

Cook, K. 2001. *Poole Harbour Reedbed Survey 2000*. Poole Harbour Study Group.

Cook, K. 2005. Physical and ecological aspects of the Poole Harbour reedbeds. In: Humphreys, J. and May, V. (eds) *The Ecology of Poole Harbour.* Elsevier, London. https://doi.org/10.1016/S1568-2692(05)80010-5

dit Durell, S.E. *et al.* 2006. Modelling the effect of environmental change on shorebirds: A case study on Poole Harbour, UK. *Biological Conservation* 131: 459–73. https://doi.org/10.1016/j.biocon.2006.02.022

Dyrynda, E.A. 1992. Incidence of abnormal shell thickening in the Pacific oyster *Crassostrea gigas* in Poole Harbour (UK), subsequent to the 1987 TBT restrictions. *Marine Pollution Bulletin* 24(3): 156–63. https://doi.org/10.1016/0025-326X(92)90244-Z

Dyrynda, P. 2005. Sub-tidal ecology of Poole Harbour – An overview. In: Humphreys, J. and May, V. (eds) *The Ecology of Poole Harbour.* Elsevier, London. https://doi.org/10.1016/S1568-2692(05)80013-0

English Nature. 2000. Poole Harbour European marine site – English nature's advice given under regulation 33(2) of the conservation (natural habitats &c.) regulations 1994. English Nature.

Ferriss, B.E., Conway-Cranos, L.L., Sanderson, B.L., and Hoberecht, L. 2019. Bivalve aquaculture and eelgrass: A global meta-analysis. *Aquaculture* 498: 254–62. https://doi.org/10.1016/j.aquaculture.2018.08.046

Harrison, A., Pinder, A., Herbert, R.J.H, O'Brien, W., Pegg, J., and Franklin, D. 2016. *Poole Park Lakes: Research and Monitoring.* BU Global Environmental Solutions (BUG) report to Borough of Poole. 96 pp.

Herbert, R.J.H., Ross, K., Hűbner, R., and Stillman, R.A. 2010. *Intertidal Invertebrates and Biotopes of Poole Harbour SSSI and Survey of Brownsea Island Lagoon.* Technical Report. Natural England, Sheffield.

Hübner R. 2009. Sediment chemistry – a case study approach. PhD, Bournemouth University.

Hübner, R., Herbert, R.J.H, and Astin, K.B. 2010. Cadmium release caused by the die-back of the salt-marsh cord grass *Spartina anglica* in Poole Harbour (UK). *Estuarine and Coastal Shelf Science* 84(4): 553–60. https://doi.org/10.1016/j.ecss.2010.02.010

Humphreys, J., Herbert, R.J.H., Harris, M., Jensen, A., Farrel, P., and Cragg, S. 2015. Introduction, dispersal and naturalisation of the Manila clam *Ruditapes philippinarum* in British estuaries 1980–2010. *Journal of the Marine Biological Association of the UK* 95(6): 1163–72. https://doi.org/10.1017/S0025315415000132

Kaiser, M.J., Edwards, D.B., and Spencer, B.E. 1996. Infaunal community changes as a result of commercial clam cultivation and harvesting. *Aquatic Living Resources* 9: 57–63. https://doi.org/10.1051/alr:1996008

Langston, W., Pope, N., Davey, M., Langston, K., O'Hara, S., Gibbs, P., and Pascoe, P. 2015. Recovery from TBT pollution in English Channel environments: A problem solved? *Marine Pollution Bulletin* 95(2): 551–64. https://doi.org/10.1016/j.marpolbul.2014.12.011

Langston, W.J, Chesman, B.S., Burt, G.R, Hawkins, S.J., Readman, J., and Worsfold, P. *Poole Harbour SPA.* Marine Biological Association, Plymouth. Report 12.

Malins, D.C., McCain, B.B., Landahl, J.T., Myers, M.S., Krahn, M.M., Brown, D.W., Chan, S.L., and Roubal, W.T. 1988. Neoplastic and other diseases in fish in relation to toxic chemicals: An overview. *Aquatic Toxicology* 11: 43–67. https://doi.org/10.1016/0166-445X(88)90006-9

Rittschoff, D. and McClellan-Green, P. 2005. Molluscs as multidisciplinary models in environment toxicology. *Marine Pollution Bulletin* 50(4): 369–73. https://doi.org/10.1016/j.marpolbul.2005.02.008

Ross, K.E. 2013. Investigating the physical and ecological drivers of change in a coastal ecosystem:

From individual to population-scale impacts. PhD Thesis. Bournemouth University. Available from: http://eprints.bournemouth.ac.uk/21351/1/Ross,Kathryn_Ph._D_2013.pdf (accessed 2 December 2019).

Thornton, A. 2016. The impact of green macroalgal mats on benthic invertebrates and over-wintering wading birds [online]. PhD Thesis. Bournemouth University. Available from: http://eprints.bournemouth.ac.uk/24874/ (accessed 29 November 2019).

Thornton, A., Herbert, R.J.H., Stillman, R.A., and Franklin, D.J. 2019. Macroalgal mats in a eutrophic estuarine marine protected area: Implications for benthic invertebrates and wading birds. In: Humphreys, J. and Clark, R.W.E. (eds) *Marine Protected Areas Science, Policy and Management*. Elsevier, Oxford, 703–64. https://doi.org/10.1016/B978-0-08-102698-4.00036-8

Underhill-Day, J.C. 2006. A condition assessment of Poole Harbour European Marine Site. Unpublished report, Footprint Ecology/Natural England. Dorset. England.

Watson, G.J., Murray, J.M., Schaefer, M., Bonner, A., and Gillingham, M. 2017. Assessing the impacts of bait collection on inter-tidal sediment and the associated macrofaunal and bird communities: The importance of appropriate spatial scales. *Marine Environmental Research* 130: 122–33. https://doi.org/10.1016/j.marenvres.2017.07.006

West, A.D., Goss-Custard, J.D., Stillman, R.A., Caldow, R.W.G., dit Durell, S.E., and McGrorty, S. 2002. Predicting the impacts of disturbance on shorebird mortality using a behaviour-based model. *Biological Conservation* 106: 319–28. https://doi.org/10.1016/S0006-3207(01)00257-9

PART II

Ecology

CHAPTER 5

The Planktonic Organisms of Poole Harbour

DANIEL J. FRANKLIN and ANDREW POWELL

Abstract

Planktonic organisms form the base of the marine food web and sustain fisheries. Poole Harbour has a rich and diverse planktonic assemblage which has been little studied. Intense blooms of the diatoms *Chaetoceros* and *Skeletonema*, and the haptophyte *Phaeocystis*, have been recorded in the harbour, and plankton production clearly helps to underpin the successful shellfish aquaculture activities of the harbour. We review the available data on the harbour plankton and provide an overview of the ways in which planktonic organisms interact with the economic activities of Poole Harbour.

Keywords: plankton, blooms, aquaculture, economics, Poole Harbour

Correspondence: dfranklin@bournemouth.ac.uk

Introduction

The term 'plankton' comes from the Greek πλαγκτός, meaning 'drifting' or 'wandering' organisms. The term 'plankton' was invented by Viktor Hensen in the 1800s and is an ecological term; it describes a lifestyle, or 'mode of living'. It carries no taxonomic information such as is denoted by 'mollusc' or 'arthropod'. Many taxonomic groups have representatives in the plankton for either all of their lifecycle ('holoplankton') or part of it ('meroplankton'). The mode of planktonic living is divided into two nutritional groups: photosynthetic, or plant-like, organisms ('phytoplankton') and non-photosynthetic, or animal, organisms ('zooplankton'). Further planktonic subdivisions are also possible such as 'tychoplankton' or 'accidental' plankton, that is, those that are brought into the planktonic realm by random disturbance. In the context of estuaries, tychoplanktonic organisms may be quite important, as many benthic organisms and normally attached organisms may be found, as well as those from the freshwater and terrestrial ecosystems linked to the estuary. Increasingly, the importance of 'mixotrophic' nutrition is studied within planktonic organisms whereby the organism combines both plant-like and animal-like feeding strategies simultaneously.

Poole Harbour is one of several ria-type estuaries in the UK, and as a temperate estuary subject to considerable nutrient inputs (and consequently designated as a 'eutrophic' and 'polluted' waterbody under environmental legislation) Poole Harbour likely shows very high rates of primary production. However, rates of primary production have never been

Daniel J. Franklin and Andrew Powell, 'The Planktonic Organisms of Poole Harbour' in: *Harbour Ecology*. Pelagic Publishing (2022). © Daniel J. Franklin and Andrew Powell. DOI: 10.53061/NXVP8477 measured in Poole Harbour, and our understanding of the Poole Harbour plankton is limited. Studies examining how plankton abundance varies over the year have demonstrated fairly typical patterns for this latitude with the spring bloom commencing around March–April and elevated/maximal abundances persisting throughout the growing season until October/ November, before declining to a winter minimum around February. As a tidal estuary subject to variable freshwater and nutrient inputs, there is likely considerable inter-annual variability between growing seasons, which could also, in part, be linked with climate variability.

In addition to a variety of short-term studies into the Poole Harbour plankton, the Environment Agency (EA) has been monitoring plankton abundance and diversity in Poole Harbour since 2007. Andrew Powell has also been assessing zooplankton abundance at one location in Poole Harbour since 2013. Poole Harbour has a long history of applied plankton research. In the 1950s a now demolished coal-fired power station on the shore of Holes Bay had an experimental facility geared towards 'sea farming': the use of flue gas and waste heat to drive the mass culture of phytoplankton in order to feed clams (e.g. Ansell *et al.* 1963).

The two objectives of this chapter are: (i) to assemble plankton taxa lists for Poole Harbour which can be used as a baseline for future monitoring, teaching and public engagement work, and (ii) to discuss the anthropogenic factors potentially at work in influencing the Poole Harbour plankton with reference to some of the principal economic activities in the harbour.

Data sources for a taxonomic assessment of the Poole Harbour plankton

For phytoplankton taxa the main data source is the EA monitoring, which commenced in 2007 in response to the EU Water Framework Directive (WFD). The EA dataset consists of monthly assessments of phytoplankton diversity at a small number of sampling locations within the harbour. Since 2007, the number of species, and the level of taxonomic discrimination for some groups, has altered within the lists that analysts use when categorising and quantifying the Poole Harbour plankton. The current version of the WFD phytoplankton taxa list (v7_Jan 2014) contains 92 diatom categories (of which 45 are species-level categories) and 102 dinoflagellate categories (of which 50 are species-level categories). Fifteen other categories are listed.

In addition to the EA dataset, since 2005 assessments have been made of a smaller number of potentially harmful algal taxa as part of the Food Standards Agency (FSA) biotoxin monitoring programme. Incidences of potentially harmful taxa in Poole Harbour from this dataset between 2001 and 2017 are summarised in Franklin *et al.* (2020). Several other short-term studies have been carried out over the years, some of which contained plankton identification and which we therefore extracted. These studies are lodged within the Poole Harbour Archive (for details, see http://www.pooleharbourstudygroup.org.uk/).

For zooplankton, the main data source is a monitoring programme carried out by Andrew Powell since 2012 at Lake Pier in Poole Harbour (Powell 2012; Powell 2013). We know of only two other published Poole Harbour zooplankton studies: Dyrynda (1989) and Barbuto *et al.* (2005).

Taxa lists for the planktonic organisms of Poole Harbour

The assembled taxonomic information on the phytoplankton of Poole Harbour is presented for the diatoms (Tables 5.1 and 5.2), dinoflagellates (Tables 5.3 and 5.4) and all other

Table 5.1 Species-level identification of diatoms in Poole Harbour

Taxa	Data source/reference	Notes
Achnanthes longipes	Harris 1983, EA	Varieties of *A. longipes* are recognised. Synonym of *Achnanthes armillaris*?
Actinocyclus octonarius	Dyrynda 1989	
Amphiprora alata	Harris 1983	Listed as '*Amphipora alata*'. *A. arenaria, A. ostrearia, A. hyalina, A. spectabilis* and *A. ventricosa* also listed but some of these are uncertain taxa
Asterionella formosa	Franklin *et al.* 2012, EA	Freshwater species from river inflow
Asterionellopsis glacialis	Franklin *et al.* 2012, EA	
Asterionellopsis kariana	EA	Synonym: *Asteroplanus karianus*
Bacillaria paxillifer	Franklin *et al.* 2012, Harris 1983	Varieties of *B. paxillifer* are recognised
Bacterosira bathyomphala	Harris 1983	Listed as '*Bacteriosira fragilis*'; *B. bathyomphala* is the currently accepted name
Biddulphia alternans	Franklin *et al.* 2012, EA	Synonym: *Trigonium alternans*
Caloneis brevis	Harris 1983	Varieties of *C. brevis* are recognised. *C. subsalina* also listed by Harris (1983)
Campyloneis grevillei	Harris 1983	Varieties of *C. grevillei* are recognised
Cerataulina pelagica	Franklin *et al.* 2012, EA	
Chaetoceros holsaticus	Drynda 1989	
Chaetoceros pseudocrinitus	Drynda 1989	Listed as *C. pseudocritinium*
Corethron hystrix	EA	
Corethron pennatum	EA	
Coscinodiscus granii	Franklin unpubl.	
Coscinodiscus wailesii	Powell unpubl.	
Dactyliosolen antarcticus	EA	
Dactyliosolen fragilissimus	Franklin *et al.* 2012, EA	
Detonula confervacea	EA	
Diatoma tenue	EA	var. *elongatum*
Ditylum brightwellii	Franklin *et al.* 2012, Drynda 1989, EA	
Eucampia zodiacus	Franklin *et al.* 2012, EA	
Grammatophora marina	EA	
Guinardia delicatula	Franklin *et al.* 2012, EA	
Guinardia flaccida	EA	
Guinardia striata	Franklin *et al.* 2012, EA	
Heliotheca tamesis	Franklin *et al.* 2012, Drynda 1989, EA	Also listed as '*Streptotheca thamensis*' (Drynda 1989)
Lauderia annulata	Franklin *et al.* 2012, EA	
Leptocylindrus danicus	Franklin *et al.* 2012, EA	EA also: cf. *danicus*
Leptocylindrus mediterraneus	EA	
Leptocylindrus minimus	Franklin *et al.* 2012, EA	EA also: cf. *minimus*
Lithodesmium undulatum	Franklin unpubl., EA	
Mediopyxis helysia	Franklin unpubl., EA	
Meridion circulare	EA	Varieties of *M. circulare* are recognised
Meuniera membranacea	EA	
Neocalyptrella robusta	EA	
Odontella aurita	Harris 1983, EA	Listed as '*Biddulphia aurita*', currently accepted name given. Varieties of *O. aurita* are recognised. *B. pulchella* also listed by Harris (1983) = *B. biddulphiana*? Regarded as a benthic taxa
Odontella mobiliensis	EA	
Odontella sinensis	Franklin unpubl., EA	
Paralia sulcata	Franklin *et al.* 2012, EA	
Podosira stelligera	EA	
Pleurosigma angulatum	Drynda 1989	Varieties of *P. angulatum* are recognised

(Continued)

Table 5.1 (Continued) Species-level identification of diatoms in Poole Harbour

Taxa	Data source/reference	Notes
Proboscia alata	EA	
Rhaphoneis amphiceros	EA	Varieties of *R. amphiceros* are recognised
Rhizosolenia imbricata	EA	
Rhizosolenia setigera	EA	Also *Rhizosolenia setigera* f. *pungens* listed (EA)
Rhizosolenia styliformis	Drynda 1989, EA	Varieties of *R. styliformis* are recognised
Skeletonema costatum	Drynda 1989, EA	
Stephanopyxis turris	EA	Varieties of *S. turris* are recognised
Striatella unipunctata	Franklin *et al.* 2012, EA	Epiphytic, stalked
Thalassionema nitzschioides	EA	Name uncertain

Environment Agency (EA) routine monitoring dataset (04/2007–05/2019) including species and taxa identified from the Marine Phytoplankton (2007–11) and also the Marine Phytoplankton truncated lists (2012–19). Food Standards Agency (FSA) datasets are also consulted – see Franklin *et al.* (2020) for a summary of this dataset. Franklin unpubl: mainly water sampling in November 2011 with identification comments from A. Kraberg (Alfred-Wegener Institut, Germany) as well as other scattered observations. T. Harris (1983) = The Holes Bay Ecological Survey, 1982/1983.

Table 5.2 Genus-level identification of diatoms in Poole Harbour

Taxa	Data source/reference	Notes
Actinoptychus	Franklin *et al.* 2012	
Asteromphalus	EA	
Bacteriastrum	EA	
Bellerochea	EA	
Chaetoceros (*phaeoceros*)	EA, Franklin *et al.* 2012	*Phaeoceros* is a subgenus. *Chaetoceros* contains many species
Chaetoceros (*hyalochaete*)	EA, Franklin *et al.* 2012	*Hyalochaete* is a subgenus. The genus *Chaetoceros* contains many species
Cyclotella	EA	
Cylindrotheca	EA	EA categories: *Ceratoneis/Nitzschia closterium/longissima* and *Cylindrotheca closterium/Nitzschia longissima* also present in datasets
Diploneis	EA	
Fragilaria	EA	
Gyrosigma/Pleurosigma	EA	
Hemiaulus	EA	
Lennoxia	EA	
Licmophora	Franklin *et al.* 2012, EA	Epiphytic, stalked
Melosira	Franklin *et al.* 2012, EA	
Navicula	EA	
Nitzschia	EA	
Plagiogrammopsis	EA	
Pseudo-nitzschia sp.	Franklin *et al.* 2012, EA, FSA	Potentially harmful (amnesic shellfish poisoning)
Rhizosolenia sp.	Franklin *et al.* 2012	
Thalassiosira sp.	Franklin *et al.* 2012, EA	

Genera are only listed in the case that a species-level identification (see Table 5.1) within these genera is not found in the other sources consulted, or additional information is made within one source (see text).

phytoplankton taxa (Table 5.5). The quality of taxonomic information within these tables is subject to variation in taxonomic ability between studies, taxonomic name changes over time and changing confidence in the level of discrimination possible within a variable group of analysts using different equipment. Nevertheless, these tables serve as a guide to all the taxonomic entities that are thought to have been encountered during the sampling of Poole Harbour. Tables 5.6, 5.7 and 5.8 present the zooplankton information from the studies of Powell. Samples were, and are, collected from Lake Pier (Hamworthy, Poole),

Table 5.3 Species-level identification of dinoflagellates in Poole Harbour.

Taxa	Data source/reference	Notes
Akashiwo sanguinea	EA	
Amphidoma caudata	EA	
Ceratium furca	EA	
Ceratium lineatum	EA	
Ceratium minutum	EA	
Ceratium tripos	EA	
Gyrodinium spirale	EA	
Heterocapsa niei	EA	
Heterocapsa triquetra	EA, Franklin *et al.* 2012	EA category: *Heterocapsa minima/Azadinium/Amphidoma* also present in datasets
Karenia mikimotoi	EA, FSA	
Noctiluca scintillans	EA	
Peridinium quinquecorne	EA	
Polykrikos schwartzii	EA	
Prorocentrum cordatum	EA, FSA	Potentially harmful. EA: *cordatum/balticum* also listed. FSA: *minimum* also listed
Prorocentrum gracile	EA	
Prorocentrum lima	EA, FSA	Potentially harmful
Prorocentrum micans	EA	Potentially harmful
Prorocentrum triestinum	EA	
Protoperidinium bipes	EA	
Protoperidinium depressum	EA	
Torodinium robustum	EA	

Table 5.4 Genus-level identification of dinoflagellates in Poole Harbour

Taxa	Data source/reference	Notes
Alexandrium	EA, FSA	Potentially harmful (paralytic shellfish poisoning)
Amphidinium	EA	
Dinophysis	EA, FSA	Potentially harmful. Category changed to Dinophysiaceace in later EA lists
Diplopsalis	EA	
Gonyaulax	EA	
Gyrodinium	EA	
Katodinium	EA	
Oxytoxum	EA	
Scrippsiella	EA	*Scrippsiella/Pentapharsodinium* also listed in EA datasets

Genera are only listed in the case that a species-level identification (see Table 5.3) within these genera is not found in the other sources consulted, or additional information is made within one source (see text).

where the high tide gives sufficient flow and depth of water to allow measured horizontal tows of a 0.3 m diameter, 250 μm plankton net. Tide profiles are obtained from United Kingdom Hydrographic Office. The length of line used, together with the calculated area of the net mouth, allows the volume of seawater filtered to be calculated. All organisms in the samples are identified to the appropriate taxonomic group using mostly live (unfixed) samples (Powell 2012; Powell 2013).

Poole Harbour plankton: Interactions with economic activities

Plankton dynamics are mainly of economic interest due to their impact on shellfish growth. In Poole Harbour, a variety of shellfish are either collected (clams and cockles, which achieved Marine Stewardship Council and Responsible Fisheries Scheme certification in 2018) or grown on aquaculture beds (oysters and, to a much lesser extent, mussels). These

Table 5.5 Species- and genus-level identifications of other photosynthetic microbes in Poole Harbour

Taxa	Data source/reference	Higher taxonomic group
Actinastrum	EA	Chlorophyte
Anabaena	EA	Cyanobacteria
Ankistrodesmus	EA	Chlorophyte
Chattonella	EA	Raphidophyte
Crucigenia tetrapedia	EA	Chlorophyte, also *Crucigenia* listed
Cryptomonas	EA	Cryptophyte
Desmodesmus	EA	Chlorophyte
Dictyocha fibula	EA	Dictyochophyte, also *Dictyocha* listed
Dictyocha speculum	EA	Dictyochophyte
Dinobryon	EA	Chrysophyte
Euglena	EA	Euglenophyte
Gonium	EA	Chlorophyte
Mesodinium rubrum	EA	Ciliate (mixotrophic)
Microcystis	EA	Cyanobacteria
Pediastrum	EA	Chlorophyte
Phaeocystis	EA, Powell/Franklin unpubl.	Haptophyte
Rhodomonas	EA	Cryptophyte
Scenedesmus	EA	Chlorophyte
Treubaria	EA	Chlorophyte

Table 5.6 Holoplanktonic zooplankton taxa recorded in Poole Harbour

Taxa	Data source/reference	Notes
Copepods		
Acartia tonsa	Dominant calanoid copepod Introduced in the 1930s (Conover 1957)	Throughout year 'Blooms' 10,000+ individuals per m³ in April and July
Caligidae		Rare
Monstrilla		Rare
Gnathia		Rare
Cladocerans		
Podon sp.		Rare
Evadne sp.		Rare
Bosmina sp.		Rare
Chaetognaths		
Parasagitta setosa		Rare/occasional throughout year
Cnidaria		
assorted small unidentified		Occasional throughout year
Aurelia ephyra		Rare
Appendicularians/Larvaceans		
Oikopleura dioica		Occasional throughout year
Ctenophores		
Pleurobrachia pileus		Occasional throughout year
cydippid larvae		Rare

shellfish are all filter feeders, which use their siphons to move water over their gills. Food (plankton) is thereby strained from the surrounding water and, using cilia, passed to the mouth. Before reaching the mouth palps are used to sort potential food particles. Oysters, clams and mussels show selectivity in particle ingestion by recognising and rejecting non-nutritious or inorganic particles through the use of their palps to sort items. Rejected particles are combined together as 'pseudofaeces' and periodically expelled. Clams and cockles, which can live fully or partially buried in the sediment, may also use their palps to directly gather food from the sediment ('deposit feeding').

Table 5.7 Meroplanktonic zooplankton taxa recorded in Poole Harbour

Taxa	Data source/reference	Notes
Cirrepede cyprids		
Austrominius modestus	Introduced in the 1940s (Bishop 1947)	Abundant February to November High numbers 100s per m³ April
Cirrepede nauplii		
Austrominius modestus	Introduced in the 1940s (Bishop 1947)	Abundant February to November High numbers 100s per m³ April
Decapod larvae		
Carcinus maenas		Zoea throughout year, megalopa only summer months
Liocarcinus sp.		Occasional
Inachus dorsettensis		Rare (seen once 29/10/18)
Polychaete larvae		
Spionidae		Most of year, March to October
Phyllodocidae *Harmothoe*		Occasional
Nereid nectochaetes		Rare
Lanice conchilega Aulophora		Occasional
Epitokous Syllidae		Occasional
Nephthys hombergi		Rare
Gastropod veligers		
Crepidula fornicata	Introduced in the late 1800s (McMillan 1938)	Very abundant April to November 1,000s per m³ June to August
Bivalve veligers		
various unidentified species		Rare
Fish eggs		Occasional February to November
Fish larvae		
various unidentified species		Occasional February to November
Tunicate larvae		
Styela clava	Davis 2007, Macleod 2016	Occasional
Bryozoan cyphonaute larvae		
Membranipora sp.		Occasional
Phoronida	Actinotrocha larva of tubicolous phoronid	
Phoronis psammophila	Adult horseshoe worm adults recorded (Dyrynda 1984)	Rare once 29/3/17

Table 5.8 Tychoplanktonic zooplankton taxa recorded in Poole Harbour.

Taxa	Data source/reference	Notes
Cumaceans		Rare
Mysids		Rare
Amphipods	First recorded Poole harbour 2004 (Arenas 2006)	Rare
Caprella mutica		
Tanaids	Dr R. Bamber pers. comm. February 2013	Rare
Tanais dulangii		
Pycnogonids	Dr R. Bamber pers. comm. October 2012	Rare
Anoplodactylus sp.		
Pelagic gastropods	Frequent throughout year	
Peringia ulvae		

A great range of both phytoplankton and zooplankton organisms will be nutritionally significant for the Poole Harbour shellfish, and the frequent very high abundances ('blooms') of certain phytoplankton taxa, such as the diatoms *Skeletonema* and *Chaetoceros* and the haptophyte *Phaeocystis* (Figure 5.1), are likely to provide exceptionally rich periods of both filter and, possibly also, deposit feeding for harbour shellfish. It is not known if the adult shellfish of Poole Harbour display feeding preferences within the great variety of phytoplankton and zooplankton organisms that they encounter. Research into shellfish feeding often concentrates on feeding in the younger life-history stages of shellfish as production of these stages can be a bottleneck in aquaculture operations. It is clear from the fact that Poole Harbour yields about one-third of the total UK oyster production that the nutritional/plankton quality here is currently good for shellfish production as are the physical conditions for oyster cultivation. Climate variability, and now anthropogenic ocean warming, drives changes in the species composition of the plankton. Around the sea regions ('ecoregions') of the British Isles some inter-regional variability is apparent. Data describing the overall diatom, dinoflagellate and copepod abundances for the sea regions adjacent to Poole Harbour are summarised in the periodic Marine Climate Change Impacts Partnership (UK government) report cards. This project uses continuous plankton record-ers on ships to monitor the composition of plankton around the British Isles.

Some of the phytoplankton species found in Poole Harbour are potentially harmful as they may contain toxins ('biotoxins'), which, if ingested by people after accumulating in shellfish, can lead to serious poisoning. *Escherichia coli* and other pathogenic bacteria, whose abundance is linked to sewage (combined sewer overflows) discharges, are also occasionally found in harbour shellfish, and local authority detection of these pathogens can trigger restrictions on fishery activities. Norovirus can also be a significant problem. Consumers can be protected from the possibility of ingesting algal toxins via both depu-ration (the cleaning of collected/harvested shellfish) and from the statutory monitoring work that is carried out by the FSA/CEFAS (UK government). This monitoring involves assessments of both the presence of specific types of phytoplankton and the presence of biotoxins in bivalve flesh. Compared to some similar UK estuarine systems the incidence of harmful species and biotoxins seems relatively low in Poole Harbour (Franklin *et al.* 2020). Significant feeding pressure must be exerted by the extensive areas of shellfish beds in Poole Harbour, and the potential capacity of aquaculture operations to mitigate harmful algal blooms is increasingly being considered in aquaculture design optimisa-tion efforts (Brown *et al.* 2019). Intensification of the UK shellfish aquaculture industry has been called for (DEFRA 2015); it would be important that if undergoing such intensi-fication, assessment is made of how the design of aquaculture operations can potentially reduce the risk of biotoxin accumulation in shellfish. This may be ever-more important given that a recent long-term study (Belgian sector of the North Sea) noted a pronounced increase in the abundance of harmful diatom and dinoflagellate genera from the 1970s to the 2000s (*Nohe et al.* 2020). Such work underlies the value of long-term datasets in understanding changes in plankton composition. At present, for Southern England, this type of work is restricted to the Western Channel Observatory, a long-term monitoring site off Plymouth.

Other economic activities such as tourism, specifically aquatic leisure activities, can also be linked to plankton dynamics if plankton blooms lead to perceptions of reduced amenity among the general public. However, there appears to be little evidence that this happens in Poole Harbour, despite occasional significant phytoplankton blooms here (and in Poole Bay; Figure 5.1). *Phaeocystis* blooms, in particular, can lead to changes in water colour, odour, as well as foam accumulations. *Phaeocystis* blooms are visible by satellite and can be obvious to beach-goers and water-users. Due to the emission of a volatile trace gas (dimethyl sulphide), there can be a sulphurous odour associated with blooms, similar to the sulphurous odour of anoxic sediments. General concerns over such blooms include potential gill clogging in captive finfish and short-term impacts on oxygen dynamics. Low dissolved oxygen can occasionally present challenges in the summer to Poole Harbour aquaculture operations (Wordsworth pers. comm.), although in a year-long monitoring effort (2011–12) dissolved oxygen was almost always above EU shellfish-directive limits in the lower part of Poole Harbour (Franklin *et al.* 2012). However, dissolved oxygen levels in some parts of Poole Harbour have been raised as a concern before (Environment Agency

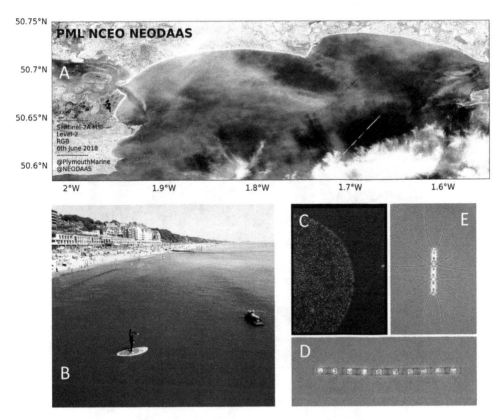

Figure 5.1 Prominent phytoplankton taxa in Poole Harbour: (A) An image from the satellite Sentinel-2 showing a *Phaeocystis* bloom in Poole Harbour (and Poole Bay) in June 2018. *Phaeocystis* blooms are a recurrent feature (Thanks to the NERC Earth Observation Data Acquisition and Analysis Service (NEODAAS)). (B) Water discolouration due to a *Phaeocystis* bloom at Bournemouth beach (Thanks to John Hourston) and (C) A microscopic image of a *Phaeocystis globosa* colony. Diatoms form recurrent and very abundant blooms in Poole Harbour, principally composed of the chain-forming diatoms *Skeletonema* (D) and *Chaetoceros* (E) (Courtesy of Robin Raine, National University of Ireland, Galway (NUIG)).

1997, 2001). The major factors influencing estuarine dissolved oxygen would be warming, nutrient inputs and river discharge (Iriarte *et al*. 2010). Knowledge of the dissolved oxygen dynamics in Poole Harbour is somewhat rudimentary and represents a good target for increased monitoring in the future in tandem with higher resolution studies of the patterns and trends in biological productivity.

Lastly, as a commercial and ferry port, and very busy leisure boating destination, Poole Harbour is now home to many introduced (non-native) species. Among these are some of the most commercially important species (e.g. Manila Clam *Ruditapes philippinarum*, Pacific Oyster *Magallana gigas*). The Slipper Limpet *Crepidula fornicata* (Figure 5.2) is also widespread and its larvae are often very dominant in zooplankton samples (Table 5.7). Other relatively recent larger arrivals include the Japanese Skeleton Shrimp *Caprella mutica*, which grows prolifically on benthic structures (Table 5.8), potentially providing a useful food source to seahorses and the seaweed *Agarophyton vermiculophylla*. The diatoms *Coscinodiscus wailesii* and *Odontella sinensis* are thought to be introduced species, as is the Copepod *Acartia (Acanthacartia) tonsa* (Table 5.6) and the barnacle *Elminius (Austrominius) modestus*. These introductions have undoubtedly modified the ecology of Poole Harbour.

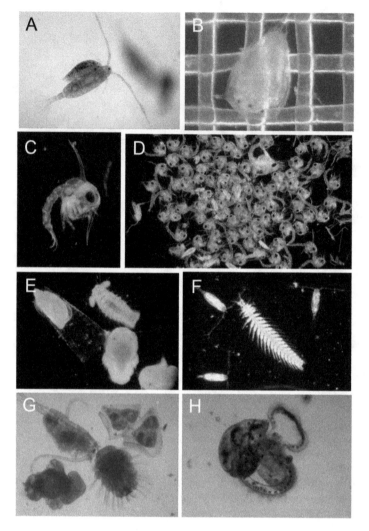

Figure 5.2 Prominent zooplankton taxa in Poole Harbour: (A) The Copepod *Acartia tonsa* with attached isopod parasite. (B) Cirrepede nauplius larva (*Austrominius modestus*) introduced to UK waters in the 1940s on 250 µm mesh (C) and (D) Decapod larvae zoea (*Carcinus maenas*) (E) Polychaete larvae including aulophora larva of *Lanice conchilega* (F) epitokous planktonic stage of Syllidae male polybostrichus (G) Bryozoan cyphonaute larvae with actinotrocha larva of *Phoronis* sp. and metatrochophore *Harmothoe* sp. (H) Gastropod veliger (*Crepidula fornicata*) introduced to UK waters in the 1880s (Andrew Powell).

The Non-native Species Secretariat (http://www.nonnativespecies.org//home/index.cfm?) maintains information on these species.

Conclusions and outlook

Plankton communities are complex and variable, and – given their importance at the base of the food web – there exists a great need for consistent long-term observations (Lombard *et al.* 2019). In Poole Harbour, where plankton production underpins a nationally important shellfish production site, there is considerable scope for expanding the monitoring of plankton production so that existing, and planned, aquaculture activities can be better designed. Important within any future monitoring activities will be the inclusion of citizen science initiatives and shellfish producers such that local people are able to gain a better understanding of the natural processes and cycles that underpin the ecological functioning of Poole Harbour and thereby support local food production.

References

Ansell, A.D., Raymont, J.E.G., Lander, K.F., Crowley, E., and Shackley, P. 1963. Studies on the mass culture of *Phaeodactylum*. II. The growth of *Phaeodactylum* and other species in outdoor tanks. *Limnology and Oceanography* 8: 184–206. https://doi.org/10.4319/lo.1963.8.2.0184

Arenas, F., Bishop, J., Carlton, J., Dyrynda, P., Farnham, W., Gonzalez, D., Jacobs, M., Lambert, C., Lambert, G., Nielsen, S., Pederson, J., Porter, J., Ward, S., and Wood, C. 2006. Alien species and other notable records from a rapid assessment survey of marinas on the south coast of England. *Journal of the Marine Biological Association of the United Kingdom* 86: 1329–37. https://doi.org/10.1017/S0025315406014354

Barbuto, P.C., Pinn, E.H., and Jensen, A.C. 2005. Summer distribution of Zooplankton in Poole Harbour. In: Humphreys, J. and May, V. (eds) *The Ecology of Poole Harbour: Proceedings in Marine Science 7*. Elsevier, pp. 131–8. https://doi.org/10.1016/S1568-2692(05)80014-2

Bishop, M. 1947. Establishment of an immigrant barnacle in British coastal waters. *Nature* 159: 501–2. https://doi.org/10.1038/159501a0

Brown, A.R., Lilley, M., Shutler, J., Lowe, C., Artioli, Y., Torres, R., Berdalet, E., and Tyler, C.R. 2019. Assessing risks and mitigating impacts of harmful algal blooms on mariculture and marine fisheries. *Reviews in Aquaculture* 12: 1663–88. https://doi.org/10.1111/raq.12403

Conover, R.J. 1957. Notes on the seasonal distribution of Zooplankton in Southampton Water with special reference to the genus *Acartia*. *Annals and Magazine of Natural History* 10: 63–7. https://doi.org/10.1080/00222935708655927

Davis, M., Boero, F., Olenin, S., Lützen, J., and Davis, M. 2007. The spread of *Styela clava* Herdman, 1882 (Tunicata, Ascidiacea) in European waters. *Aquatic Invasions* 2: 378–90. https://doi.org/10.3391/ai.2007.2.4.6

DEFRA. 2015. United Kingdom multiannual national plan for the development of sustainable aquaculture. Available from: https://assets.publishing.service.gov.uk/government/uploads/system/uploads/attachment_data/file/480928/sustainable-aquaculture-manp-uk-2015.pdf (accessed 1 August 2021)

Dyrynda, P. 1984. Poole harbour subtidal survey – Southern sector. Report to Nature Conservancy Council.

Dyrynda, P. 1989. Marine biological survey of the bed and waters of Holes Bay, Poole Harbour Dorset 1988 report to Dorset county council survey report: Plankton sampling section H.

Environment Agency. 1997. Candidate sensitive area (eutrophic) and polluted waters (eutrophic). Poole Harbour. Environment Agency, SW region.

Environment Agency. 2001. Candidate sensitive area (eutrophic) and polluted waters (eutrophic). Poole Harbour. Environment Agency, SW region.

Franklin, D.J., Herbert, R.J.H., Chapman, I., Willcocks, A., Humphreys, J., and Purdie, D.A. 2020. Consequences of nitrate enrichment in a temperate estuarine marine protected area; response of the microbial primary producers and consequences for management. In: Humphreys, J. and Clark, R. (eds) *Marine Protected Areas: Science, Policy and Management*. Elsevier, pp. 685–702. https://doi.org/10.1016/B978-0-08-102698-4.00035-6

Franklin, D.J., Humphreys, J., Harris, M., Jensen, A.C., Herbert, R.J.H., and Purdie, D.A. 2012. An investigation into the annual cycle of phytoplankton abundance in Poole Harbour and its relationship with Manila clam nutrition. Report to the Marine Management Organisation.

Harris, T. 1983. The Holes Bay ecological survey. Report to the nature conservancy council.

Iriarte, A., Aravena, G., Villate, F., Uriarte, I., Ibanez, B., Llope, N., and Stenseth, N.C. 2010. Dissolved oxygen in contrasting estuaries of the Bay of Biscay: Effects of temperature, river discharge and chlorophyll a. *Marine Ecology-Progress Series* 418: 57–71. https://doi.org/10.3354/meps08812

Lombard, F., Boss, E., Waite, A.M., Vogt, M., Uitz, J., Stemmann, L., Sosik, H.M., Schulz, J., Romagnan, J-B., Picheral, M., Pearlman, J., Ohman, M.D., Niehoff, B., Moller, K.O., Miloslavich, P., Lara-Lpez, A., Kudela, R., Lopes, R.M., Kiko, R., Karp-Boss, L., Jaffe, J.S., Iversen, M.H., Irisson, J-O., Fennel, K., Hauss, H., Guidi, L., Gorsky, G., Giering, S.L.C., Gaube, P., Gallager, S., Dubelaar, G., Cowen, R.K., Carlotti, F., Briseno-Avena, C., Berline, L., Benoit-Bird, K., Bax, N., Batten, S., Ayata, S.D., Artigas, L.F., and Appeltans, W. 2019. Globally consistent quantitative observations of planktonic ecosystems. *Frontiers in Marine Science*. https://doi.org/10.3389/fmars.2019.00196

Macleod, A., Cook, E.J., Hughes, D., and Allan, C. 2016. Investigating the impacts of marine invasive non-native species. A report by Scottish

Association for Marine Science Research Services Ltd for Natural England & Natural Resources Wales, p. 59. Natural England Commissioned Reports, Number223.

McMillan, N.F. 1938. Early records of *Crepidula fornicata* in English waters. *Proceedings of the Malacological Society of London* 23: 236. https://doi .org/10.1093/oxfordjournals.mollus.a064357

Nohe, A., Goffin, A., Tyberghein, L., Lagring, R., De Cauwer, K., Vyverman, W., and Sabbe, K. 2020. Marked changes in diatom and dinoflagellate biomass, composition and seasonality in the Belgian Part of the North Sea between the 1970s and 2000s. *Science of the Total Environment* 716: 136316. https://doi.org/10.1016/j.scitotenv.2019 .136316

Powell, A. 2012. *Seasonal Changes in Abundance and Diversity of Mesozooplankton in Poole Harbour*. MSc dissertation. Bournemouth University.

Powell, A. 2013. Seasonal variation of zooplankton in Poole Harbour. *Porcupine Marine Natural History Society Bulletin* 34: 55–60.

CHAPTER 6

Intertidal and Lagoon Macrofauna and Macroflora of Poole Harbour

ROGER J.H. HERBERT, RICHARD A. STILLMAN, KATHRYN E. ROSS, ANN THORNTON, ALICE E. HALL, JESSICA BONE, LEO CLARKE, ELENA CANTARELLO and PHILIP PICKERING

Abstract

Poole Harbour, on the south coast of England, is one of Europe's largest estuaries and is an important economic hub and centre for waterborne recreation. Here we describe intertidal and lagoon habitats, document an inventory (2002–21) of macrofauna and flora and make comparisons with earlier reviews. Since 2002, over 300 species have been recorded within intertidal mudflats, sandflats, lagoon habitats and upon hard substrata, approximately 9% of which are not native to the British Isles. These local stocks and assets comprised of living things represent part of the harbour's 'Natural Capital' and provide ecosystem services. Time-series data since the 1970s are presented that indicate the range of population variability of important species, with some trends apparent. These can help managers determine the likelihood of 'tipping points' within the ecosystem. Robust and consistent benthic monitoring protocols combined with high resolution imagery of the harbour will improve the quality of data and aid the interpretation of future surveys.

Keywords: intertidal, lagoon, time series, monitoring, benthic, estuary, English Channel

Correspondence: rherbert@bournemouth.ac.uk

Introduction

In an era of significant environmental change, it is important to collate and document biological records to which future assessments and evaluations may be compared. As we understand more clearly the interdependence of biodiversity and the benefits of ecosystems to human wealth, health and wellbeing, the concept of 'Natural Capital' is helpful. In the context of Poole Harbour this represents local stocks and assets comprised of living things. Within intertidal areas and lagoons, these 'stocks' are at an increasing risk of decline as a result of sea level rise, development pressures and pollution. Over the past two decades, to comply with statutory monitoring objectives, there have been several major surveys of the intertidal and lagoon biodiversity of Poole Harbour, yet project requirements have not extended to determine which species are 'new' to the harbour and which have not been seen for some years. Therefore, the purpose of this study is to document a current

Roger J.H. Herbert, Richard A. Stillman, Kathryn Ross, Ann Thornton, Alice E. Hall, Jessica Bone, Leo Clarke, Elena Cantarello and Philip Pickering, 'Intertidal and Lagoon Macrofauna and Flora of Poole Harbour' in: Harbour Ecology. Pelagic Publishing (2022). © Roger J.H. Herbert, Richard A. Stillman, Kathryn Ross, Ann Thornton, Alice E. Hall, Jessica Bone, Leo Clarke, Elena Cantarelloand Philip Pickering. DOI: 10.53061/OPZF8277 inventory (2002–21) of intertidal and lagoon macrofauna and flora and where possible to make some comparisons with earlier work. Earlier reviews of the intertidal macrofauna (defined here as organisms greater than 0.5 mm) are from 1972 to 2002 (Caldow *et al.* 2005) and specifically for Holes Bay from 1991 to 2002 (Bowles and English 2005). Subtidal fauna was usefully described by Dyrynda (2005) and fisheries by Jensen *et al.* (2005). A list of the flora and fauna of the Poole Harbour brackish lagoons was incorporated within Herbert *et al.* (2019), but records are also included here due to their proximity to intertidal habitats and relative ease of access. To ensure that the species inventory is as comprehensive as possible we have also sought information from local recorders and natural history groups who are referred to within the Acknowledgements section.

Poole Harbour

At high tide, Poole Harbour has an area of approximately 3,600 ha, making it one of Europe's largest lowland estuaries (Humphreys and May 2005). The harbour comprises extensive intertidal and subtidal habitats, saltmarshes, saline lagoons, seagrass beds and several islands, of which Brownsea Island is the largest (Figure 6.1). Several small rivers and streams discharge into the south and west of the harbour near the town of Wareham, the navigable limit. From 1980 to 2009, the area of intertidal sediments (excluding saltmarsh) was found to be relatively constant at ~1000–1367 ha (Herbert *et al.* 2010).

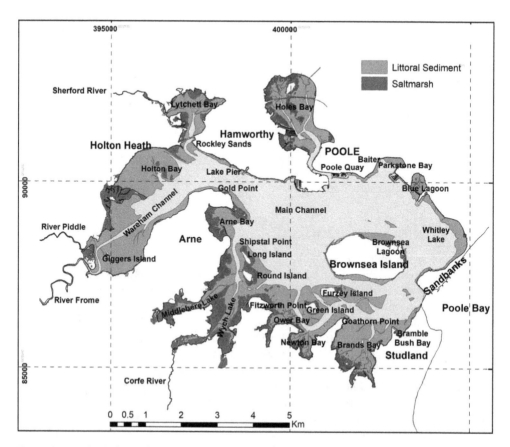

Figure 6.1 Poole Harbour showing the extent of intertidal sediment and saltmarsh habitats
© Crown Copyright and database right (2010) Ordnance Survey Licence Number 1000022021.
Saltmarsh & Sediment data from East Dorset Habitat map © Environment Agency.

The harbour consists of one main basin and two smaller basins in the north, Lytchett Bay and Holes Bay, and is classified as microtidal, having a spring tide range of 1.8 m and a neap tide range of 0.6 m. There is a double high water in the harbour which produces a relatively long stand at high tide. The entrance is particularly narrow, with fast tidal streams, and despite the low tidal range about 45% of the water leaves the harbour on a spring ebb tide (Humphreys 2005). Above the intertidal areas are four saline lagoons; all are artificial with the largest being Brownsea Island Lagoon (Herbert *et al.* 2010; 2019). Water is retained in the lagoons at low tide and is generally of much more variable salinity compared to the main harbour.

Poole Harbour is of international importance for nature conservation, and areas are designated as Sites of Special Scientific Interest (SSSI), a Special Protection Area (SPA) under the EU Birds Directive and also, as an important wetland under the Ramsar Convention. The harbour supports an internationally important assemblage of over 20,000 waterfowl that feed within intertidal and lagoon habitats, including nationally or internationally important populations of Shelduck, Black-tailed Godwit, Avocet, Redshank, Little Egret and Spoonbill (see chapter in this volume by Ross *et al.* for further details).

The catchment of the harbour is mainly agricultural, and the chalk rivers Frome and Piddle enter the navigable harbour at Wareham. The growing town of Poole lies immediately to the north of the harbour, where there is a wide range of marine industries including boat building and several marinas. Oil and gas are extracted from the Wytch Farm oil field (Perenco UK Ltd.), and continental ferries and small freight vessels operate from Poole Port. Maintenance dredging is managed by Poole Harbour Commissioners, with the last major capital dredge carried out during the winter of 2005/2006. Within the harbour there is an important fishery for Manila Clam *Ruditapes philippinarum* (Jensen *et al.* 2005; Birchenough *et al.* 2019) and a small bait-dragging fishery for King Ragworm *Alitta virens* (Birchenough 2013). Shellfish lease beds occur to the west of Brownsea Island, where Mussels *Mytilus edulis* and Pacific Oysters *Magallana gigas* are cultivated. There is also considerable recreational use of the harbour by small vessels including dinghy sailing, kayaking, kite surfing and wind surfing.

Methods – Biological survey

Data has been extracted from surveys and projects (Table 6.1) between the navigable limit at Wareham and the chain ferry at the harbour entrance. These include surveys of soft sediments, lagoons and artificial hard substrata. The major surveys focused on the sediments in the harbour, which make up the bulk of intertidal habitat. These surveys were conducted between the extreme high water spring tide and the extreme low water spring tide mark and included brackish lagoon habitats but excluded saltmarshes. Assessing the condition of intertidal sediment habitats within the SSSI was the principal objective of the 2002, 2009 and 2017 surveys commissioned by Natural England. The 2002 and 2009 surveys also calculated the biomass of macrofauna available to feeding species of waterfowl in different parts of the harbour. The surveys in 2002 and 2009 utilised an 80-point 500 × 500 m sampling grid (detailed in Thomas *et al.* 2004; Caldow *et al.* 2005; Herbert *et al.* 2010).

Benthic core samples from intertidal soft sediments were washed over a 500 μm sieve, and invertebrates were 'picked' under a low power stereo binocular microscope. Quality assurance procedures included checking by an experienced taxonomist to ensure adequate removal of fauna, including juvenile polychaetes and oligochaetes. All fauna were identified to species level where possible. Faunal counts within core samples were converted to densities per square metre. Additional core samples were taken for sediment particle size analysis. Organic matter content was determined through loss on ignition. Between

Table 6.1 Sources of data and information on intertidal and lagoon surveys of Poole Harbour, 2002–2021

Survey no.	Location /project	Fieldwork dates	No. benthic core sites	Cores per site	Core size	Core depth (cm)	Notes and references
1	SSSI condition assessment	September–October 2002	80	1	0.01 m²	30	Thomas et al. (2004) Report to Natural England; Caldow et al. (2005)
2	EU WFD monitoring	April 2008	31	3	0.01 m² ×3 combined	15	Environment Agency https://data.gov.uk/dataset/76963b49-a391-4370-a02f-ae3dc8445915/marine-benthic-invertebrate-species#licence-info
3	SSSI condition assessment	September 2009–March 2010	80	5	0.01 m²	15	Herbert et al. (2010) Report to Natural England
4	Middlebere Creek	September 2010–April 2011	15	3	0.01 m²	15	Ross (2013) Bournemouth University PhD
5	EU WFD monitoring	April 2011	25	3	0.01 m² × 3 combined	15	Environment Agency https://data.gov.uk/dataset/76963b49-a391-4370-a02f-ae3dc8445915/marine-benthic-invertebrate-species#licence-info
6	Welly zone	2012–14	6	36	0.01 m²	15	Dorset Wildlife Trust
7	Ower Bay	September 2013, December 2013,	6	36	0.01 m²	15	Thornton (2016) Bournemouth University PhD
	Brands Bay	September 2014	6	36	0.01 m²	15	
	Holes Bay		6	36	0.01 m²	15	
8	EU WFD monitoring	April 2015	20	3	0.01 m² ×3 combined	15	Environment Agency https://data.gov.uk/dataset/76963b49-a391-4370-a02f-ae3dc8445915/marine-benthic-invertebrate-species#licence-info
9	Pottery Pier	June 2015					National Trust BioBlitz
10	Seagrass (Zostera) survey	July 2015					Envision (2015)
11	Wytch Channel	June and November 2015	72	1	0.01 m²	30	Clarke (2017) Bournemouth University PhD
12	Pottery Pier, Brownsea Island southern shore, Castle Pier, Sandbanks pier and wall	May 2016					R.J.H. Herbert

Survey no.	Location /project	Fieldwork dates	No. benthic core sites	Cores per site	Core size	Core depth (cm)	Notes and references
13	SSSI condition assessment	November 2017	12	1	0.01 m²	15	Herbert et al. (2018) Natural England
14	Baiter Point	April 2017					Conchological Society of GB and Ireland S. Trewhella
15	Baiter Point and Holes Bay	October 2018					
16	Baiter Point	October 2018					D. Fenwick
17	Banks Road sea wall, Sandbanks	November 2019 and 2020					J Bone, A. Lamb, R.J.H. Herbert
18	Brownsea Lagoon	November 2009 2010–11 2017	6 15	5 3	0.01 m² 0.01 m²	15 15	Herbert et al. (2009) Ross (2013) Bournemouth University PhD Harrison et al. (2018)
19	Arne Lagoon	August 2012, 2013, 2017	4	1	0.01 m²	15	Herbert et al. (2019)
20	Poole Park Lagoon	2015–16 2017	2	1	0.01 m²	15	Harrison et al. (2016); Herbert et al. (2019) Bone (2017)
21	Brownsea Island Lagoon SSSI con- dition assessment	September 2015	12	3	0.01 m²	15	Thomas and Worsfold (2016) for Natural England
22	Seymers Lagoon	November 2016, April 2017	10	1	0.01 m²	15	Bone (2017); Herbert et al. (2019)
23	Brownsea Island and Sandbanks	March 2020					National Trust
24	Banks Road sea wall, Sandbanks	December 2020					J Bone
25	Arne Moors, Wareham	April 2021	30	3	0.01 m²	15	Cesar (2021)
26	Holes Bay	January–June 2021. Algal mat samples and core samples	18 in March and June	6 at three tidal levels	0.01 m²	30	MarineScope Taxonomy (EU Interreg Project RanTrans)
27	Baiter Point	November 2021					R.J.H. Herbert
28	Sandbanks chain ferry slipway	November 2021					R.J.H. Herbert
29	Blue Lagoon	November 2021					R.J.H. Herbert

Table 6.2 Poole Harbour salinity (ppt), January 2011–November 2017 (Poole Bay shellfish water, 2011–13)

Site name	Grid ref	Mean	Max	Min
Poole Harbour 1 Wareham Channel Buoy 82	SY96308909	23.5	32.6	1.2
Poole Harbour 3 (Hutchins Buoy 71)	SY99438929	29.5	34.4	11.1
Poole Harbour 6 (near buoy no 36)	SZ03518920	32.1	35.2	25.7
Poole Harbour 12 (South Deep)	SZ01628664	31.3	34.6	17.3
Poole Bay shellfish water	SZ09508900	34.3	35.1	32.5

Source: Environment Agency.

January 2011 and November 2017, monthly surface salinity data were collected by the Environment Agency at four long-term monitoring sites in the harbour and one within Poole Bay (Table 6.2). Observations on harder substrata such as the clay shards at Pottery Pier at the western end of Brownsea Island and on structures at Castle Pier and Sandbanks were quantified using an abundance scale based on Crisp and Southward (1958).

Using data from the 2009 survey (Herbert *et al.* 2010), a biotope classification was assigned to each site based on its similarity with those defined in the Marine Habitat Classification System of Britain and Ireland v0405 (Conner *et al.* 2004). This hierarchical system combines both physical habitat characteristics with the observed biological community. Biotopes were assigned using assemblage nMDS data output from PRIMER (Clarke and Gorely 2015), sediment analysis and expert judgement. If an appropriate biotope could not be reasonably assigned, a conservative approach was adopted by moving up a tier in the hierarchy.

Data were also extracted from PhD theses and known consultancy reports. In addition to authors' own surveys, records were also provided by Dorset Wildlife Trust, members of the Conchological Society of Great Britain and Ireland, local naturalists and recorders. Several verified records were extracted from the National Biodiversity Network database. Taxonomy and nomenclature followed that of the World Register of Marine Species (WoRMS: www.marinespecies.org). Brief details of other surveys, including those on hard substrata, and sampling locations are indicated in Table 6.1 and Figure 6.2.

Results

Salinity and temperature

There is a gradient in surface salinity along the length of the harbour, with lowest mean values in the Wareham Channel and the highest values at the harbour entrance (Table 6.2). Reduced salinities can occur at any time of year but are most extreme in the winter months, for example in January 2013 when a salinity of 1.2 ppt was recorded in the Wareham Channel. The greatest range in surface salinity is also recorded in the Wareham Channel, and this decreases towards the harbour entrance. A gradient in mud interstitial salinity data obtained in 2011 showed lowest values in the Wareham Channel (min 12 ppt, mean 25 ppt) and highest at the seaward end around Newtons Bay (30–35 ppt).

General patterns and associations

Since 2002, 317 species taxa have been recorded within intertidal and lagoon habitats in Poole Harbour. These comprise 44 plants and algae, 258 invertebrates and 15 fish. Of the invertebrates, polychaete worms are the most numerous group, which is not surprising

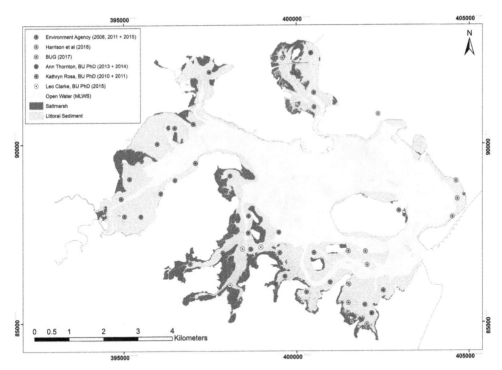

Figure 6.2 Locations of sampling points within intertidal soft sediment habitats and lagoons (Environment Agency and Bournemouth University (2008–17). Not all sites were sampled by the Environment Agency each year. For the large-scale surveys in 2002 and 2009 (Table 6.1), see Caldow *et al.* (2005) and Herbert *et al.* (2010) respectively.

considering the prevalence of soft sediment habitats. More fish are likely to forage over intertidal habitats at high water than have been recorded in samples or pools at low tide. The list of seaweeds should be regarded as provisional as to the best of knowledge this group has not been previously documented and requires further survey and research. Overall, species distribution correlates well with particle size of harbour sediments (Figure 6.3). Species characteristic of sands, such as the annelid worm *Scoloplos armiger*, Lugworm *Arenicola marina* and amphipod *Urothoe* spp., are prevalent near the harbour entrance and around some of the island shores. Polychaete and bivalve assemblages are common throughout the harbour and are primarily differentiated by proximity to areas of reduced salinity (Figure 6.4). The most diverse sites occurred south of Baiter and Parkstone Bay and in the Wareham Channel, whereas areas of lower diversity occurred on the sandflats and on shores south of Brownsea Island (Figure 6.5).

In sheltered creeks where periods of reduced salinity are frequent, such as upper Wareham Channel, upper Holes Bay, Brands Bay and Middlebere Creek, the isopod *Cyathura carinata*, Mud Scud *Corophium volutator* and chironomid larvae can achieve their highest densities. The most ubiquitous species is the Common Ragworm *Hediste diversicolor* that was prevalent in all muddy sediments. An association with the small bivalve *Macoma balthica* characterises the majority of intertidal mudflat habitat in the Wareham Channel, though densities are relatively low.

Sites characterised by an assemblage consisting of the Catworm *Nephtys hombergii*, high densities of oligochaetes and the polychaete *Streblospio shrubsolii* were found to the south of Brownsea Island and in the vicinity of Brands Bay (Figure 6.4). These areas have tentatively been assigned the sub-biotope LS.LMu.UEst.NhomStr (*Nephtys hombergii* and *Streblospio shrubsolii* in intertidal mud). Because of an imprecise match, separation

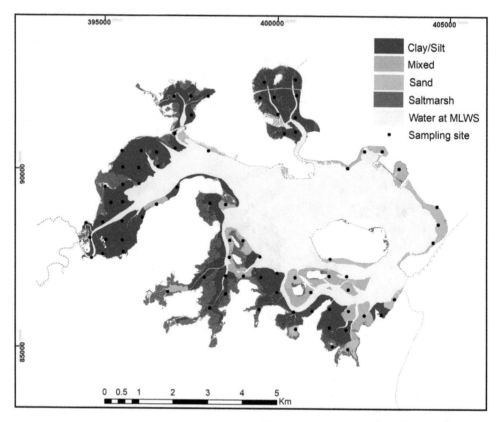

Figure 6.3 Distribution of sediments (Herbert *et al.* 2010). Clay/Silt: >45% of sample weight, <63 μm; Sand: >45% of sample weight, >125 μm; Mixed: >45% of sample weight, 63–125 μm (Crown Copyright and database right (2010) Ordnance Survey Licence No. 100022021. Sediment data from East Dorset Habitat map © Environment Agency, 2010).

of these assemblages from other upper estuarine groups is really only valid in a local context. These biotopes are sometimes indicative of sub-optimal or stressful conditions, perhaps associated with high nutrient levels or macroalgae, and the area is worth continued monitoring.

Mixed-sediment habitats have not been well captured by the systematic surveys. The area just below Baiter in Parkstone Bay is particularly diverse (Figure 6.5) and so is the shore at Pottery Pier on the western end of Brownsea Island (Figure 6.6). The substratum of this shore, which consists of clay pottery shards, must be unique in the British Isles and in many ways is similar to that found on some rocky shores. In surveys in 2015 and 2016, moderate densities of the barnacles *Semibalanus balanoides* and *Austrominius modestus* were recorded on the surface of the shards, while on the undersides there were exceptionally large numbers of Edible Periwinkle *Littorina littorea*, Shore Crabs *Carcinus maenas* and Shanny fish *Lipophrys pholis*. A light settlement of Pacific Oyster occurred on the shards and occasional specimens of the protected Native Oyster *Ostrea edulis* were recorded at extreme low water. The locally distributed southern nudibranch *Aeolidiella alderi* may also be found beneath stones and clay shards.

True intertidal hard substrata in the harbour are limited to port infrastructure and sea defences, which have not yet been thoroughly investigated. Around Castle Pier on Brownsea Island, the limpet *Patella vulgata* and periwinkles *Melarhaphe neritoides* and *Littorina saxatilis* may be found, along with red seaweeds *Catanella caespitosa* and *Chondrus crispus*. Since

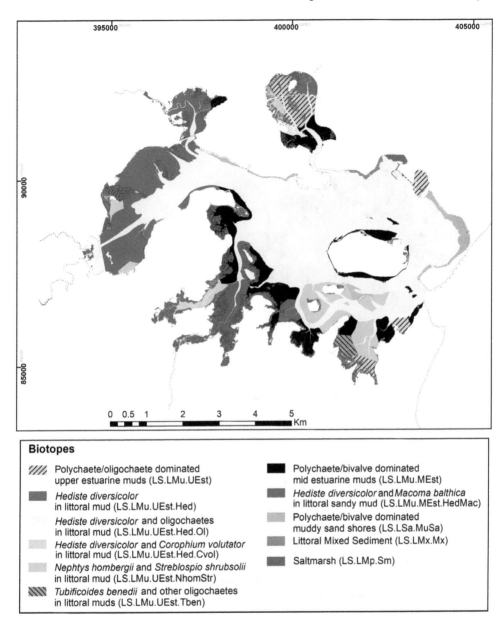

Biotopes

Polychaete/oligochaete dominated upper estuarine muds (LS.LMu.UEst)

Hediste diversicolor in littoral mud (LS.LMu.UEst.Hed)

Hediste diversicolor and oligochaetes in littoral mud (LS.LMu.UEst.Hed.Ol)

Hediste diversicolor and *Corophium volutator* in littoral mud (LS.LMu.UEst.Hed.Cvol)

Nephtys hombergii and *Streblospio shrubsolii* in littoral mud (LS.LMu.UEst.NhomStr)

Tubificoides benedii and other oligochaetes in littoral muds (LS.LMu.UEst.Tben)

Polychaete/bivalve dominated mid estuarine muds (LS.LMu.MEst)

Hediste diversicolor and *Macoma balthica* in littoral sandy mud (LS.LMu.MEst.HedMac)

Polychaete/bivalve dominated muddy sand shores (LS.LSa.MuSa)

Littoral Mixed Sediment (LS.LMx.Mx)

Saltmarsh (LS.LMp.Sm)

Figure 6.4 Intertidal biotopes of Poole Harbour (from Herbert *et al*. 2010). Biotopes classified according to *The Marine Habitat Classification for Britain and Ireland. v04 05* (Connor *et al*. 2004). Areas of seagrass (*Zostera marina*) have been excluded (see Envision (2015) for details) (© Crown Copyright and database right (2010) Ordnance Survey Licence Number 1000022021).

around the turn of the millennium, Pacific Oysters have naturalised on dock walls near the RNLI buildings and on other structures. Brown seaweeds, including Spiral Wrack *Fucus spiralis* and Bladder Wrack *F. vesiculosus*, occur on sea walls including the outer wall of Brownsea Island Lagoon. A particularly rich biodiversity hotspot is the walled remains of an old swimming pool (Figure 6.6) and associated debris on the shore at Baiter, where numerous species have been recorded including a variety of tunicates, the limpet *P. depressa*, barnacle *Chthamalus montagui*, anemones *Actinia equina* and *Anemonia viridis* and a persistent population of the large chiton Velvety Mail-shell *Acanthochitona fascicularis* (Figure 6.7).

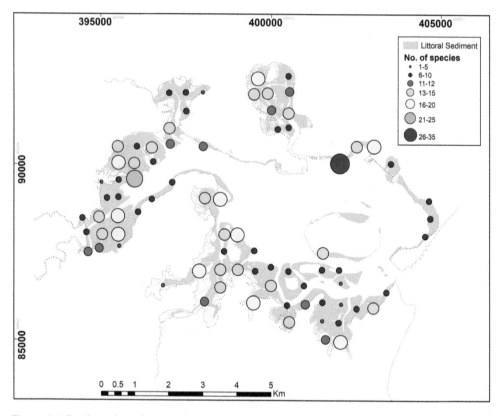

Figure 6.5 Total number of species found at each sampling station in 2009 (Herbert *et al.* 2010).

The fauna of the artificial lagoons on Brownsea Island, Arne and Poole Park Lake is variable yet characterised by site-specific assemblages that include British lagoonal specialists, such as the Starlet Sea Anemone *Nematostella vectensis*, amphipods *Monocorophium insidiosum* (Figure 6.8) and Spire Shell *Ecrobia ventrosa* (Herbert *et al.* 2019). Improved sluice flushing at Poole Park Lake may increase the biodiversity of the lagoon, which has both Tassleweed species *Rupia cirrhosa* and *R. maritima*.

Species associations within the Eelgrass *Zostera marina* habitats present at Whitley Lake and the Salterns area, which extend into subtidal areas (Envision 2015), have not been investigated and this would be worthy of further study.

Time series

A time series for intertidal macrofauna was first presented in Caldow *et al.* (2005) and is extended here in Table 6.3 and specifically for Holes Bay (Table 6.4). Here follows a brief comment on some of the more important species and groups from data obtained in 2009 and subsequent surveys.

Common Ragworm *Hediste diversicolor*

The mean abundance of ragworm has been remarkably stable over the years. However, in 2009 there was a significantly higher density in the southern and eastern bays compared to the previous survey in 2002. Although ubiquitous on mudflats, the largest populations occur in the Wareham Channel where densities can exceed 4000 ind. m^{-2}.

Figure 6.6 (a) Remains of the old seawater swimming pool at Baiter, now a wonderful rock pool! 12 September 2018 (Photo D. Fenwick), (b) Benthic survey with the RNLI, October 2009, (c) Pump-scoop fishing disturbance on mudflats in Arne Bay. Circular scars 5–15 m across (Fearnley *et al.* 2013; Clarke 2017), (d) The shore at Pottery Pier Brownsea Island, a unique substratum. May 2016 (Photo R.J.H. Herbert).

Figure 6.7 (a) Strawberry Worm *Eupolymnia nebulosa* beneath stones on lower shore at Baiter (Photo D. Fenwick), (b) Polychaete worm *Amblyosyllis spectabilis*, (c) King Ragworm *Alitta virens* (Photo S. Birchenough), (d) Large chiton *Acanthochitona fascicularis* beneath rocks in old swimming pool at Baiter, 12 September 2018 (Photo D. Fenwick).

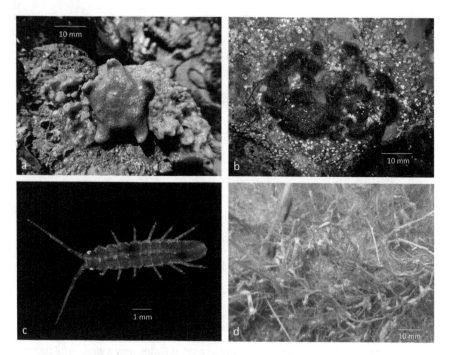

Figure 6.8 (a) Cushion Star *Asterina gibbosa* on the western shore of Brownsea Island, April 2020 (Photo P. Pickering), (b) Non-native Red Ripple Bryozoan *Watersipora subatra* on shore at Baiter, 12 September 2018 (Photo D. Fenwick), (c) *Idotea chelipes* (Photo S. Trewhella), (d) Epifauna upon *Agarophyton vermiculophyllum* within Brownsea Lagoon in August 2017: Starlet Sea Anemone *Nematostella vectensis* with tubes of amphipod *Monocoropium insidiosum*. A few plants of tassleweed *Ruppia* sp. are also visible (Photo R.J.H. Herbert).

Table 6.3 Time-series comparison of mean density (ind. m⁻²) of key species and major groups of macroinvertebrates in intertidal flats of Poole Harbour, 1972–2015

Species/group	Year						
	1972	1987	2002	2008	2009	2011	2015
Alitta virens			81	16	8	28	5
Hediste diversicolor			615	1,437	650	841	565
Nephtys hombergii			47	72	75	60	63
Sedentary polychaetes	2,023	6,570	4,909		925	3,073	4,568
Oligochaeta (*Tubificoides* spp.) and Nemertea	6	1,884	1,841	13,584	1,466	7,414	3,325
Abra tenuis			254	821*	113	2,945	3,973
Cerastoderma spp.			30	33	24	28	22
Macoma balthica	42	10	1	12	9	23	12
Mya arenaria			10	8	29	44	85
Peringia (*Hydrobia*) *ulvae*	214	135	756	1,846	490	2,337	3,055
Ruditapes philippinarum	0	0	5	26	13	23	28
Scrobicularia plana	53	239	8	78	5	5	28
Corophium volutator	1,540	2,882	374	92	468	48	4
All crustaceans (excl. Ostracoda, Copepoda)	1,577	3,059	1,804	583	800	204	137

*Data provided refers to juv. *S. plana,* but this is most likely to be *A. tenuis* and is shown here.

Data pre-2008 from Table 5 in Caldow *et al.* (2005); data for 2008, 2011, 2015 from Environment Agency surveys collected from approximately 30 stations using similar protocol; data from 2009 Natural England survey (Herbert *et al.* 2010, Appendix 4 and Table 7).

Table 6.4 Time-series comparison of mean density (ind. m^{-2}) of key species and major groups of macroinvertebrates on intertidal flats in Holes Bay, Poole Harbour

Species/group	Year						
	1972	1987	1991	2002	2009	2011	2015
Errant polychaetes	619	123	2,750	992	168	2,555	467
Sedentary polychaetes	5,008	177	582	3,491	64	4,722	1,517
All worms (inc. oligochaeta)	5,627	506	3,908	5,093	836	15,865	2,666
Macoma balthica	77	1	1	0	6	11	33
Scrobicularia plana	32	0	4	6	0	0	0
All bivalves	146	5	785	300	27	1,833	2,883
Peringia ulvae	31	0	253	1,013	647	2,322	850
Corophium volutator	2,457	50	623	13	647	67	0
All crustaceans	2,265	53	1,135	848	144	122	0
All individuals	8,069	563	6,081	7,254	414	40,751	6,799
Number of samples	21	22	31	19	45	9	6

Data pre-2009 from Caldow *et al.* 2005); data from 2009 in Appendix 4 and Table 7 (Herbert *et al.* 2010); data for 2011 and 2015 from Environment Agency surveys (Sites 13, 14, 15).

Data excludes Ostracoda, Copepoda and Nematoda.

Catworm *Nephtys hombergii*

Highest densities of Catworm are found mainly in the southern and eastern inlets around Newton, Ower and Brands bays. Mean densities are remarkably stable at around 60–70 ind. m^{-2}.

King Ragworm *Alitta virens*

Although there were higher densities of the King Ragworm (Figure 6.7) recorded in 2011 (28 ind. m^{-2}), numbers remain relatively low compared to 2002. However, as the burrowing capability of the species is significant, it is possible that densities could have been underestimated due to sampling methods used.

Worms – Oligochaetes

Densities of oligochaetes, largely comprised of the genus *Tubificoides*, were much higher in 2011 and 2015 than they were in the previous survey in 2009 but not as great as in 2008. Highest densities were recorded in the Wareham Channel, the entrance to Holes Bay and the southern inlets around Brands Bay.

Worms – Sedentary polychaetes

The density of sedentary polychaete worms has increased significantly since 2009 when values had fallen. These comprise mainly small Cirratulidae and Spionidae. In addition to the much higher densities it should be noted that the Environment Agency data refers to the most abundant species as *Tharyx* sp., whereas the earlier survey identified this as *Aphelochaeta marioni*. There remains some uncertainty in the separation of these species among taxonomists (Thomas and Worsfold 2016).

Laver Spire Shell *Peringia ulvae*

Mean densities have increased markedly since the earlier surveys. Formerly known as *Hydrobia ulvae*, this gastropod snail is widespread and is now very characteristic of intertidal mudflats throughout the harbour. Highest densities were recorded in the southern inlets around Brands Bay.

White Furrow Shell *Abra tenuis*

In both 2011 and 2015, recorded densities of this small bivalve were significantly greater than all previous studies. However, it is possible that smaller individuals may have been overlooked in the past. The species is widespread with highest densities recorded in the Wareham Channel.

Cockles *Cerastoderma* spp.

Densities of cockles have remained fairly constant throughout all previous surveys (~20–30 ind. m^{-2}). Distribution is patchy with highest values at the entrance of Lytchett Bay and Holes Bay. Although the Lagoon Cockle *C. glaucum* has been recorded in the harbour, the two species Cockle *C. edule* and *C. glaucum* have not been differentiated in the time series. However, numbers of undetermined Cardidae are referred to separately in Environment Agency data and it is presumed, though not confirmed, that these are juvenile/spat.

Peppery Furrow Shell *Scrobicularia plana*

Mean densities of this clam were significantly higher in 2015 than in 2011. These values may indicate a continued recovery following a decline as a result of persistent residues of tributyltin (TBT) antifouling paints.

Sand Gaper *Mya arenaria*

The density of this clam continues to rise in the harbour. The species is widespread, with abundance greatest within the Wareham Channel. There is no indication of clam size in Environment Agency data and therefore overall biomass may be low, as reported by Herbert *et al.* (2010).

Manila Clam *Ruditapes philippinarum*

The mean density of Manila Clam has been relatively stable from 2011 to 2015 at approximately 25 ind. m^{-2}. This is higher than measured in 2009, when the surveys included a larger part of the harbour. Highest densities were found in the Wareham Channel and Arne Bay.

Baltic Tellin *Macoma balthica*

There has been a slight increase in mean density of this clam; however, values most likely fall within the range of natural variability. The species is most common in the Wareham Channel, where densities of 60–100 ind. m^{-2} were recorded.

European Mud Scud *Corophium volutator*

Mean densities of this burrowing amphipod continue to fall. In 2009, highest densities were found in Middlebere Creek, Holes Bay and in the upper part of the Wareham Channel. The density in the upper part of Middlebere Creek, obtained in April 2011 by Ross (2013), remained as high. Although lower densities have been recorded, the overall distribution is largely unchanged in the main harbour; however, none were found in Holes Bay in 2015.

Other crustacea

Apart from *C. volutator*, the only other native macroinvertebrate of any numerical significance is the isopod *Cyathura carinata*. This species is characteristic of upper sections of creeks and most notably in the Wareham Channel. The mean harbour density in both 2011 and 2015 was very similar and comparable with 2009 (82 ind. m^{-2}).

Non-native species

Twenty-seven species classified as non-native (GBNNS 2018) were recorded on the harbour's shores and lagoons during the period 2002–21, approximately 9% of benthic species (Table 6.5). Some of these were found in sediment core samples while others were observed during ad hoc field visits over the period. The assessment of change in abundance is based on a comparison of data obtained from core samples, casual observation and referenced studies. The most notable recent record is the red alga *Agaraphyton vermiculophyllum* (Figures 6.8 and 6.9), native to the Pacific, which is now present in Brownsea Island Lagoon and elsewhere in the harbour. This was previously recorded as *Gracilariopsis longissima* (Herbert *et al.* 2010); however, its non-native status, which was suspected by Thomas and Worsfold (2016), has now been confirmed by genetic analysis (Krueger-Hadfield *et al.* 2017). These records, together with the extensive population in Christchurch Harbour and specimens from the Kingsbridge Estuary in Devon, represented the first observations along the English coast (Krueger-Hadfield *et al.* 2017). The protected Starlet Sea Anemone found in the saline lagoons, is now regarded as non-native (GBNSS 2020), following genetic analysis (Sheader *et al.* 1997; Reitzel *et al.* 2008). However, its inclusion on the IUCN Red List of vulnerable species is currently unchanged (MarLin 2020). The ribbon worm *Cephalothrix simula*, which can produce high levels of the marine neurotoxin Tetrodotoxin (Turner *et al.* 2018), was found beneath rocks at Baiter in 2018. This is only the second record for this species in the UK and like many non-natives its Pacific origins may suggest importation via container vessels or from oysters; hence it may have been in the harbour unnoticed for decades. Wild populations of the Pacific Oyster are currently small; however, there has been some more dense settlement in the Blue Lagoon at the eastern end of the harbour (Mills 2016).

Naturalisation of the Manila Clam on intertidal sediment has continued (Humphreys *et al.* 2015; Clarke *et al.* 2019a) and densities continue to increase. The status of non-native species known to be present on floating platforms and pontoons was not assessed, but see Arenas *et al.* 2006. Some of these species, such as the brown kelp *Undaria pinnatifida*, initially established on structures prior to colonising natural habitats. The non-native status of some taxa is cryptogenic (e.g. *Polydora cornuta* and *Tharyx* 'species A'), and confirmation of the identity of polychaetes in the genus *Streblospio* would be helpful (Thomas and Worsfold 2016).

Discussion

The composition and condition of intertidal sediment assemblages have generally remained similar over the past two decades. The habitat is dominated by annelid worms supplemented by lower densities of molluscs and crustaceans. The updated time series of mean densities of the more common species reveals substantial variability, as expected, especially for some annelid taxa. As concluded by Caldow *et al.* (2005), 'change is the norm' in Poole Harbour. Of course, there have been very significant changes in sampling

Table 6.5 Non-native species recorded on shores and lagoons in Poole Harbour, 2002–21

Species	Details	Change since 2002
Agarophyton vermiculophyllum (red alga)	Established in Brownsea Island Lagoon and now confirmed in Holes Bay and elsewhere in the harbour	Increasing. Thomas and Worsfold (2016); Krueger-Hadfield *et al.* (2017); Harrison *et al.* 2018; Environment Agency
Ammothea hilgendorfi (sea spider)	Common on mixed-sediment shores, particularly Pottery Pier on Brownsea Island	Unknown
Amphibalanus improvisus (Bay Barnacle)	Generally rare, on hard substrata: Wareham Channel and Sandbanks	No significant change
Austrominius modestus (barnacle)	Common on shells and stones at mixed-sediment shores, for example, Pottery Pier	No significant change
Botrylloides diegensis (sea squirt)	Beneath rocks in old swimming pool at Baiter Point	Unknown
Botrylloides violaceus (sea squirt)	Beneath rocks in old swimming pool at Baiter Point	Unknown
Bugula neritina (sea mat)	Beneath rocks in old swimming pool at Baiter Point	Unknown
Caulacanthus okamurae (red alga)	On Banks Road sea wall, Sandbanks	Unknown
Cephalothrix simula (ribbon worm)	Beneath rocks at Baiter Point	First recorded by David Fenwick in September 2018. Identification confirmed by CEFAS from DNA analysis
Colpomenia peregrina (Oyster Thief, brown alga)	On foreshore in Blue Lagoon	Unknown
Corella euymota (sea squirt)	Beneath rocks in old swimming pool at Baiter Point	Unknown
Crepidula fornicata (Slipper Limpet)	Mainly subtidal but can be common on mixed-sediment shores at extreme low water mark	Stable/unknown
Desdemona ornata (polychaete worm)	Common in cores from southern inlets and Brownsea Island Lagoon	Possibly increasing
Eusarsiella zostericola (ostracod)	Prevalent in core samples in western harbour and Holes Bay	Unknown
Ficopomatus enigmaticus (tube worm)	Poole Park Lagoon, can be abundant	Harrison *et al.* 2016; Herbert *et al.* 2018
Grateloupia turuturu (red alga)	Found commonly where there is mixed sediment, particularly Pottery Pier on Brownsea Island and at Baiter Point	Possibly increasing
Magallana gigas (Pacific Oyster)	Light, wild settlement on mixed sediment shore at Pottery Pier and on sea walls. Higher densities on mudflats at low water inside entrance to Blue Lagoon	Stable/possibly increasing. Mills (2016); R.J.H. Herbert personal observations
Mercenaria mercenaria (American Hard Clam)	Found occasionally at Pottery Pier	Stable, unknown
Mya arenaria (Sand Gaper Clam)	Naturalised across harbour	Stable/possibly increasing
Nematostella vectensis (Starlet Sea Anemone)	Abundant in Brownsea Island Lagoon, Poole Park Lagoon. Also found in Seymers Lagoon on Brownsea Island	Unknown Bone 2017, 2019; Herbert *et al.* 2010, 2019
Perophora japonica (sea squirt)	Found beneath rocks in old swimming pool at Baiter Point	Unknown (D. Fenwick pers.com)
Pomatopyrgus antipodarum (New Zealand Mudsnail)	In Brownsea Island Lagoon	Unknown (Thomas and Worsfold 2016)

(Continued)

Table 6.5 (Continued) Non-native species recorded on shores and lagoons in Poole Harbour, 2002–21

Species	Details	Change since 2002
Ruditapes philippinarum (Manila Clam)	Naturalised across harbour	Stable/possibly increasing (Humphreys *et al.* 2015; Clarke *et al.* 2019a)
Styela clava (Leathery Sea Squirt)	Occasional on mixed-sediment shores	Stable/unknown
Sargassum muticum (Wireweed, brown alga)	Commonly where there is mixed sediment and pools, Pottery Pier on Brownsea Island	No significant change
Undaria pinnatifida (Wakame, brown alga)	Single specimen naturalised on Baiter foreshore in 2009. Three specimens found naturalised on south shore (mixed sediments) of Brownsea Island (5/5/2016). Baiter Point (2/11/2021)	Possibly increasing
Watersipora subatra (Red Ripple Bryozoan)	Beneath rock in old swimming pool at Baiter Point (12/9/2018) (D. Fenwick), on sea wall at Banks Road and chain ferry	Unknown, first recorded in Poole marina (Arenas *et al.* 2006)

Most species classified as non-native are listed in GBNNS (2020). Note that species on floating docks and pontoons have been excluded (see Arenas *et al.* 2006).

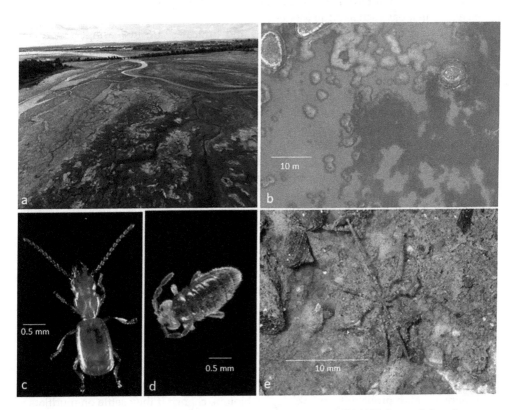

Figure 6.9 (a) Holes Bay algal mats, August 2017, AUV (A. Harrison). (b) AUV photo of algae in Brownsea Lagoon: the green alga *Chaetomorpha linum* with red clumps of *Agarophyton vermiculophyllum*, August 2017 (Photo D. Harrison), (c) Beetle *Aepus marinus* (Photo S. Trewhella), (d) Springtail *Axelsonia littoralis* (Photo S.Trewhella), (e) Sea spider *Ammothea hilgnendorfi* at Baiter 12 September 2018 (Photo D. Fenwick).

protocols and methods. Despite this, the range of 'natural variability' is becoming more apparent for some taxa, and departures from long-term trends are becoming easier to identify. Densities of most bivalves are stable or increasing, although numbers of *Scrobicularia plana* have not returned to levels observed in the 1970s–1980s. Measured densities of the small bivalve *Abra tenuis* have risen remarkably since previous surveys, the reason for which is unknown. It is possible that this is due to different methods of sample processing and this should be investigated prior to future surveys. The declining abundance of the amphipod *Corophium volutator* is a particular concern considering its importance as prey for wading birds and should be a focus for future monitoring.

Poole Harbour itself has many physical lagoonal characteristics (Humphreys 2005) and in the sediments there are benthic invertebrates which are typically found in lagoons such as the crustaceans *Monocorophium insidiosum*, *Microdeutopus gryllotalpa* and *Idotea chelipes* (Figure 6.8) and the Lagoon Cockle *C. glaucum*, which was most common in Poole Park Lagoon where it was present in the 1970s (Boyden and Russell 1972). With projected sea level rise it is possible that the distribution of artificial lagoonal habitat will change as sites revert to intertidal areas when sea walls are overtopped and fall into disrepair, whereas new lagoons might form in regions that become flooded with seawater (Herbert *et al.* 2019). The Starlet Sea Anemone, Tentacled Lagoon Worm *Alkmaria romijni* and Lagoon Sand Shrimp *Gammarus insensibilis* are typically found in brackish lagoonal habitats and are each protected under Schedule 5 of the Wildlife and Countryside Act. However, as all were also sampled in intertidal mudflats, they are likely to quickly colonise new lagoons to be created in areas such as Arne Moors.

Non-native species, particularly seaweeds and tunicates, are increasingly abundant on artificial structures and within mixed-sediment habitats, and new arrivals are to be expected in future years. Currently, most appear ecologically benign, though some require more intensive monitoring and research, notably the alga *A. vermiculophyllum*, Pacific Oyster and nemertean worm *C. simula*. Within mudflats, the naturalised Manila Clam now supports an economically valuable fishery, which has been awarded Marine Stewardship Council status. Modelling suggests that the expansion of Manila Clam has been facilitated by high levels of larval retention in the harbour, which is influenced by the attraction of larvae to regions of reduced salinity, particularly the Wareham Channel (Herbert *et al.* 2012). Although the abundance of the native Palourde Clam *Ruditapes decussatus* appears to have declined, there is no evidence yet to indicate that this is due to the increasing densities of *R. phillipinarum*. However, there have been concerns that the harvesting of clams by pump-scoop dredging, which leaves prominent circular-shaped scars on the mud surface (Fearnley *et al.* 2013; Figure 6.6), has had a negative impact on intertidal assemblages (Parker and Pinn 2005). This was investigated by Clarke *et al.* (2017) at a newly opened harvesting site in the Wytch area in 2015. Following pump-scoop clam dredging, the site experienced an increase in species richness and abundance of annelid worms *H. diversicolor* and *Aphelochaeta marioni*, while the density of the bivalve mollusc *A. tenuis* was reduced. Both molluscs *A. tenuis* and *P. ulvae* declined significantly within the dredged area. Yet, although the changes were attributed to pump-scoop dredging they were not substantial enough to change the biotope classifications for the site. It was concluded that most of the invertebrates are too small to be affected by the dredge. The numbers of wild Pacific Oysters are slowly increasing on sea walls and other hard substrata, and there is a high probability of habitat and biotope change if they settle on mudflats in large numbers and form reefs; yet disturbed areas of mixed sediment are more likely to be colonised initially.

Green macroalgal mats have been a feature of creeks in Holes Bay (Figure 6.9) since the 1970s (Holme and Bishop 1980). Extensive mats may now colonise mudflats throughout the harbour from spring to autumn (Pinn and Jones 2005), although mat thickness and duration vary from year to year (Thornton 2016; Thornton *et al.* 2019). The density of smaller species of invertebrate infauna beneath the mat increases with rising algal biomass, and mat coverage is the strongest predictor of variation in overall invertebrate community assemblage on the mudflats (Thornton 2016). It is possible that the increasing abundance of the mudsnail *P. ulvae* is linked to the growth of green algae (*Ulva* spp.). In some bird species, evidence was found of changes in feeding behaviour and the ability to obtain fauna entrained within the algae, including Shore Crabs *Carcinus maenas* (Thornton 2016). With eutrophication unlikely to reduce significantly in medium term, the impact of green algae on the health of intertidal habitats must remain a principal concern. Efforts to differentiate and map the distribution of algal species associated with intertidal soft sediments would be extremely valuable.

The important legacy of inorganic pollution in Poole Harbour should not be overlooked in attempting to interpret environmental change in intertidal habitats. Highest values of sediment metal concentrations (chromium, iron, nickel, copper, zinc, arsenic, silver, cadmium, selenium, lead) were found in northern Holes Bay, Arne Bay and the Wareham Channel (Hübner 2009; Hübner *et al.* 2010; Oaten 2017; Oaten *et al.* 2017). The high metal pollution in Holes Bay is believed to be the legacy of a coal-fired power station that closed in 1993 and the MERC chemical plant which closed in 1998, both of which discharged effluent into the bay. Combined sewer outfalls and sewage treatment works also discharge into the north-east corner of the bay. However, the magnitude of metal pollution in Poole Harbour is currently not severe according to EC Regulation 1881/2006, as the highest concentrations of cadmium found in clams were below safe human consumption limits. Yet as of 2011, tributyltin (TBT), which is known to cause significant harm to invertebrates, was still found in the waters of Lytchett Bay, Brands Bay, Wytch Farm and by the old power station site (Langston *et al.* 2015).

With a predicted increase in storm frequency and rise in sea level, the shape of Poole Harbour and the size and type of intertidal habitats are likely to change significantly in future decades. New set-back schemes near Wareham on the Arne Peninsula may enable the formation of lagoons, saltmarsh and intertidal mudflats. There may also be the construction of new sea defences, breakwaters and other coastal infrastructure that could see an increase in species more typical of hard substrata. In addition to the arrival of non-native species, some warm-adapted taxa that are native to the southern biogeographic realm (Lusitanian province) are responding to rising sea temperatures and expanding their range eastwards along rocky shores on the south coast of England (Hawkins *et al.* 2009). Relatively little is known about range expansion of soft-sediment fauna; however, the Purple Topshell *S. umbilicalis*, which has become more common on the Dorset coast and Solent region since 2000 (Herbert *et al.* personal observations), is now well established and abundant on the mixed-sediment shore at Pottery Pier on Brownsea Island. The Toothed Topshell *Phorcus lineatus*, which is also expanding its range eastwards, was first recorded in the harbour in 2018 on the shore at Baiter.

Improved efforts to identify lesser known species and groups of the extreme upper shore would be valuable to increase our knowledge and understanding of their role within the ecosystem, such as the intertidal beetle *Aepus marinus* and springtail *Axelonia littoralis* (Shaw and Trewhella 2019; Figure 6.9). Recording efforts to improve knowledge of seaweeds and seagrass in the harbour would be very useful as this has been generally overlooked.

The maintenance of time-series data as presented is important as this can help managers determine the likelihood of 'tipping points' in ecological systems, as exemplified in Poole Harbour by Watson *et al.* (2018). More robust and consistent benthic monitoring protocols combined with high resolution imagery of the harbour, perhaps through the use of unmanned aerial vehicles (UAV) (Clarke *et al.* 2019b), automation of environmental monitoring and molecular techniques such as the use of eDNA and metabarcoding, will improve the quality of data and aid the interpretation of future surveys.

Acknowledgements

We are grateful to the following individuals and organisations:

Poole Harbour Commissioners, Poole Harbour Study Group, National Trust, Dorset Wildlife Trust, Natural England, RSPB and RNLI, Prof. John Humphreys, skipper of *Sea Rush*, Dr Sue Burton of Natural England, Sarah Birchenough of SIFCA, Suzy Witt from the Environment Agency, Dr Dan Franklin, Francis Grandfield, Ralf Hübner, Dr Annesia Lamb, Prof. Christine Maggs at Bournemouth University, David Fenwick, Steve Trewhella and Peter Tinsley. Data from the Environment Agency is provided under the Open Government Licence. Bournemouth University acknowledges support from the EcoSal Atlantis EU Interreg IVB (Atlantic) programme, Marineff EU Interreg project VA (Channel) programme and RaNTrans EU Interreg project VA (Channel) programme.

Appendix 6.1

Species Recorded in Intertidal and Lagoon Habitats within the Poole Harbour, 2002–21.

Phylum TRACHEOPHYTA (Flowering plants)

Lemna sp. 21

Ruppia cirrhosa 20

Ruppia maritima 20

Ruppia sp. 18

Zostera marina 10

Phylum CHLOROPHYTA (Green algae)

Blidingia marginata 17

Chaetomorpha linum 18, 20, 21

Cladophora rupestris 17, 28

Ulva clathrata 7

Ulva compressa 7, 17

Ulva fenestrata 17

Ulva intestinalis 3, 6, 7, 9, 12, 21

Ulva lactuca 3, 6, 9, 12, 18

Ulva linza 17

Ulva rigida 7

Ulva sp. 18, 21, 22, 27–29

Phylum OCHROPHYTA (Brown algae)

Ascophyllum nodosum 3, 9

Colpomenia peregrina 29

Ectocarpus siliculosus 9

Fucus spiralis 3, 9, 12, 17, 27–29

Fucus serratus 9, 12, 17, 27

Fucus vesiculosus 3, 6, 9, 27, 29

Pelvetia canaliculata 6

Saccharina latissima 6

Sargassum muticum 3, 12, 18, 22, 27–29

Undaria pinnatifida 3, 12, 27

Phylum RHODOPHYTA (Red algae)

Agarophyton vermiculophyllum 18, 21

Aglaothamnion sp. 18

Carradoriella denudata 18

Carradoriella elongata 18

Catenella caespitosa 9, 12, 17

Caulacanthus okamurae 17

Ceramium pallidum 18

Ceramium secundatum 18

Ceramium sp. 12

Chondria coerulescens 16

Chondrus crispus 6, 9, 12, 17, 27, 28

Corallina officinalis 28

Corallinales sp. 23

Gelidium pusillum 17

Grateloupia turuturu 3, 16, 27

Mastocarpus stellatus 9, 27, 28

Osmundea hybrida 9

Osmundea pinnatifida 9, 17, 27, 28

Palmaria palmata 28

Porphyra linearis 3, 17

Spyridia griffithsiana 17

Phylum PORIFERA (Sponges)

Amphilectus fucorum 6

Halichondria panacea 3, 9, 12, 17, 27

Hymeniacidon perlevis 6

Sycon ciliatum 9, 16, 27

Phylum CNIDARIA

Anthozoa (Sea anemones)

Actiniaria 1, 2, 4, 5, 7, 8, 11, 13, 26

Actinia equina 9, 12, 16, 27, 28

Anemonia viridis 3, 6, 12, 22, 27

Cereus pedunculatus 3, 9

Edwardsiidae indet. 2, 5

Nematostella vectensis 18, 20, 21, 22, 25, 26

Phylum PLATYHELMINTHES (Flatworms)

Platyhelminthes sp. 2, 5, 21

Phylum NEMATODA (Roundworms)

Nematoda indet. 1–5, 8, 13, 18, 21, 22, 25, 26

Phylum NEMERTEA (Ribbon worms)

Cephalothrix simula 16

Cerebratulus sp. 5

Lineus viridis 3

Nemertea indet. 1–3, 5, 7, 8, 21, 26

Phylum ANNELIDA (Segmented worms)

Polychaeta (Bristleworms)

Alitta virens 1, 2, 3, 5, 8, 11

Alkmaria romijni 2

Amblyosyllis spectabilis 16

Ampharete acutifrons 7

Ampharete baltica 3

Ampharete grubei 1

Ampharete sp. 2, 5, 8, 11

Aonides oxycephala 3

Aphelochaeta filiformis 1

Aphelochaeta marioni 3, 4, 5, 7, 11, 18

Arenicola marina 1, 3, 5–7, 9, 12, 13, 22

Arenicolidae 8

Brania pusilla 2

Capitella sp. 2, 5, 7, 8, 11, 19, 26

Capitella capitata 1, 3, 4, 11, 13, 18, 21, 22

Chaetozone christiei 3

Chaetozone gibber 8

Chaetozone zetlandica 3, 18

Chaetozone sp. 7

Cirratulidae (indet.) 3, 7, 11, 13

Cirriformia tentaculata 1, 3

Cossura spp. 5, 8

Cossura longocirrata 3

Cossura pygodactylata 2

Dipolydora sp. 26

Dipolydora quadrilobata 2

Desdemona ornata 2–5, 7, 8, 11, 18

Eteone longa 1–5, 7, 11

Euclymene oerstedii 8

Eumida punctifera 3

Eumida cf. *sanguinea* 3

Eupolymnia nebulosa 16

Exogone naidina 2, 5, 8

Ficopomatus enigmaticus 20

Galathowenia oculata 2, 8

Gattyana cirrhosa 2

Glycera alba 2

Glycera tridactyla 1, 3, 7, 11

Harmothoe spp. 1

Hediste diversicolor 1–8, 11, 13, 18–22, 26

Heteromastus filiformis 1, 11

Hypereteone foliosa 3

Hypereteone lighti 25

Hypereteone sp. 8

Janua heterostropha 3

Lanice conchilega 6, 12

Malacoceros sp. 26

Malacoceros fuliginosus 1, 2

Malacoceros tetracerus 3

Maldanidae sp. 5

Manayunkia aestuarina 2, 5, 8, 21, 25

Marphysa sanguinea 11

Mediomastus fragilis 3, 21

Melinna palmata 2–5, 7, 8, 11

Microphthalmus cf. *similis* 3

Microspio mecznikowianus 5

Microspio sp. 2, 8

Nais sp. 21

Neanthes fucata 11

Neoamphitrite figulus 2, 3

Nephtys hombergii 1–5, 7, 8, 11, 13

Nephtys kersivalensis 3

Nephtys sp. 5

Nereididae sp. 5, 6, 8, 26

Notomastus latericeus 3, 8

Opheliidae sp. 21

Ophryotrocha sp. 2, 5

Parapionosyllis minuta 3

Paranais litoralis 5, 21, 26

Phyllodoce mucosa 1–3, 7, 8

Phyllodocidae sp. 8, 11, 13

Polycirrus sp. 2, 3

Polycirrus caliendrum 1

Polydora cornuta 2–5, 8, 18, 20, 21

Polydora sp. 3, 7, 11

Prosphaerosyllis tetralix 2

Pseudopolydora paucibranchiata 3, 8

Pygospio elegans 1–5, 7, 8, 18, 21

Sabella pavonina 12

Scolelepis foliosa 1

Scolelepis squamata 1

Scolelepis sp. 3, 7, 13

Scoloplos armiger 1, 3, 7

Serpulidae sp. 3

Sphaerosyllis taylori 2

Spio armata 2

Spio martinensis 2, 3

Spionidae sp. 3, 11, 19, 22

Spirobranchus lamarcki 3, 6, 9, 12, 17, 27, 28

Spirorbis (Spirorbis) spirorbis 9, 12, 27

Streblospio shrubsolii 2–5, 7, 11, 18, 25

Streblospio sp. 8, 11, 21

Terebellidae sp. 13

Tharyx sp. (inc. species A) 2, 5, 8, 21

Tharyx killariensis 8

Oligochaeta (Worms with few bristles)

Baltidrilus costatus 2, 21, 22, 26

Enchytraeidae sp. 2, 18, 21, 26

Tubificoides amplivasatus 2, 8

Tubificoides benedii 2–5, 8, 18, 19, 21, 25, 26

Tubificoides pseudogaster 2–5, 8, 18, 21, 25

Tubificoides sp. 1, 3, 5, 7, 9, 11, 13, 20

PHYLUM ARTHROPODA

Subphylum CRUSTACEA

Cirripedia (Barnacles)

Amphibalanus improvisus 2, 3

Austrominius modestus 3, 6, 9, 11, 12, 16, 17, 27

Chthamalus montagui 27, 28

Perforatus perforatus 12, 28

Semibalanus balanoides 9, 12, 17, 27

Sacculina carcini 16

Copepoda (Copepods)

Copepoda sp. 2, 5, 6, 8

Harpacticoida sp. 3

Ostracoda (Seed shrimps)

Cyprideis torosa 25

Eusarsiella zostericola 2, 3, 5, 8, 11

Ostracoda sp. 3, 4, 8, 11, 18, 21

Podocopida sp. 2, 5

Mysida (Opossum shrimps)

Mesopodopsis slabberi 1

Neomysis integer 1

Praunus flexuosus 1, 5

Praunus inermis 18

Mysidae spp. 1, 6, 20

Amphipoda (Sideswimmers)

Aoridae sp. 2, 5, 8, 26

Ampelisca brevicornis 2, 3

Aora gracilis 2

Apocorophium acutum 8

Apohyale prevostii 26

Bathyporeia sarsi 1

Bathyporeia indet. 3

Corophium arenarium 1

Corophium volutator 1–5, 7, 8, 18, 21, 22, 25

Crassicorophium bonellii 8

Gammaropsis palmata 1

Gammarus sp. 7, 9, 11, 13

Gammarus insensibilis 26

Gammarus locusta 1–3, 7, 18, 20

Gammarus salinus 2

Gammaridae sp. 2, 5

Melita palmata 3, 7, 15, 18, 20, 21, 22, 26

Microdeutopus anomalus 5, 8, 26

Microdeutopus gryllotalpa 1–3, 7, 8, 20, 22, 26

Microdeutopus sp. 8

Microprotopus maculatus 2, 3, 5, 7, 8, 13

Monocorophium insidiosum 2, 18, 20, 21, 22

Orchestia gammarellus 21

Pariambus typicus 8

Phtisica marina 3

Stenothoidae sp. 5

Talitrus saltator 1

Urothoe poseidonis 1

Urothoe pulchella 3

Urothoe sp. 13

Isopoda (Sea lice)

Cyathura carinata 1–5, 7, 8, 11, 25

Dynamene bidentata 7

Idotea balthica 1, 3, 7

Idotea chelipes 2, 3, 15, 18, 21, 22, 26

Idotea granulosa 12, 21

Idotea neglecta 1

Idotea pelagica 1

Isopoda sp. 5

Janira maculosa 16

Lekanesphaera hookeri 21

Lekanesphaera rugicauda 18

Lekanesphaera sp. 7

Ligia oceanica 17

Tanaidacea (Tanaids)

Apseudopsis latreillii 3

Heterotanais oerstedii 25

Tanais dulongii 15

Decapoda (Large crustaceans)

Brachyura sp. 8

Carcinus maenas 1–3, 5–9, 11–13, 16, 17, 19, 20, 22, 26, 27, 28

Crangon crangon 1, 3

Liocarcinus navigator 3

Palaemon longirostris 1

Palaemon serratus 1

Palaemon sp. 16

Palaemon varians 4, 18, 19, 20, 21, 22

Pisidia longicornis 3, 16, 27

Porcellana platycheles 12, 16, 27, 28

Subphylum CHELICERATA

Acari (Mites)

Acariformes spp. 2, 5, 7, 8, 21

Pycnogonida (Sea spiders)

Ammothea hilgendorfi 6, 9, 12, 15, 16

Pycnogonum litorale 12

Subphylum MYRIAPODA (Sea centipedes)

Strigamia maritima 15

Subphylum HEXAPODA (Insects and springtails)

Collembola 5

Axelsonia littoralis 15

Lepidoptera

Pyralidae sp. 21

Hemiptera 2

Gerris odontogaster 21

Philaenus sp. 21

Diptera

Chironomidae 1–5, 7, 11, 13, 18, 21, 22, 26

Chironomus salinus 20

Dolichopodidae 7, 13, 18, 21, 26

Coleoptera

Carabidae sp. 11

Aepus marinus 15

Hexapoda indet. 3, 11, 13, 18

Phylum MOLLUSCA (Sea snails and slugs)

Abra sp. 11

Abra tenuis 1–5, 7, 8, 11, 13, 18, 19, 22, 26

Acanthochitona crinita 14

Acanthochitona fascicularis 14–16, 27

Aeolidiella alderi 12

Aeolidiidae sp. 8

Akera bullata 2, 5, 8

Alderia modesta 3, 5, 8

Berthella plumula 3, 5, 8

Buccinum undatum 14

Cerastoderma sp. 19, 22

Cerastoderma edule 1–4, 6–8, 11–13, 16, 26

Cerastoderma glaucum 3, 4, 5, 8, 9, 11, 18, 20

Cerithiopsis tubercularis 14

Crepidula fornicata 1, 3, 6, 9, 12, 14, 18, 22

Doris pseudoargus 14

Ecrobia ventrosa 18, 20, 21

Elysia viridis 14

Eubranchus exiguus 16

Haminoea navicula 1, 3

Heteranomia squamula 16

Hydobiidae sp. 11

Lamellaria latens 14

Leptochiton asellus 11

Lepidochitona cinerea 2, 3, 7, 9, 12

Littorina spp. 1, 6

Littorina fabalis 6

Littorina littorea 1, 3, 6, 7, 9, 12, 14, 16, 17, 26, 27, 28

Littorina obtusata 1, 3, 9, 17, 27, 28

Littorina saxatilis 1, 3, 11, 12, 17, 26

Limapontia depressa 2, 5, 8, 15, 26

Lucinoma borealis 3, 8, 11, 13

Macoma balthica 1–5, 7, 8, 11

Magallana gigas 3, 9, 12, 17, 27–29

Melarhaphe neritoides 12

Mercenaria mercenaria 9

Mimachlamys varia 14, 16, 27, 28

Musculus subpictus 16

Mya arenaria 1–5, 7–9, 11, 14, 19, 20, 26

Myosotella myosotis 14, 26

Mysia undata 2

Mytilus edulis 3, 5, 8, 12, 14, 17, 27, 28

Nucella lapillus 12, 28

Ocenebra erinaceus 12, 14, 27, 28

Ostrea edulis 6, 9, 12, 14, 27, 29

Parvicardium exiguum 3

Patella depressa 27, 28

Patella ulyssiponensis 28

Patella vulgata 3, 9, 12, 14, 17, 27–29

Peringia ulvae 1–8, 11, 13, 14, 17–19, 20, 22, 25, 26

Phorcus lineatus 16, 27

Potamopyrgus antipodarum 21

Retusa obtusa 2–5, 11

Ruditapes decussatus 2, 3, 11, 16

Ruditapes philippinarum 1–5, 7, 8, 9, 11–14, 18, 26

Scrobicularia plana 1–5, 8, 11, 25

Solen marginatus 3

Steromphala cineraria 12, 14, 16, 27, 28

Steromphala umbilicalis 1, 3, 7, 9, 12, 14, 16, 26, 27, 28

Tectura virginea 14

Thraciidae sp. 26

Timoclea ovata 5, 8

Tricolia pullus 16

Tritia incrassata 14

Tritia reticulata 1, 3, 6, 14, 16

Trivia monacha 14

Turtonia minuta 26

Venerupis corrugata 3

Veneridae sp. 5, 8

Phylum BRYOZOA (Sea mats)

Amathia lendigera 2

Bryozoa indet. 2, 17

Bugula sp. 9, 27

Bugula neritina 16, 27

Conopeum reticulum 21

Cradoscrupocellaria reptans 16

Crisia sp. 21

Einhornia crustulenta 2

Electra pilosa 9

Membranipora membranacea 9

Scruparia chelata 16

Watersipora subatra 16, 17, 27, 28

Phylum PHORONIDA (Horseshoe worms)

Phoronis sp. 8

Phylum ECHINODERMATA (Starfish and urchins)

Amphipholis squamata 2, 3, 9, 12, 16

Amphiuridae sp. 2

Asterias rubens 16

Asterina gibbosa 23

Ophiothrix fragilis 15

Psammechinus miliaris 15, 27, 29

Phylum HEMICHORDATA (inc. Acorn worms)

Enteropneusta 2

Phylum CHORDATA
TUNICATA (Sea Squirts)

Ascidia spp. 1, 9, 12

Ascidiella aspersa 3, 6, 16, 27–29

Botryllus schlosseri 9, 12, 16, 27, 28

Botrylloides diegensis 6, 16, 27

Botrylloides leachii 3, 9, 16, 27

Botrylloides violaceus 6, 16

Botrylloides sp. 12

Ciona intestinalis 6, 16, 27

Corella eumyota 16

Didemnum maculosum 16

Molgula sp. 6, 16

Perophora japonica 16

Polycarpa scuba 16, 27

Styela clava 3, 6, 12, 16, 27

PICES (Fish)

Anguilla anguilla (European Eel) 19

Atherina presbyter (Sand Smelt) 19, 20

Chelon labrosus (Thick-lipped Grey Mullet) 18–20

Clupea harengus (Atlantic Herring) 19, 20

Coryphoblennius galerita (Montagu's Blenny) 24

Dicentrarchus labrax (European Bass) 20

Gasterosteus aculeatus (Three-spined Stickleback) 18–20

Labridae sp. (wrasse) 19

Limanda limanda (Dab) 19

Lipophrys pholis (Shanny) 3, 9, 12, 20, 28

Nerophis lumbriciformis (Worm Pipefish) 16

Platichthys flesus (European Flounder) 19, 20

Pomatoschistus microps (Common Goby) 2, 18, 20, 22

Pomatoschistus sp. (goby) 3, 4, 6, 18

Symphodus melops (Corkwing Wrasse) 20

(Numbers refer to data sources and references in Table 6.1).

References

Arenas, F., Bishop, J.D.D., Carlton, J.T., Dyrynda, P.J., Farnham, W.F., Gonzalez, D.J., Jacobs, M.W., Lambert, C., Lambert. G., Nielsen, S.E., Pederson, J.A., Porter, J.S., Ward, S., and Wood, C.A. 2006. Alien species and other notable records from a rapid assessment survey of marinas on the south coast of England. *Journal of the Marine Biological Association of the United Kingdom* 86: 1329–37. https://doi.org/10.1017/S0025315406014354

Birchenough, S. 2013. Impact of Bait Collecting in Poole Harbour and Other Estuaries within the Southern IFCA District. Marine Management Organisation Fisheries Challenge Fund Project FES 286.

Birchenough, S.E., Clark, R.W.E., Pengelly, S., and Humphreys, J. 2019. Managing a dredge fishery within a marine protected area: Resolving environmental and socio-economic objectives. In: Humphreys, J. and Clark, R. (eds) *Marine Protected Areas: Science, Policy and Management.* Elsevier, pp. 459–73. https://doi.org/10.1016/B978-0-08-102698-4.00023-X

Bone, J. 2017. *A Comparison of the Benthic Fauna and Habitat Health of Three Poole Harbour Lagoons and Their Potential as Refugia.* Unpublished BSc Thesis. Bournemouth University.

Bone, J. 2019. Behind anemone lines: Determining the environmental drivers influencing lagoonal benthic communities, with special reference to the anemone *Nematostella vectensis.* MRes Thesis. Bournemouth University, Department of Life and Environmental Sciences.

Bowles, F. and English, P. 2005. Sediment quality and benthic invertebrates in Holes bay. In: Humphreys, J. and May, V. (eds) *The Ecology of* *Poole Harbour: Proceedings in Marine Science 7.* Elsevier, Amsterdam, pp. 223–9. https://doi.org/10.1016/S1568-2692(05)80024-5

Boyden, C.R. and Russell, P.J.C. 1972. The distribution and habitat range of the brackish water cockle (*Cardium (Cerastoderma) glaucum*) in the British Isles. *Journal of Animal Ecology* 41: 719–34. https://doi.org/10.2307/3205

Caldow, R., McGrorty, S., West, A., Lev dit Durell, S.E., Stillman, R., and Anderson, S. 2005. Macroinvertebrate fauna in the intertidal mudflats. In: Humphreys, J. and May, V. (eds) *The Ecology of Poole Harbour: Proceedings in Marine Science 7.* Elsevier, Amsterdam, pp. 91–108. https://doi.org/10.1016/S1568-2692(05)80012-9

Cesar, C.P. 2021. Wareham coastal change project: Intertidal benthic ecology survey. Project Code: 5193244, Environment Agency.

Clarke, K.R. and Gorley, R.N. 2015. PRIMER v7: User manual/tutorial. PRIMER-E, Plymouth, 296 pp.

Clarke, L. 2017. Ecosystem Impacts of Intertidal Invertebrate Harvesting: From Benthic Habitats to Bird Predators. PhD Thesis. Bournemouth University.

Clarke, L.J., Esteves, L.S., Stillman, R.A., and Herbert, R.J.H. 2017. Impacts of a novel shellfishing gear on macrobenthos in a marine protected area: Pump-scoop dredging in Poole Harbour, UK. *Aquatic Living Resources* 31(5). https://doi.org/10.1051/alr/2017044

Clarke, L.J., Esteves, L.S., Stillman, R.A., and Herbert, R.J.H. 2019a. Population dynamics of a commercially harvested, non-native bivalve in an area protected for shorebirds: *Ruditapes philippinarum* in Poole Harbour, UK. *Aquatic Living Resources* 32(10). https://doi.org/10.1051/alr/2019008

Clarke, L.J., Hill, R.A., Ford, A., Herbert, R.J.H., Esteves, L.S., and Stillman, R.A. 2019b. Using remote sensing to quantify fishing effort and predict shorebird conflicts in an intertidal fishery. *Ecological Informatics* 50: 136–48. https://doi.org/10.1016/j.ecoinf.2019.01.011

Connor, D.W., Allen, J.H., Golding, N., Howell, K.L., Lieberknecht, L.M., Norther, K.O, and Recker, J.B. 2004. *The Marine Habitat Classification for Britain and Ireland. v04 05.* JNCC, Peterborough. Obtainable from www.jncc.gov.uk/Marine Habitat Classification.

Crisp, D.J. and Southward, A.J. 1958. The distribution of intertidal organisms along the coasts of the English Channel. *Journal of the Marine Biological Association of the United Kingdom* 37: 157–208. https://doi.org/10.1017/S0025315400014909

Dyrynda, P. 2005. Sub-tidal ecology of Poole Harbour – An overview. In: Humphreys, J. and May, V. (eds) *The Ecology of Poole Harbour: Proceedings in Marine Science 7.* Elsevier, Amsterdam, pp. 35–47. https://doi.org/10.1016/S1568-2692(05)80013-0

Envision. 2015. Poole Harbour SPA seagrass assessment. Report to Natural England. Envision Mapping Ltd. Newcastle, Northumberland.

Fearnley, H., Cruickshanks, K., Lake, S., and Lily, D. 2013. The effects of bait harvesting on bird distribution and foraging behaviour in Poole Harbour SPA: Footprint Ecology report for Natural England.

GBNSS. 2020 Great Britain Non-native species secretariat. Available from: http://www.nonnativespecies.org/home/index.cfm (accessed 21 January 2020).

Harrison, A., Pinder, A., Herbert, R.J.H, O'Brien, W., Pegg, J., and Franklin, D. 2016. Poole Park Lakes: Research and monitoring. Bournemouth University Global Environmental Solutions (BUG) report to Borough of Poole, 96 pp.

Harrison, A., Hill, R., Ford, A., Herbert, RJ.H., and Pinder, A. 2018. Aerial surveys of opportunistic green algae and *Gracilaria vermiculophylla* in Poole Harbour. Bournemouth University Global Environmental Solutions (BUG) report (BUG2777) to Natural England, 12 pp.

Hawkins, S.J., Sugden, H.H., Mieszkowska, N., Moore, P.J., Poloczanska, E., Leaper, R., Herbert, R.J.H., Thompson, R.C., Jenkins, S.R., Southward, A.J, and Burrows, M.T. 2009. Consequences of climate-driven biodiversity changes for ecosystem functioning of North European rocky shores. *Marine Ecology Progress Series* 396: 245–59. https://doi.org/10.3354/meps08378

Herbert, R.J.H., Ross, K., Huebner, R., and Stillman, R.A. 2010. *Intertidal Invertebrates and Biotopes of Poole Harbour SSSI and Survey of Brownsea Island Lagoon.* Report to Natural England. Bournemouth University.

Herbert, R.J.H., Willis, J., Jones, E., Ross, K., Huebner, R., Humphreys, J., Jensen, A., and Baugh, J. 2012. Invasion in tidal zones on complex coastlines: Modelling larvae of the non-native Manila clam, *Ruditapes philippinarum*, in the UK. *Journal of Biogeography* 39(3): 585–99. https://doi.org/10.1111/j.1365-2699.2011.02626.x

Herbert, R.J.H., Grandfield, F., Redford, M., Rivers, E., Hall, A.E., and Harrison, A. 2018. Littoral sediment condition of Poole Harbour SSSI. Bournemouth University Global Environmental Solutions (BUG) report (BUG2777) to Natural England, 34 pp.

Herbert, R.J.H., Ross, K., Whetter, T., and Bone, J. 2019. Maintaining ecological resilience on a regional scale: Coastal saline lagoons in a Northern European marine protected area. In: Humphreys J. and Clark, R. (eds) *Marine Protected Areas: Science, Policy and Management.* Elsevier, pp. 631–47. https://doi.org/10.1016/B978-0-08-102698-4.00032-0

Holme, N.A. and Bishop, G.M. 1980. Report on the sediment shores of Dorset, Hampshire and the Isle of Wight. Report to the Nature Conservancy Council. No. 280.

Hübner, R. 2009. Sediment Chemistry: A Case Study Approach. PhD Thesis. Bournemouth University.

Hübner, R., Herbert, R.J.H., and Astin, K.B. 2010. Cadmium release caused by the die-back of the salt-marsh cord grass *Spartina anglica* in Poole Harbour (UK). *Estuarine and Coastal Shelf Science* 84: 553–60. https://doi.org/10.1016/j.ecss.2010.02.010

Humphreys, J. 2005. Salinity and tides in Poole Harbour: Estuary or lagoon? In: Humphreys, J. and May, V. (eds) *The Ecology of Poole Harbour: Proceedings in Marine Science 7.* Elsevier, Amsterdam, pp. 35–47. https://doi.org/10.1016/S1568-2692(05)80008-7

Humphreys, J. and May, V. (eds). 2005. *The Ecology of Poole Harbour: Proceedings in Marine Science 7.* Elsevier, Amsterdam, p. 282.

Humphreys, J., Harris, M., Herbert, R.J.H., Farrel, P., Jensen, A.C., and Cragg, S.M. 2015. Introduction, dispersal and naturalization of the Manila clam *Ruditapes philippinarum* in British estuaries, 1980–2010. *Journal of the Marine Biological Association of the UK* 95(6): 1163–72. https://doi.org/10.1017/S0025315415000132

Jensen, A., Humphreys, J., Caldow, R., and Cesar, C. 2005. The Manila clam in Poole Harbour. In: Humphreys, J. and May, V. (eds) *The Ecology of Poole Harbour: Proceedings in Marine Science 7.* Elsevier, Amsterdam, pp. 163–73. https://doi.org/10.1016/S1568-2692(05)80018-X

Kruegger-Hadfield, S., Magill, C., Bunker, F., Mieszkowska, N., Sotka, E., and Maggs, C. 2017. When invaders go unnoticed: The case of *Gracilaria vermiculophylla* in the British Isles. *Cryptogamie, Algologie* 38(4): 379–400. https://doi.org/10.7872/crya/v38.iss4.2017.379

Langston, W., Pope, N., Davey, M., Langston, K., O'Hara, S., Gibbs, P., and Pascoe, P. 2015. Recovery from TBT Pollution in English Channel environments: A problem solved? *Marine Pollution Bulletin* 95: 551–64. https://doi.org/10.1016/j.marpolbul.2014.12.011

MarLIN. 2022. The marine life information network. Available from: https://www.marlin.ac.uk/ (accessed 4 April 2022)

Mills, S. 2016. Population Structure and Ecology of Wild *Crassostrea Gigas* (Thunberg, 1793) on the South Coast of England. PhD Thesis. Southampton University.

Oaten, J.F.P. 2017. Sediment metal concentrations, Poole Harbour. University of Southampton [Dataset]. https://doi.org/10.5258/SOTON/D0113

Oaten, J.F.P., Hudson, M.D., Jensen, A.C., and Williams, I.D. 2017. Seasonal effects to metallothionein responses to metal exposure in a naturalised population of *Ruditapes philippinarum* in a semi-enclosed estuarine environment. *Science of the Total Environment* 575: 1279–90. https://doi.org/10.1016/j.scitotenv.2016.09.202

Parker, L. and Pinn, E. 2005. Ecological effects of pump-scoop dredging for cockles on the intertidal benthic community. In: Humphreys, J. and May, V. (eds) *The Ecology of Poole Harbour: Proceedings in Marine Science 7*. Elsevier, Amsterdam, pp. 205–18. https://doi.org/10.1016/S1568-2692(05)80022-1

Pinn, E. and Jones, M. 2005. Macroalgal mat development and associated changes in infaunal biodiversity. In: Humphreys, J. and May, V. (eds) *The Ecology of Poole Harbour: Proceedings in Marine Science 7*. Elsevier, Amsterdam, pp. 231–7. https://doi.org/10.1016/S1568-2692(05)80025-7

Reitzel, A.M., Darling, J.A., Sullivan, J.C., and Finnerty, J.R. 2008. Global population genetic structure of the starlet anemone *Nematostella vectensis*: Multiple introductions and implications for conservation policy. *Biological Invasions* 10: 1197–213. https://doi.org/10.1007/s10530-007-9196-8

Ross, K. 2013. *Investigating the Physical and Ecological Drivers of Change in a Coastal Ecosystem: From Individual to Population-scale Impacts*. PhD Thesis. Bournemouth University.

Shaw, P. and Trewhella, S. 2019. Recent unusual UK collembola records – Entomobryomorpha and Poduromorpha. *British Journal of Entomology and Natural History* 32: 217–30.

Sheader, M., Suwailem, A.M., and Rowe, G.A. 1997. The anemone, *Nematostella vectensis*, in Britain: Considerations for conservation management. *Aquatic Conservation: Marine and Freshwater Ecosystems* 7: 13–25. https://doi.org/10.1002/(SICI)1099-0755(199703)7:1<13::AID-AQC210>3.0.CO;2-Y

Thomas, N.S., Caldow, R., McGrorty, S., dit Durell, S.E.A.L.V., West, A.D., and Stillman, R.A. 2004. *Bird Prey Availability in Poole Harbour*. Poole Harbour Study Group Publication No. 5, Wareham.

Thomas, P. and Worsfold, T. 2016. Brownsea Island Lagoon condition assessment – Interpretative survey report natural England. APEM Ref: 414247.

Thornton, A. 2016. The Impact of Green Macroalgal Mats on Benthic Invertebrates and Overwintering Wading Birds. PhD Thesis. Bournemouth University.

Thornton, A., Herbert, R.J.H., Stillman, R.A., and Franklin, D.J. 2019. Macroalgal mats in a eutrophic estuarine marine protected area: Implications for benthic invertebrates and wading birds In: Humphreys, J. and Clark, R. (eds) *Marine Protected Areas: Science, Policy and Management*. Elsevier, pp. 703–27. https://doi.org/10.1016/B978-0-08-102698-4.00036-8

Turner, A.D., Fenwick, D., Powell, A., Dhanji-Rapkova, M., Ford, C., Hatfield, R.G., Santos, A., Martinez-Urtaza, J., Bean, T.P., Baker-Austin, C., and Stebbing, P. 2018. New invasive nemertean species (*Cephalothrix simula*) in England with high levels of Tetrodotoxin and a microbiome linked to toxin metabolism. *Marine Drugs* 16: 452. https://doi.org/10.3390/md16110452

Watson, S., Herbert, R.J.H., Grandfield, F., and Newton, A. 2018. Detecting ecological thresholds and tipping points in the natural capital assets of a protected coastal ecosystem. *Estuarine and Coastal Shelf Science* 215: 112–23. https://doi.org/10.1016/j.ecss.2018.10.006

WoRMS. World register of marine species. Available from: www.marinespecies.org (accessed January 2018).

CHAPTER 7

The Fishes of Poole Harbour

PHILIP PICKERING and ROGER J.H. HERBERT

Abstract

This chapter and the accompanying list describe the fish species recorded in Poole Harbour over five decades (1970–2020). It also describes, where known, why these fish occur in the harbour, what affects their presence, their distribution and what period of the year they are seen. It is accepted by the authors, due to the difficulties of observing fish as explained later in this chapter, that there may be significant omissions in the information presented here.

Keywords: fish, distribution, Poole Harbour, species list, English Channel

Correspondence: philip.pickering@nationaltrust.org.uk

An overview

Observations in this chapter are mostly not from scientific research, a number being anecdotal, but they are largely backed up by confirmation from more than one source, not all of whom wished to be credited in references. They are believed to be quite accurate as the sources are either observations made by the authors theselves or those of knowledgeable individuals from within the local community of commercial fishing, diving and angling.

To understand the presence and abundance of fish living in this part of the English Channel, it is useful to draw some parallels between the behaviours of birds and fish. Birds and fish both have resident populations, migratory populations, summer or winter preferences, and predators and prey species, and include rare and occasional visitors, as well as have large fluctuations in population due to factors such as food sources, breeding site availability and human interference. Birds, however, are readily observed, whereas fish are not, and yet they are both exposed to a similar range of negative pressures. In Poole Harbour these include:

- habitat reclamation and development;
- a relative disregard when planning marine civil engineering developments;
- vulnerability to pollution;
- exploitation by humans on a commercial scale;
- changes to the seabed due to aquaculture and fisheries;
- regular disturbance of the seabed due to dredging for shellfish or dragging for ragworm. The dumping of maintenance dredging spoil in Brownsea Roads with the resultant instant change in local turbidity and consequent smothering.

Philip Pickering and Roger J.H. Herbert, 'The Fishes of Poole Harbour' in: *Harbour Ecology*. Pelagic Publishing (2022). © Philip Pickering and Roger J.H. Herbert. DOI: 10.53061/SBDK9176

Some of these pressures would be strongly resisted by conservation organisations and the public should they be applied to very visible populations of birds yet are barely registered when these same pressures are applied to fish.

Development

Old maps of Poole show the degree of reclamation that has taken place along the northern shore in the last 150 years and how the land-sea interface has changed from one of salt-marsh, gentle shelving mud and sandflats to many more abrupt human-made quaysides or limestone block sea defence structures, both of which encourage different species from those that previously occurred.

Following a full Environmental Impact Assessment, the current 'Middle Channel' was last straightened and deepened in 2005/6, to a minimum depth of 7.5m below Chart Datum. However, with this and previous large-scale capital dredges in the 1980s, there was relatively little information available on the impact on the subtidal ecology.

Pollution

Sources of pollution include agricultural fertilisers in freshwater drainage, that is, from rivers and surface water drainage from urban areas including roads containing vehicle oil, exhaust particulates and tyre compounds. Industrial pollutants from local industrial estates such as electro-plating (including chromium and cadmium) and chemical effluent from British Drug Houses (MERC) discharging for decades into Holes Bay. Sewage effluent including detergents, disinfectants and even drug residues in human waste. There is also a significant contribution from antifouling eroding from the many thousand private boats on mooring and in marinas.

Exploitation

There is a significant commercial fishery and sport angling community within the harbour, although currently there is a general awareness of the need to protect the marine ecology of the harbour to ensure healthy populations of fish. Details of current fisheries and bylaws are available from the Southern Inshore Fisheries and Conservation Authority (SIFCA, 2022)

Observation

Unfortunately, unlike birds, which can be studied with relative ease to the extent that one rare visitor is rarely missed, it is really very difficult to get a truly accurate picture of the fish ecology of the harbour as most of the time they are hidden beneath the surface and seen only when captured, glimpsed from above or spotted on occasions when the water clarity makes diving worthwhile. Many of the observations in this chapter are sourced from local commercial activities. However, as local byelaws limit fishing with fixed nets from the beginning of November to the end of March – apart from mullet ring nets, sand eel trawl and research beach seine – these byelaws also significantly limit the recording opportunities in the warmer months when there are far more species and in greater numbers than the colder months. Therefore, any list of species can only be an overview of the general picture, and there is likely to be many omissions in the species listed here, their distribution and frequency of occurrence.

Identification and detection of the presence of fish including species and population sizes by the sampling of environmental DNA (eDNA) is a possibility for the future, although the dispersion of DNA in seawater is large and salinity negatively influences DNA preservation.

Geography

Situated on the south coast of England, approximately halfway up the English Channel, Poole Harbour is in a good location for supporting a relatively large proportion of the total number of inshore species of fish in comparison to other UK locations such as the harbours and estuaries of the east coast. This is because it includes temperate species which are commonly associated with more southerly latitudes such as Trigger Fish and Red Mullet, which are at or near the most northerly limit of their range, but also species usually associated with northerly cold waters like Cod and Coalfish, which are at or near the most southerly limit of their range. However, as the depth of water in the harbour averages less than 1 m outside of the marked channels and even within the channels has a maximum depth of only around 17 m and in the English Channel depths rarely exceed 50 m, it can be safely assumed that we are never, or only accidentally, visited by the deep-water species that occur over the continental shelf and beyond.

The varied, largely estuarine or lagoonal geography of the harbour creates a range of habitats that support many different species of fish, with most of the north European families represented in either the resident populations or the seasonal migratory and occasional visitor. This variation in habitat ranges from:

- The relatively deep (17 m) saline, fast flowing (up to 4 kts) waters of the harbour entrance and Brownsea Roads.
- The principal channels of Middle Ship Channel, Wych Channel, South Deep Channel, North Channel and Wareham Channel are still relatively deep and at times fast flowing.
- The shallow, muddy bays and 'lakes' that comprise the largest area.
- The three virtually landlocked Holes Bay, Lytchett Bay and Blue Lagoon.
- The rivers Frome, Piddle, Sherford and Corfe.
- Brackish lagoons on Brownsea Island, Poole and Arne.
- The human-made structures of Poole Quay, various jetties and limestone block marinas. These features being a relatively recent and quite unique habitat and one that has been quickly colonised by a significant number of our harbour species such as Wrasse and juvenile Pollock, which use the shelter offered among the algae, hollows and holes.

Much of the harbour seabed consists of sandy or muddy sediments with some shingle stone and shell, with the sand component tending to give way to an increasingly muddy bottom further into the harbour and out of the scoured channels; in the warmer seasons, the bottom may also have quite a covering of marine algae. There are also significant areas of managed shellfish beds comprising Pacific Oyster *Magallana gigas* and Mussels *Mytilus edulis*. In addition to these various habitats are the less immediately obvious wide-ranging factors which dictate suitability for a particular species including tidal coverage, temperature, turbidity, oxygen and salinity (there is a significant salinity gradient between the harbour entrance and the upper reaches of the Wareham Channel).

Some species, such as the Shanny *Lipophrys pholis*, are able to cope with extremes. This little fish of the Blenny family is very common on the south shore of Brownsea, living in the intertidal zone among the shards left over from the pottery works last operational in the late nineteenth century. In order to thrive here it must sometimes endure being out of the water for several hours a day, suffer a rapid temperature variation of well over 10°C due to being trapped under damp tiles in the sun and then being immersed by a fresh tide. Low oxygen levels can occur in warm water and reduced salinity in rainy conditions. At the other end of the spectrum would be one of the shoaling pelagic species, perhaps the Mackerel *Scomber scombrus* or Scad *Trachurus trachurus*, both late summer, early autumn visitors and reliant on clear, fully saline, highly oxygenated deep water from the harbour entrance up to Poole Quay (although they do occasionally visit less typical locations such as Poole Yacht Club marina). Both these species quickly succumb if not kept in quite a narrow range of conditions.

Species list

Depending on the source consulted, it is considered that there are around 17,000 to 20,000 species of fish in the world's oceans; in some tropical reef ecosystems, there can be several thousand in a relatively small area (Census of Marine Life 2003). The total number of species recorded in the list accompanying this chapter is 87 (Table 7.1). Assuming the likelihood of a number of species being present that are unrecorded (especially gobies), perhaps another 15 altogether, that still makes the tally around 100, making the piscatorial diversity of the harbour quite poor in a global context. However, the numbers are quite good in comparison with other British harbours and coastal regions.

The list was first compiled in 2013. Made up originally from the authors' (PP) angling and diving notes and local knowledge it has since been added to from more recent catches and observations, as well as other trusted sources, including Southern Inshore Fisheries and Conservation Authority (SIFCA), the Environment Agency, local angling charter skippers and commercial fishermen. It's arranged in families so all the cod (gadiforms), for instance, are together. As are all the wrasses and gobies. Where possible, date and location details have also been included. As mentioned elsewhere in this chapter, the list isn't closed or complete.

All the locations referenced in this chapter can be found on the Admiralty chart of Poole Harbour 2611.

Notes of interest

Round fish

Moronidae (bass)

The harbour is a significant nursery area for European bass, and they are present throughout the harbour all year round as juveniles and mature adults, although there is a significant increase in the population in the warmer months coinciding with the arrival of prey species such as Lesser Sand Eel. Fish in excess of 1 kg have been recorded in the drainage channel running from Fleetsbridge into Holes Bay. Unfortunately, there has been a decline in numbers over the last few decades that has been attributed to commercial pressures on the offshore populations; bass have gone from being a relatively obscure species valued chiefly by anglers to being a highly prized food fish. However, they are a much sought-after sport fish and have in recent years gained a little protection in Poole from catch and release byelaws and a no-take nursery area, west of a line drawn between Jerry's Point and Salterns.

Table 7.1 The fish species of Poole Harbour, 1970–2020

Species	Common name	Location or distribution	Abundance	Notes
Blennies and Gobies				
Parablennius gattorugine	Tompot Blenny	Town pier/Castle pier	Common	Fish traps
Lipophrys pholis	Shanny	Shard point to Pot pier	Frequent	Beneath tiles
Coryphoblennius galerita	Montagu's Blenny	Whitley Lake	One	B.U. vertipool 14/12/20 (J. Bone)
Pomatoschistus microps	Common Goby	Town pier/Castle pier	Frequent	Fish traps
Gobius paganellus	Rock Goby	Town pier/Castle pier	Frequent	Diving
Pomatoschistus minutus	Sand Goby	B'sea castle beach	Abundant	Diving/seine net
Pomatoschistus pictus	Painted Goby	Wych Channel	Unknown	Diving
Gobiusculus flavescens	Two-spotted Goby	Town pier/Castle pier	Common	Diving/cast net
Gobius niger	Black Goby	Town pier/Castle pier	Unknown	Diving
Thorogobius ephippiatus	Leopard Spotted Goby	Town pier/Castle pier	Unknown	Diving
Aphia minuta	Transparent Goby	Middle Ship Channel	Unknown	EA Trawl
Pipefish and Seahorse				
Nerophis lumbriciformis	Worm Pipefish	B'sea south shore	Locally frequent	Beneath tiles
Hippocampus ramulosus	Long-snouted Seahorse	Widespread in harbour	Occasional	By-catch in nets
Hippocampus hippocampus	Short-snouted Seahorse	Widespread in harbour	Rare	By-catch in nets
Syngnathus acus	Greater Pipefish	Widespread in harbour	Common	By-catch in nets
Syngnathus rostellatus	Lesser Pipefish	Main Channel	Unknown	EA Trawl
Entelurus aequoreus	Snake Pipefish	Nr. Sandbanks Yacht Company	Unknown	Seine net
Syngnathus typhle	Deep Snouted Pipefish	Whitley Lake	Unknown	Seine net
Mullet				
Liza ramada	Thin-lipped Mullet	Widespread	Common April–October	Ring net
Chelon labrosus	Grey Mullet (Thick-lipped)	Widespread	Common April–October	Ring net
Liza aurata	Golden Grey Mullet	Large shoals, shallows, Whitley L.	Frequent April–October	Ring net
Bream				
Sparus auratus	Gilt-head Bream	Widespread	Increasingly common	Angling, Fixed nets
Spondyliosoma cantharus	Black Bream	Widespread	Common	Angling
Pagrus pagrus	Couch's Bream	B. Roads	Rare	Fixed net, angling
Flatfish				
Pleuronectes platessa	Plaice	Widespread	Common	Angling
Platichthys flesus	Flounder	Widespread	Common	Angling
Limanda limanda	Dab	Main Channel	Unknown	EA Trawl
Arnoglossus laterna	Scaldfish	Main Channel	Unknown	EA Trawl
Solea solea	Dover Sole	Widespread	Frequent	Nets, Angling

Solea lascaris	Sand Sole	Shallow sands	Infrequent	Cockle dredge
Buglossidium luteum	Solenette	Brownsea south shore	Several	Seine net, south shore
Zeugopterus punctatus	Topknot	Main Channel	Rare	Nets
Scophthalmus rhombus	Brill	B. Roads	Small/juvenile	Angling, nets
Psetta maxima	Turbot	B. Roads	Small/juvenile	Angling, nets
Cod				
Gadus morhua	Cod	B. Roads and deeper channels	Adult-rare autumn	Nets
Trisopterus luscus	Pout Whiting, Bib, Pouting	B. Roads	Frequent, Summer, Autumn	Diving
Pollachius pollachius	Pollack	Widespread in deeper water	Frequent	Angling, diving
Pollachius virens	Coalfish, Saithe	Main Channel	Rare, winter	Nets
Trisopterus minutus	Poor Cod	B. Roads	Occasional, Summer	Diving
Merlangius merlangus	Channel Whiting	B. Roads	Infrequent Late Autumn, Winter	Cormorant eating, under Rhizostoma
Wrasse				
Labrus bergylta	Ballan Wrasse	Town pier/Castle pier	Common, Summer, Early Autumn	Angling, diving
Crenilabrus melops	Corkwing Wrasse	Town pier/Castle pier	Common	Angling, fish traps, diving
Ctenolabrus rupestris	Goldsinny	Town pier/Castle pier	Infrequent, Summer	Diving
Labrus mixtus	Cuckoo Wrasse	Town pier/Castle pier	infrequent/summer	Diving
Symphodus bailloni	Baillion's Wrasse	South Deep	Unknown	Diving
Herring				
Clupea harengus	Herring	Widespread	Common, Autumn, Winter	Nets, Angling
Engraulis encrasicolus	Anchovy	Castle beach	Occasional	SIFCA seine net, ring net
Sprattus sprattus	Sprat	Widespread	Common	Nets, angling
Alosa fallax	Twaite Shad	Widespread	Occasional	Angling, ring net
Alosa alosa	Allis Shad	Widespread	Occasional	Ring net
Cartilaginous fish				
Mustelus asterias	Starry Smooth-Hound	Widespread in deeper water	Frequent summer	Angling, nets
Scyliorhinus stellaris	Nursehound	B. Roads	Infrequent	Angling
Raja montagui	Spotted Ray	Wych Channel	Unknown	Diving
Raja undulata	Undulate Ray	Wych Channel	Unknown	Diving
Others				
Dicentrarchus labrax	Bass	Widespread in harbour	Common	Angling, diving
Balistes capriscus	Grey Triggerfish	Town pier/Castle pier	Unknown	Fish trap
Scomber scombrus	Mackerel	Widespread in deeper water	Common Summer, Autumn	Angling

(Continued)

Table 7.1 (Continued) The fish species of Poole Harbour, 1970–2020

Species	Common name	Location or distribution	Abundance	Notes
Thunnus sp.	tuna	One in Brownsea Road	16/10/2020	Jumping after prey
Trachurus trachurus	Scad	Widespread in deeper water	Common Summer, Autumn	Angling
Belone belone	Garfish	Widespread in deeper water	Common summer, Autumn	Angling
Callionymus lyra	Common Dragonet	Town pier/Castle pier	Common	Diving
Callionymus reticulata	Reticulated Dragonet	Brownsea Roads, Wych Channel	Unknown	Diving
Echiichthys vipera	Lesser Weaver	B. Roads	Common some years	Angling/sand eel trawl
Trachinus draco	Greater Weaver	Middle Mud, Aunt Betty	Occasional	Sand eel trawl
Taurulus bubalis	Long-spined Sea Scorpion	Town pier/Castle pier	Frequent	Fish traps
Agonus cataphractus	Pogge	B. Roads	Unknown	Diving
Gaidropsaurus mediterraneus	Shore Rockling	Town pier/Castle pier	Unknown	Fish trap
Ciliata mustela	Five-bearded Rockling	B. Roads	Common, Late Autumn, Winter	Angling
Atherina presbyter	Sand Smelt	Widespread	Spring to Autumn	Seen from surface, angling
Ammodytes tobianus	Sand Eel	Widespread	Common, Summer, Early Autumn	Seen from surface
Hyperoplus lanceolatus	Launce, Greater Sand Eel	B. Roads	Frequent, Summer, Autumn	Seen from surface
Aspitrigla cuculus	Red Gurnard	Main channels	Occasional	Nets
Trigla lucerna	Tub Gurnard	Main channels	Frequent, Summer, Autumn	Angling
Gasterosteus aculeatus	Three-spined Stickleback	Lagoon sluice	Common	Hand net
Spinachia spinachia	Fifteen-spined Stickleback	Town pier/Castle pier	Occasional	Diving, fish trap
Pungitius pungitius	Nine-spined Stickleback	Reedbeds, DWT	Common	Seen from surface
Mullus surmuletus	Red Mullet	B. Roads	Common, Summer, Autumn	Angling, nets
Anguilla anguilla	European Eel	Widespread	Increasingly rare	Angling
Conger conger	Conger Eel	B. Roads	Frequent	Fish trap/angling/diving
Cyclopterus lumpus	Channel Lumpsucker	B. Roads, castle beach	Occasional	Fixed nets, beach seine
Zeus faber	John Dory	Eelgrass patch, B. Roads	Unknown	Diving
Salmo trutta	Sea Trout	Passing through harbour, Castle jetty	Increasingly rare	Seen in harbour rivers
Salmo salar	Salmon	Passing through harbour	Increasingly rare	Seen in harbour rivers
Acipenser sp.	sturgeon	Off Ham common beach	Two	Ring net
Petromyzon marinus	Sea Lamprey	Widespread	Occasional	On larger fish in nets

Gadidae (cod family)

Of the Cod-like species, **Pollock** *Pollachius pollachius* are probably the most numerous and are present in significant numbers throughout the warmer months. Although generally they are mainly juvenile fish of less than 30 cm in length there are occasional fish of 2–3 kg caught in the main channels during the colder months. Most inhabit the deeper and more saline waters of the main channels and can be observed when diving, hanging in the tide in shoals of several thousand between the Brownsea jetties. Very small juveniles of about 5 cm inhabit weedy rocky shallows wherever they occur.

Cod *Gadus morhua* of around 3 kg are regularly recorded in the colder months as far up as the Wareham Channel.

Whiting *Merlangius merlangus*, although abundant in Poole Bay, are strangely rare in the harbour. However, they do visit as very small juveniles within the mantle of Barrel Jellyfish, and the author observed a fish of around 1 kg being tackled in Brownsea Roads by a cormorant.

Pout *Trisopterus luscus* **and Poor Cod** *Trisopterus minutus* are frequent summer visitors and are regularly seen between the Brownsea Island jetties – the Pout in large shoals of small fish around 100 g, the Poor Cod as individuals among the rock armour.

Coalfish *Pollachius virens* is a close relative of the Pollock and an occasional winter visitor. This fish is on the southernmost limit of its range as it's normally associated with cold waters right up to the Arctic. It will be interesting to note whether observations of it decline in the coming years as sea temperatures are predicted to increase.

Clupidae (herring, sprats, anchovies)

Herring *Clupea harengus* are found throughout the harbour up to the mouth of the river Frome, arriving early in October and remaining until April. Once entering the harbour, they seem to come and go, preferring a spring tide. In the daytime they shoal up in the channels or deeper water. At night they disperse on to the shallows at high water, even more so when there is moonlight. Some years they are abundant and other years almost non-existent. The seasons of 2018 and 2019 were poor, with 2011–17 being exceptionally good. All records are for adult fish, male and female, as the net sizes allow the passage of juveniles, so it's not clear whether they are also present. Local spawning is likely as by February 90% of the fish are spent (their gonads have released sperm or eggs).

Sprats *Sprattus sprattus* are also abundant in the harbour although it's uncertain how widespread they are, perhaps being more common in the eastern parts. They are easy enough to find off Poole Town Quay and can be relatively easily caught using tiny fish skin lures. In the late summer of some years they occur in huge shoals in Brownsea Roads.

Anchovy *Engraulis encrasicolus* is possibly as widespread as sprats but currently thought to be less frequent visitors. One was caught with a large quantity of sprat in a SIFCA beach seine in the summer of 2018. However, they are caught in small quantities as by-catch in mullet ring nets every year. There is some evidence that they were present in quantities worth commercially fishing off the Hamworthy shore within living memory.

Shad, Twaite and Allis, *Alosa fallax* **and** *A. alosa* are only occasionally caught as by-catch in the harbour, although they are considered common as by-catch during the autumn and winter Sprat season in Poole Bay, Twaite being by far the predominant species. It is unclear whether either species spawn in the rivers of Poole Harbour.

Mugilidae (mullet)

All three species of mullet (Mugilidae), the **Thick-lipped Grey Mullet** *Chelon labrosus*, the **Thin-lipped Grey Mullet** *Liza ramada* and the **Golden Grey Mullet** *Liza aurata* found in UK waters, occur in significant numbers and frequency in the harbour, enough to sustain a small commercial fishery. UK waters represent the northern limit of their range and as such they are very much a fish of the warmer months. Depending on sea temperature they may start arriving in March and leave again for deeper offshore waters by November. Thick-lipped Mullet are found throughout the eastern end of the harbour up to Rockley and Lychett Bay, but they avoid the low salinity of the river Frome and upper reaches of the Wareham Channel. They also frequent the very shallow muddy tidal creeks of all the harbour's lakes and bays, feeding on organic matter and small invertebrates that they filtered out, and unwanted material is expelled in puffs from their gills. The Thick-lipped Mullet especially like rasping off algae from human-made structures such as boat hulls and concrete slipways, leaving very distinctive tracks where they have visited. It is this species which is most commonly observed in Poole Harbour marinas.

Thin lipped have a much greater tolerance to low salinities and can be found all the way up to the tidal limits of the river Frome.

The Golden Grey frequents the higher salinity areas over similar grounds to the Thick-lipped, although it is not quite as common.

All three species spawn in deeper offshore waters in the winter months.

Sparidae (bream)

Gilt-head Bream *Sparus auratus* are widespread in the harbour up to the Wareham Channel. Preferring channels or deeper water in the daylight, they can be seen at night feeding in extremely shallow water of less than 30 cm deep. Like golden grey mullet at night, they will jump out of the water if disturbed by a spotlight. Once considered a rarity they are now (as of 2020) a common and increasing, year-round species in the harbour. They are usually to be found swimming in small schools or even shoaled up with bass or mullet. The average size fish is about 400–700 g. It's breeding status within the harbour is not certain as few juvenile fish have been seen.

Black Bream *Spondyliosoma cantharus* A summer visitor of large numbers of mainly juvenile fish which can found in much of the harbour including most of the Wareham Channel, with occasionally larger fish occurring in the deeper waters of the eastern parts of the harbour or off town quay.

Couch's Bream *Pagrus pagrus* infrequent visitor that has been recorded in sand eel trawls on Soldier bank and angling in Brownsea Roads. This species is protected – Poole Rocks MCZ

Salmonidae (Salmon Salmo salar and Sea Trout Salmo trutta)

Although Salmon and Sea Trout should be among the frequent records, unfortunately they are not. The author has only three recent records of a sea trout, one caught in a SIFCA seine net off Brownsea Castle beach in November 2020, one photographed alongside Brownsea Castle jetty in 2018 and the other caught accidentally near Upper Wych. Although these species are still recorded in the harbour's rivers and therefore must pass through the harbour at some stage, the numbers are very low. They are reported as by-catch in commercial nets or occasional angling reports. Although traditionally there has been a reluctance to report accidental landings of these species, it's quite difficult to make an accurate assessment of the true population. Data from the harbour river monitoring would give a more accurate picture of the populations.

Flatfish

Plaice *Pleuronectes platessa* in the summer, **Flounder** *Platichthys flesus* in the winter has been the general understanding of the flatfishes of Poole Harbour. However, the frequency and distribution are more nuanced than that, with both species present in varying numbers for most of the year and several other species of flatfish also present albeit in smaller numbers.

Plaice increase in numbers in the spring and move out again in the autumn. They are more usually found on the harder, sandy bottom and tend to stay below the low water mark, while Flounder increase in numbers in the autumn and tend to depart for deeper waters again in late January to February. They occur right up to the edge of the tide and often favour a muddier substrate. They are well able to survive stranding for a tide with only a thin covering of watery mud. The numbers of Flounder have reduced significantly in recent decades whereas Plaice have increased, although not to the levels seen in the 1960s.

Dover Sole *Solea solea* is probably the next most common flatfish and is present year-round. Distribution in the harbour must be widespread as they have been recorded from the Wareham Channel, Wych Channel, Little Channel and Holes Bay as well as around the more usual locations at the eastern end of the harbour. Also, not uncommon are tiny juveniles from Bramblebush Bay and the south shore of Brownsea caught when using a push net while looking for brown shrimp.

Sand Sole *Solea lascaris* are infrequently captured both around Middle Sands and on White Ground Lake every year. They are always juveniles between 100 and 200 mm. Usually caught in the autumn as by-catch when using a cockle dredge.

There is a single record of a **Solenette** *Buglossidium luteum* capture in a beach seine on Brownsea south shore, but this species is probably also present in reasonable numbers.

Brill *Scophthalmus rombus* and **Turbot** *Psetta maxima* are found at the eastern end of the harbour, preferring the deeper water. Usually found in Brownsea Roads. They often take sand eel baits intended for Bass and also occur in the area to the northern side of the Middle Channel where it shallows up to the Middle Sands. Frequently caught year-round. Very rarely are adults seen, usually juveniles up to 30 cm but more commonly around 20 cm caught as by-catch in mullet ring nets and cockle dredges. The numbers seen within the harbour remain unchanged in recent decades.

There are two locations within the harbour where **Topknot** *Zeugopterus punctatus* are occasionally caught as by-catch in commercial nets, one being the main channel north of Stone Island and the other being near Bell Buoy. They are usually 100–125 mm in length.

Dab *Limanda limanda* are surprisingly rare in the harbour as they are quite common off the beaches of Poole Bay. It has been difficult to obtain reliable records, the only ones shown in this chapter's list being from Environment Agency sample trawls in the main channel; interestingly, however, there is a reference to them being frequently caught in Brands Bay in Aflalo's late nineteenth-century sea angling book (Aflalo 1901).

Pipefish and seahorses

Long-snouted Seahorse *Hippocampus ramulosus* (a protected species) is found in most of the harbour as far up as Rockley, except in shallow mudflats. A few are found among Eelgrass *Zostera marina* near Salterns Marina, but preferred areas are bends in deeper water (less tidal) channels as weed congregates here. They are also found on mooring chains, especially those west of the Royal Marine's Hard near Ham Common. They are resident all year round, and by-catch numbers have remained stable over the last few decades. Luckily, it's more the case of the seahorse holding on to the net rather than the net catching the seahorse, so most are returned unharmed. They are also occasionally caught in mullet ring nets during the summer months.

Short-snouted Seahorse *Hippocampus hippocampus* (a protected species) is found in the same locations as the spiny seahorse except they are more tolerant to freshwater so can be found closer to freshwater sources. Ring net by-catch numbers have been stable over the last few decades. Although they are lower in numbers in the harbour than the long-snouted seahorse, they are considered common and long-standing residents in Poole Bay by the commercial fishing community.

Pipefish are well represented and are common and widespread in the harbour. These include the large **Greater Pipefish** *Syngnathus acus* and the tiny **Worm Pipefish** *Nerophis lumbriciformis*. A close relative of the seahorse, the pipefishes are more abundant but less well known. They are very frequently seen writhing to escape the beak of a cormorant on the surface. Also recorded are the **Lesser Pipefish** *Syngnathus rostellatus*, the **Snake Pipefish** *Entelurus aequoreus* and the **Broadnosed Pipefish** *Syngnathus typhle*.

Cartilaginous fish (sharks and rays)

These species are poorly represented in the harbour, although they probably visit more often than commonly known. Most will only be so close inshore in the warmer months and therefore outside of the fixed net season and furthermore not so likely to occur in shallow areas that are ring netted for Mullet (with the exception of Smooth-Hound). Tope are not on the list although anecdotally they were irregularly caught by rod and line in the harbour entrance as recently as the 1970s.

Small-spotted Catshark *Scyliorhinus canicula* These are occasionally caught at the eastern limit of the harbour and usually over towards Stone Island.

Starry Smooth-hound *Mustelus asterias* are found at the eastern end of the harbour (South Deep is favoured). First arrivals appear in late April and are normally gone by late September. Their population numbers appear in the thousands especially in South Deep as it is thought to be a breeding area and comprises mainly large to very large adult fish. Most of the specimens seen in Poole are the Starry Smooth-Hound rather than the Common Smooth-Hound, although some recent discussion may be found to indicate that these species are one and the same.

Adult **Undulate Ray** *Raja undulata* and **Small-eyed Ray** *Raja microocellata* have both been seen while diving in the east Wych Channel within the last five years.

Others

Mackerel *Scomber scombrus* are a warm-water visitor and usually arrive in the harbour from July and leave by late October, although the timing and especially the abundance vary quite considerably from year to year. They tend to stick to the deeper waters of the eastern areas, although they can sometimes be found as far up as the Wareham Channel and commonly in the turning basin and between the town quays. They seem to be mostly adult specimens that are generally present in association with their prey of sand eel and sprat.

Scad or **Horse Mackerel** *Trachurus trachurus* keep roughly the same arrival and departure seasons as Mackerel, although their feeding habits differ as they have very large eyes, making them effective night hunters of small fish. They have been observed at night driving large shoals of baitfish out of the water and onto the rocks in the Brownsea Roads area. They are the only species of 'Jack' (Carangidae) to be commonly found in UK waters, a family usually associated with more warm temperate zones.

Labridae (wrasse) are generally present throughout the year with much larger populations in the warmer months. They are not tolerant of very cold weather or low salinity following periods of very heavy rain, so probably move out to deeper waters at these times. They prefer to inhabit rocky terrain or areas providing plenty of cover; this includes most of the human-made structures within the harbour and also the eelgrass beds. Small specimens occur as far up as Rockley. Some very large **Ballan Wrasse** *Labrus bergylta* inhabit the rock armour between the Brownsea jetties. **Ballan** and **Corkwing** *Crenilabrus melops* are common whereas **Goldsinny** *Ctenolabrus rupestris*, **Cuckoo** *Labrus mixtus* and **Baillion's** *Crenilabrus bailloni* are occasionally seen.

European Eel *Anguilla Anguilla* In the 1960 and 1970s, eels were so abundant that it was almost impossible to fish in many parts of the harbour for any other species, as the eels would get to the bait before anything else. One of the authors (PP) can also recall paddling in the water at Whitecliff just by the railway bridge into Poole Park before the reclamation of land known as Baiter Park and the water being thick with elvers. Unfortunately, the European Eel population has reduced by over 95% in many areas of its former range (Jacoby and Gollock 2014). That said it is still possible to find eels over much of the harbour right up to and beyond its tidal limits.

The **Conger Eel** *Conger conger* is a frequent visitor in summer and autumn; relatively small specimens of up to 10 kg are caught at night by angling in the Brownsea roads area

Lamprey (sp.) The following is an interesting commercial fishing record: 'They can occur anywhere in harbour, normally seen October/November, around half dozen a year (1980–2019) attached to a bigger Bass or Pollock. Strangely we would have them all over one or two nights. If we caught on a particular night other boats fishing in different areas would also report the same. Sizes: 4–10 inches.' The colouration was described as blue-black back to white on the undersides, so these could be either juvenile, sea or river lamprey. The author (PP) can also report that what were thought to be Brook Lamprey were once (early 1970s) locally common in the stream that drains from Upton Heath along Watery Lane and into Lychett Bay.

Other than the (in)famous Atlantic **Sturgeon** *Acipenser sturio* killed in the river Frome in 1911, there are two other recent records of Sturgeon of over a metre in length caught and released from mullet ring nets off Ham common in the last 10 years, caught 4 or 5 years

Figure 7.1 (a) Tompot Blenny (b) Shore Rockling, (c) Anchovy and (d) Corkwing Wrasse.

apart and in roughly the same place it could have been the same animal. The Atlantic Sturgeon is considered extinct over much of its range with the nearest breeding location being the Gironde in France. It is a possibility that these were captive-bred specimens of a species not native to the UK that were released into the harbour. **Sterlet** *Acipenser ruthenus* would be a likelihood as although this is a freshwater species they are readily available as lesser pond fish and apparently do have some tolerance to the lower salinity conditions found in this area, although it would be wonderful to think that the Atlantic Sturgeon could make a return to UK waters in Poole Harbour.

Blennies and Gobies. There are over 20 species of blenny and over 60 species of goby occurring in European waters and differentiating between the species (especially some of the gobies) can be quite difficult. The list is therefore quite conservative at the time of writing and only those with fairly positive identification have been included in the list accompanying this chapter. These two families are likely to provide the largest contribution in the expansion of the list in future years. The two commonest gobies are probably the **Sand Goby** *Pomatoschistus minutus* and the **Two-spotted Goby** *Gobiusculus flavescens*. Interestingly, it's been observed that the Two Spotted goby as a mid-water species make a significant contribution to the diet of breeding terns in years when the sand eel and sprat are late in arriving. The commonest Blenny is the **Shanny** *Lipophrys pholis* and probably followed by the **Tompot Blenny** *Parablennius gattorugine*.

Sand Smelt *Atherina presbyter* are abundant and widespread throughout the warmer months. Casual observation would indicate that the spring influx is related to water temperature and although the numbers drop dramatically in November they are probably present all year round. Very large shoals occur around the jetties of Brownsea Roads during the summer and early autumn, especially in moderate tidal streams where they maintain station while feeding on plankton. They are the only representative of the 'Silversides' family to be found in UK waters and like the sand eel are a very important prey species for predatory fish and birds alike.

Lesser Sand Eel *Ammodytes tobianus* has played an important part in the local ecology as a prey species for both the predatory fish of the harbour like bass and birds especially terns. Unfortunately, their numbers, although fluctuating a little, have been in steady decline over the last few decades, and their current numbers are only 10–20% of what was once considered normal. They usually occur through the harbour entrance and over the sand bars to the sides of the main channels.

Garfish *Belone belone* are present at similar times to Mackerel, a fish associated with summer and early autumn. They are common and widespread at the eastern end of the harbour especially in the deeper waters of the harbour entrance, Brownsea Roads and main channel and their range extend at least as far as the turning basin and between the town quays. They have also featured in SIFCA beach seine nettings from Brownsea Castle beach. Garfish juveniles of not more than 50 mm have been seen close to the surface by the Brownsea National Trust jetty. It was also here that a group of around six fish were observed apparently cooperating in hunting prey (in this case small sand smelt). In the autumn of 2019 the author observed a mature Garfish of over 50 cm being taken while very much alive by a Greater Black-backed Gull in the waters off the lagoon wall of Brownsea.

Juvenile **John Dory** *Zeus faber* have been occasionally seen while diving in the small eelgrass patches between the Brownsea Island jetties.

On 16 October 2020 a **tuna** was seen close to 'Brownsea' cardinal buoy in Brownsea Roads leaping from the water in pursuit of prey (Mackerel) by two reliable witnesses. This fish was in excess of a metre in length and is therefore quite likely to have been a **Bluefin** *Thunnus thynnus* as its size precludes some of the other possibilities, and there have been several other Bluefins recorded quite far up the English Channel during 2020.

Readers are encouraged to submit their records to the Poole Harbour Study Group.

References

Aflalo, F.G. 1901. *Sea and Coast Fishing*. Grant Richards.

Census of Marine Life. 2003. How many fish in the sea? Census of marine life launches first report. *Science Daily*. Available from: www.science-daily.com/releases/2003/10/031024064333.htm (Accessed 24 October 2003).

Jacoby, D. and Gollock, M. 2014. *Anguilla anguilla*. The IUCN Red List of Threatened Species 2014: e.T60344A45833138. Available from: http://dx.doi.org/10.2305/IUCN.UK.2014-1.RLTS.T60344A45833138.en

SIFCA. 2022. Southern Inshore Fisheries and Conservation Authority. https://www.southern-ifca.gov.uk/ (Accessed 17 June 2022).

Acknowledgements

The following people and organisations contributed to the observations in this work: Organisations: SIFCA, Sea search, Environment Agency, Dorset Wildlife Trust, National Trust. People: Mike Bailey (local fisherman), Louise Schmitt (local fisherman), Paul Trowbridge (local fisherman), Chris Mowlem (local fisherman) and several others who do not wish to be named. Finally, apologies to all those who know more but have not been consulted.

CHAPTER 8

Waterbirds of Poole Harbour: Figures and Trends

KATHRYN E. ROSS, KATHARINE M. BOWGEN, NIALL H.K. BURTON,
ROGER J.H. HERBERT and RICHARD A. STILLMAN

Abstract

Poole Harbour supports a diverse and numerous waterbird assemblage. This assemblage includes internationally important numbers of non-breeding waterbirds, including Shelduck *Tadorna tadorna* and Black-tailed Godwit *Limosa limosa*, and nationally important numbers of Avocet *Recurvirostra avosetta*, Eurasian Spoonbill *Platalea leucorodia* and Little Egret *Egretta garzetta*. Important breeding populations include Common Tern *Sterna hirundo*, Sandwich Tern *Sterna sandvicensis*, Mediterranean Gull *Larus melanocephalus* and Black-headed Gull *Chroicocephalus ridibundus*. Here, we report waterbird numbers in Poole Harbour based on the Wetland Bird Survey (WeBS), Seabird Monitoring Programme (SMP) and various other ad hoc surveys. In addition, we report notable population trends and recent research.

Keywords: waterbirds, WeBS, Poole Harbour, Special Protection Area, population trends

Correspondence: Kathryn.Ross@toiohomai.ac.nz

Introduction

Poole Harbour, with its diverse range of wetland habitats, including mudflats, sandflats, lagoons, saltmarsh and coastal grasslands, supports an abundant assemblage of bird life, particularly waterbirds. Waterbirds are useful indicators of ecosystem health and environmental change (Ogden *et al.* 2014). They provide many ecosystem services, including ecological and cultural values such as birdwatching and ecotourism (Green and Elmberg 2014). The status of waterbirds in Poole Harbour has been examined previously by Pickess and Underhill-Day (2002) and Pickess (2007). In this chapter, we provide updated numbers of waterbirds recorded in Poole Harbour and notable population trends following on from these previous accounts. We focus mainly on non-breeding waterbirds. Furthermore, we present a summary of recent ornithological research conducted in Poole Harbour in the decade prior to this publication and highlight a few conservation management issues.

Protection and designations

Poole Harbour was classified as a Special Protection Area (SPA) in 1999 in accordance with the European Council Directive 2009/147/EC on the conservation of wild birds, also

Kathryn E. Ross, Katharine M. Bowgen, Niall H.K. Burton, Roger J.H. Herbert and Richard A. Stillman, 'Waterbirds of Poole Harbour: Figures and Trends' in: *Harbour Ecology*. Pelagic Publishing (2022). © Kathryn E. Ross, Katharine M. Bowgen, Niall H.K. Burton, Roger J.H. Herbert and Richard A. Stillman. DOI: 10.53061/STXT1634 known as the Birds Directive. This directive protects rare and vulnerable birds (listed in Annex 1 of the Directive) and regularly occurring migratory species. Following the UK's exit from the European Union, UK SPAs have retained their protected status as part of the UK national site network under UK legislation, including the Wildlife & Countryside Act 1981 and the Conservation of Habitats and Species Regulations 2010. To qualify as an SPA interest feature, a species must regularly occur in numbers exceeding 1% of the national or biogeographic (international) population, or as part of an assemblage of >20,000 individuals. The original SPA designation was made based on the presence of qualifying numbers of non-breeding Pied Avocet *Recurvirostra avosetta*, Shelduck *Tadorna tadorna* and Black-tailed Godwit *Limosa limosa*, and a waterfowl assemblage including Dark-bellied Brent Goose *Branta bernicla bernicla*, Cormorant *Phalacracorax carbo*, Teal *Anas crecca*, Goldeneye *Bucephala clangula*, Pochard *Aythya farina*, Red-breasted Merganser *Mergus serrator*, Dunlin *Calidris alpina*, Curlew *Numenius arquata*, Redshank *Tringa totanus*, Spotted Redshank *Tringa erythropus*, Greenshank *Tringa nebularia*, and Black-headed Gull *Chroicocephalus ridibundus*, and breeding populations of Common Tern *Sterna hirundo* and Mediterranean Gull *Larus melanocephalus* (JNCC 2017). Poole Harbour also qualifies as a Ramsar site due to its internationally important waterbird species and assemblage.

In 2017, the seaward boundary of the Poole Harbour SPA was extended to include subtidal areas (Figure 8.1), due to several of the waterbird features of interest utilising areas outside of the existing SPA boundary. The landward boundary was also extended to

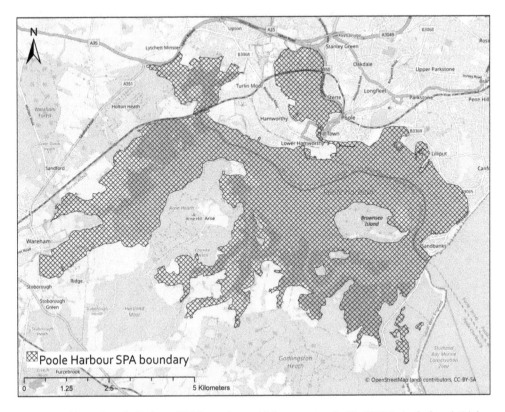

Figure 8.1 Map of Poole Harbour SPA boundary, which was extended in 2017 to include subtidal areas (Base map and data from OpenStreetMap and OpenStreetMap Foundation © OpenStreetMap contributors. Boundary contains public-sector data from © JNCC/NE/NRW/SNH/DOENI 2019. Contains OS data © Crown Copyright and database right 2016).

include the north-west of Lychett Bay, which is now regularly inundated by the tide and used for foraging and roosting by several species, including Shelduck and Black-tailed Godwit (Natural England 2016). In addition, three species were added to the list of qualifying features on account of their increased numbers within the SPA: non-breeding Little Egret *Egretta garzetta* and Eurasian Spoonbill *Platalea leucorodia* and breeding Sandwich Tern *Sterna sandvicensis*. A summary of the SPA qualifying ornithological features is presented in Table 8.1.

Data sources

Data on non-breeding waterbirds were taken from the Wetland Bird Survey (WeBS) Core Counts. WeBS is the UK monitoring scheme for non-breeding waterbirds and is a partnership of the British Trust for Ornithology (BTO), the Royal Society for the Protection of Birds (RSPB) and the Joint Nature Conservation Committee (JNCC), in association with the Wildfowl & Wetlands Trust (WWT). Core Counts is the principal WeBS scheme, providing information to assess the status and trends of non-breeding waterbird populations from national to site levels (Frost *et al.* 2020; Woodward *et al.* 2019). The associated WeBS

Table 8.1 A summary of qualifying ornithological features for Poole Harbour SPA and qualifying counts

Species	Breeding (B) or Non-breeding (NB)	International (I) or national (N) importance	Year recognised	Type	Qualifying count and importance at time of qualification
Black-tailed Godwit *Limosa limosa*	NB	I	1999	Regularly occurring migrant	1,576 individuals[1] 2.3% of biogeographic population[3]
Shelduck *Tadorna tadorna*	NB	I	1999	Regularly occurring migrant	3,569 individuals[1] 1.2% of biogeographic population[3]
Waterbird assemblage	NB	I	1999	Non-breeding waterfowl assemblage	23,498 individuals[1]
Pied Avocet *Recurvirostra avosetta*	NB	N	1999	Annex 1	459 Individuals[1] 36.1% of GB population[3]
Eurasian Spoonbill *Platalea leucorodia*	NB	N	2017	Annex 1	20 individuals[2] 45% of GB population[4]
Little Egret *Egretta garzetta*	NB	N	2017	Annex 1	114 individuals[2] 2.5% of GB population[5]
Common Tern *Sterna hirundo*	B	N	1999	Annex 1	155 pairs[3] 1.3% of GB population[3]
Mediterranean Gull *Larus melanocephalus*	B	N	1999	Annex 1	5 pairs[3] 22.7–38.5% of GB population[3]
Sandwich Tern *Sterna sandvicensis*	B	N	2017	Annex 1	181 pairs[6] 1.6% of GB population[5]

Sources:

[1] WeBS 1992/93–1996/97 mean peak.

[2] WeBS 2009/10–2013/14 mean peak.

[3] From SPA citation (March 1999) held on Register of European Marine Sites for Great Britain, based on five year mean, 1993–7.

[4] GB count is based on Holt *et al.* (2015).

[5] GB population estimates from Musgrove *et al.* (2013).

[6] Seabird Monitoring Programme (SMP) Brownsea Island breeding seabird data, 2010–14 mean.

Low Tide Counts scheme provides information on the distributions of foraging water-birds on estuaries. Core Counts have been regularly conducted in Poole Harbour since the mid-1960s and are carried out monthly, typically at high tide, by skilled volunteer bird-watchers. Unless otherwise stated, the WeBS figures quoted here refer to the mean peak counts for the most recent five winters for which data were available (2014/15–2018/19).

Data on breeding seabirds were compiled from several sources. Counts of breeding terns and gulls are conducted annually at several sites within the harbour as part of the partnership Seabird Monitoring Programme (SMP), coordinated by JNCC. Counts of apparently occupied nests (AONs) were recorded following standard protocols (Walsh 1995). AONs are counted regularly on Brownsea Island by Dorset Wildlife Trust, while additional ad hoc gull counts have been made on the *Spartina* saltmarsh islands within the Wareham Channel (Chown 2015; Hopper 2016).

Data from other ad hoc surveys reported on the Birds of Poole Harbour website (https://www.birdsofpooleharbour.co.uk/) have also been included: breeding season surveys of Redshank (Archer and Branston 2014) and Water Rail *Rallus aquaticus* (Hopper 2013), a winter survey of Woodcock *Scolopax rusticola* (Hopper 2014) and a year-round survey of the harbour's Ardeidae (Hopper 2018).

Overwintering populations

The ornithological interest of Poole Harbour is greatest during the winter months, when the SPA supports nationally and internationally important numbers of non-breeding waterbirds. Table 8.2 shows the five-year mean peak counts for species recorded by WeBS between 2014/15 and 2018/19 and the month in which peak counts were recorded. The most abundant species (those comprising more than 5% of the total harbour assemblage) were Wigeon *Mareca penelope* (12%), Teal (10%), Black-tailed Godwit (7%), Dunlin (7%), Avocet (5%), Lapwing *Vanellus vanellus* (5%), Herring Gull *Larus argentatus* (5%) and Common Gull *Larus canus* (5%); mean peak counts of each of these species exceeded 1,400. Some species occur in relatively high numbers but, due to their national abundance, are not considered nationally important populations (e.g. Dunlin, Lapwing, Curlew and Oystercatcher *Haematopus ostralegus*). Conversely, some species occurring in much lower numbers are considered nationally important due to smaller national populations (e.g. Greenshank and Green Sandpiper *Tringa ochropus*).

Up to 2.1% of the biogeographic population of Black-tailed Godwit is currently sup-ported by Poole Harbour, making it the species of greatest international importance. Until recently, the species of greatest national importance here was Avocet. During the 2014/15–2018/19 period, Poole Harbour held up to 17.6% of the Great Britain (GB) winter popula-tion of Avocet, making it the fourth most important site for this species (Frost *et al.* 2020). However, since 2006, Spoonbill has become proportionally more important with Poole Harbour supporting up to 33% of national numbers in 2018/19.

Shelduck numbers in Poole Harbour currently represent 2.6% of the GB population; however, they no longer exceed the threshold for international importance. Other impor-tant wildfowl species include the Siberian-breeding Dark-bellied Brent Geese *Branta berni-cla bernicla,* with numbers consistently in the thousands, comprising 1.8% of the national population. The Nearctic-breeding Light-bellied Brent Geese *Branta bernicla hrota* is a more sporadic visitor, with only a single count of 95 individuals recorded in the last five years (although this was sufficient to exceed the threshold for national importance).

Several species of gull overwinter in the harbour in large numbers, peaking in January/February – Black-headed Gull being the most common (10% of the harbour assemblage), and Herring and Common Gull (both 6%). However, it is the less abundant Mediterranean Gull that occurs in nationally important numbers during the winter (1.4% of the GB population).

Table 8.2 Non-breeding waterbird numbers based on five-year (2014/15–2018/19) mean peak counts for Poole Harbour from Wetland Bird Survey (WeBS) Core Counts and the month of the peak count in the most recent year with a count (after Frost *et al.* 2020)

Common name	Latin name	Five-year mean peak count	Peak month	% of GB population*	% of biogeographic population**
Wigeon	*Mareca penelope*	3,916	Dec	0.9	0.3
Teal	*Anas crecca*	3,265	Dec	0.8	0.7
Black-headed Gull	*Chroicocephalus ridibundus*	3,258	Feb	0.1	0.2
Black-tailed Godwit	***Limosa limosa***	**2,336**	**Mar**	**6.0**	**2.1**
Dunlin	*Calidris alpina*	2,283	Dec	0.7	0.2
Brent Goose (Dark-bellied)	*Branta bernicla bernicla*	1,761	Feb	1.7	0.8
Avocet	***Recurvirostra avosetta***	**1,535**	**Jan**	**17.6**	**1.6**
Lapwing	*Vanellus vanellus*	1,467	Jan	0.2	0.1
Herring Gull	*Larus argentatus*	1,437	Jan	0.2	0.1
Common Gull	*Larus canus*	1,436	Feb	0.2	0.1
Shelduck	***Tadorna tadorna***	**1,225**	**Feb**	**2.6**	**0.5**
Curlew	*Numenius arquata*	1,108	Feb	0.9	0.1
Redshank	*Tringa totanus*	1,059	Nov	1.1	0.4
Oystercatcher	*Haematopus ostralegus*	996	Nov	0.3	0.1
Cormorant	*Phalacrocorax carbo*	664	Nov	1.1	0.6
Canada Goose†	*Branta canadensis*	475	Dec	0.3	7.9
Mallard	*Anas platyrhynchos*	353	Sep	0.1	<0.1
Pintail	*Anas acuta*	263	Feb	1.3	0.4
Coot	*Fulica atra*	222	Jan	0.1	<0.1
Bar-tailed Godwit	*Limosa lapponica*	207	Mar	0.4	0.1
Red-breasted Merganser	*Mergus serrator*	207	Dec	2.0	0.2
Little Egret	***Egretta garzetta***	**192**	**Sep**	**1.7**	**0.2**
Mute Swan	*Cygnus olor*	185	Sep	0.4	0.4
Grey Plover	*Pluvialis squatarola*	164	Feb	0.5	0.1
Gadwall	*Mareca strepera*	159	Dec	0.5	0.1
Shoveler	*Spatula clypeata*	140	Jan	0.7	0.2
Great Crested Grebe	*Podiceps cristatus*	139	Jan	0.8	<0.1
Greylag Goose	*Anser anser*	96	Jan	<0.1	<0.1
Great Black-backed Gull	*Larus marinus*	86	Feb	0.1	<0.1
Goldeneye	*Bucephala clangula*	79	Jan	0.4	<0.1
Tufted Duck	*Aythya fuligula*	74	Dec	0.1	<0.1
Turnstone	*Arenaria interpres*	70	Mar	0.2	0.1
Mediterranean Gull	***Ichthyaetus melanocephalus***	**55**	**Mar**	**1.4**	**<0.1**
Spoonbill	***Platalea leucorodia***	**52**	**Oct**	**33.3**	**0.3**
Knot	*Calidris canutus*	49	Jan	<0.1	<0.1
Snipe	*Gallinago gallinago*	46	Nov	<0.1	<0.1
Sanderling	*Calidris alba*	46	Jan	0.2	<0.1
Grey Heron	*Ardea cinerea*	45	Oct	0.1	<0.1
Ringed Plover	*Charadrius hiaticula*	44	Nov	0.1	0.1
Moorhen	*Gallinula chloropus*	35	Sep	<0.1	<0.1
Sandwich Tern	***Thalasseus sandvicensis***	**28**	**Sep**	**52.8**	**<0.1**
Little Grebe	*Tachybaptus ruficollis*	27	Nov	0.2	<0.1
Greenshank	*Tringa nebularia*	25	Sep	3.1	<0.1
Lesser Black-backed Gull	*Larus fuscus*	24	Oct	<0.1	<0.1
Brent Goose (Light-bellied)	*Branta bernicla hrota*	20	Nov	1.3	0.1
Black-necked Grebe	*Podiceps nigricollis*	16	Dec	13.9	<0.1
Spotted Redshank	*Tringa erythropus*	12	Oct	17.9	<0.1
Pochard	*Aythya ferina*	10	Feb	<0.1	<0.1
Shag	*Phalacrocorax aristotelis*	9	Dec	<0.1	<0.1
Yellow-legged Gull	*Larus michahellis*	7	Sep	0.8	<0.1
Barnacle Goose	*Branta leucopsis*	6	Feb	<0.1	<0.1
Whimbrel	*Numenius phaeopus*	5	Sep	5.0	<0.1

(Continued)

Table 8.2 (Continued)

Common name	Latin name	Five-year mean peak count	Peak month	% of GB population*	% of biogeographic population**
Common Scoter	*Melanitta nigra*	5	Nov	<0.1	<0.1
Kingfisher	*Alcedo atthis*	5	Sep	0.1	<0.1
Green Sandpiper	*Tringa ochropus*	5	Sep	1.7	<0.1
Common Sandpiper	*Actitis hypoleucos*	5	Sep	9.6	<0.1
Water Rail	*Rallus aquaticus*	5	Nov	0.1	<0.1
Common Tern	*Sterna hirundo*	4	Sep	<0.1	<0.1
Golden Plover	*Pluvialis apricaria*	4	Mar	<0.1	<0.1
Great Northern Diver	*Gavia immer*	3	Dec	0.1	0.1
Ruff	*Calidris pugnax*	2	Oct	0.2	<0.1
Curlew sandpiper	*Calidris ferruginea*	2	Oct	4.0	<0.1
Goosander	*Mergus merganser*	2	Nov	<0.1	<0.1
Great White Egret	*Ardea alba*	2	Sep	2.8	<0.1
Slavonian Grebe	*Podiceps auritus*	2	Jan	0.2	<0.1
Little Stint	*Calidris minuta*	2	Oct	2.0	<0.1
Jack Snipe	*Lymnocryptes minimus*	1	Jan	<0.1	<0.1
Red-throated Diver	*Gavia stellata*	1	Jan/Feb	<0.1	<0.1
Eider	*Somateria mollissima*	1	Nov	<0.1	<0.1
Black Swan[†]	*Cygnus atratus*	1	Feb	2.1	<0.1

*GB population estimates from Woodward *et al.* (2020), except for Spoonbill which is based on the maximum monthly WeBs count for 2018/19.

**Biogeographic population estimates from BirdLife International (2015).

[†]Non-native/introduced species.

Species exceeding thresholds for international importance are highlighted in dark grey and national importance in light grey. Recognised interest features in bold.

The number of Woodcock wintering in the pastures and grazed floodplain of Poole Harbour was estimated as 644 birds in 2014, with the highest densities of 72 birds/km² recorded in the south-west of the harbour (Hopper 2014).

Population trends

Here we examine recent changes in the numbers of non-breeding waterbirds in the Poole Harbour SPA, drawing from Woodward *et al.* (2019) and providing a comparison with regional (Southwest England) and national trends. Figure 8.2 shows long-term trends for the five non-breeding waterbird interest features of the Poole Harbour SPA and that for the overall assemblage. With the exception of Shelduck, each species has undergone population increases since SPA classification. Table 8.3 shows the percentage changes in population numbers in the short (5-year), medium (10-year) and long (25-year) term for all species with 5-year mean peak counts of >25 individuals.

While numbers of Black-tailed Godwit have been steadily increasing in Poole Harbour in the long term, the trends for GB and the Southwest region have been climbing at a faster rate (Woodward *et al.* 2019). Thus, the proportion of the overwintering population supported by Poole Harbour is decreasing, suggesting the site is close to carrying capacity for this species.

Avocet numbers in Poole Harbour have increased steadily from a few individuals in the early 1980s, after the Dorset Wildlife Trust began managing water levels at Brownsea Island Lagoon, creating suitable habitat. Since 2004, numbers have remained relatively stable, although the proportion of the GB population in Poole Harbour has dropped in this period (Woodward *et al.* 2019). As numbers increase at other sites in the Southwest

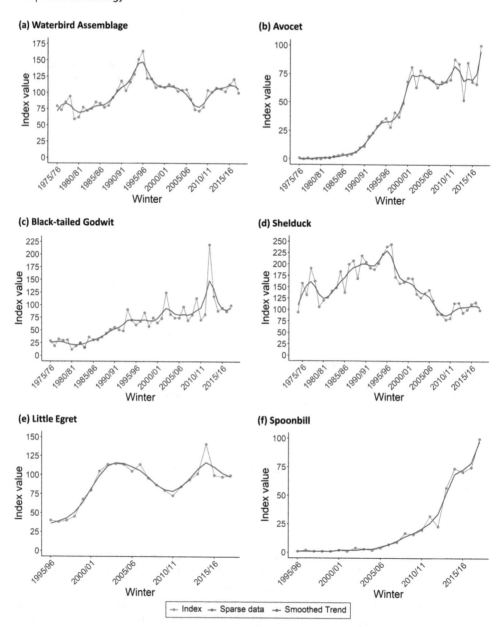

Figure 8.2 Trends in five-year mean peak counts for Poole Harbour SPA waterbird assemblage and interest feature species (based on Woodward *et al.* 2019). Values reported as an index relative to most recent year. Note different y-axis scales for each species and different x-axis scales for (e) and (f).

region and nationally, the relative importance of Poole Harbour for Avocets is dropping. This suggests that Poole Harbour is at carrying capacity for this species.

Little Egret numbers increased substantially between the mid-1990s and 2004, after which a drop in numbers occurred. Since 2010, however, Little Egret numbers have been steadily rising again. This pattern appears to generally track the Southwest regional and national trends for this species (Woodward *et al.* 2019), suggesting that the environmental conditions within Poole Harbour remain relatively favourable for Little Egret.

Spoonbill was a rare visitor to Poole Harbour prior to 2004, with only one or two individuals appearing each winter. Their numbers have risen steadily since then, and

Table 8.3 Changes in non-breeding waterbird populations in the short (5-year), medium (10-year) and long (25-year) term

Common name	Latin name	Short term % Δ	Medium term % Δ	Long term % Δ
Wigeon	*Mareca penelope*	53	87	370
Teal	*Anas crecca*	11	111	173
Black-tailed Godwit	*Limosa limosa*	–23	11	63
Dunlin	*Calidris alpina*	14	–12	–54
Brent Goose (Dark-bellied)	*Branta bernicla bernicla*	43	59	–2
Avocet	*Recurvirostra avosetta*	–9	10	317
Lapwing	*Vanellus vanellus*	19	27	–63
Shelduck	*Tadorna tadorna*	9	–5	–45
Curlew	*Numenius arquata*	5	–33	–36
Redshank	*Tringa totanus*	19	34	–41
Oystercatcher	*Haematopus ostralegus*	–6	–17	–38
Cormorant	*Phalacrocorax carbo*	18	32	13
Mallard	*Anas platyrhynchos*	–25	–12	-14
Pintail	*Anas acuta*	47	–13	33
Coot	*Fulica atra*	100	100	100
Red-breasted Merganser	*Mergus serrator*	–36	–26	-55
Bar-tailed Godwit	*Limosa lapponica*	12	96	-31
Little Egret	*Egretta garzetta*	21	–3	164
Mute Swan	*Cygnus olor*	8	–13	56
Grey Plover	*Pluvialis squatarola*	19	50	–41
Gadwall	*Mareca strepera*	–11	32	146
Shoveler	*Spatula clypeata*	–35	–24	–52
Great Crested Grebe	*Podiceps cristatus*	5	56	318
Goldeneye	*Bucephala clangula*	–44	–37	–54
Tufted Duck	*Aythya fuligula*	5	33	–53
Turnstone	*Arenaria interpres*	7	50	185
Spoonbill	*Platalea leucorodia*	204	1029	7800
Knot	*Calidris canutus*	–8	42	–2
Sanderling	*Calidris alba*	182	1450	1450
Grey Heron	*Ardea cinerea*	11	15	818
Ringed Plover	*Charadrius hiaticula*	26	1	–66
Little Grebe	*Tachybaptus ruficollis*	45	126	41
Greenshank	*Tringa nebularia*	–6	35	33
Waterbird assemblage		10	26	–1

Based on a reference winter of 2016/17.

Increases of 25% or more in bold, declines of >25% highlighted in light grey and >50% in dark grey (after Woodward *et al.* 2019).

Only species with 5-year mean peak counts of 25 or above were included.

Poole Harbour now represents the second-most important wintering site in the UK for Spoonbill, after the North Norfolk Coast. A peak count of 71 birds was recorded in October 2017. Numbers of Spoonbill overwintering in the UK have been increasing, but a growing proportion of the national and regional population overwinter in Poole Harbour (Woodward *et al.* 2019), suggesting conditions at this site are favourable for this species. Ringing data have indicated that many of Poole Harbour's wintering Spoonbills originate in the Netherlands, but a bird from Denmark has also been reported (Hopper 2018).

Shelduck numbers have been relatively stable over the past decade, following a period of significant decline between the mid-1990s and 2010. While this long-term trend some-what mirrors the national trend, it contrasts with the Southwest regional trend (Woodward *et al.* 2019). A declining proportion of regional numbers are supported by Poole Harbour, suggesting that site-specific pressures may be affecting Shelduck.

Several other species of wildfowl have experienced notable long-term declines in Poole Harbour (Woodward *et al.* 2019). Shoveler *Spatula clypeata* numbers have declined 52%

over 25 years, in contrast to increasing regional and national trends. Several diving ducks have also experienced long-term declines in numbers, including Tufted Duck *Aythya fuligula*, which has declined by 53% despite increasing regional and national trends over this period. Notable declines have also been observed in Goldeneye (–54%) and Pochard (–97%), though these declines follow the regional and national trends so may not indicate a site-specific issue.

Wader species that have experienced long-term declines in Poole Harbour include Dunlin (–54%), Oystercatcher (–38%), Lapwing (–63%) and Grey Plover *Pluvialis squatarola* (–41%). While numbers are still lower than the peak observed in the early 1990s, Bar-tailed Godwit *Limosa lapponica* numbers have been climbing steadily over the past decade. The Turnstone *Arenaria interpres* population has also increased markedly over the past 25 years (+180%).

Notable long-term population increases have occurred in Wigeon (+370%) and Teal (+173%). Great Crested Grebe *Podiceps cristatus*, Little Grebe *Tachybaptus ruficollis* and Coot *Fulica atra* numbers have also increased over the past decade (+56%, +126%, +100% respectively).

Apart from a dip in numbers between 2006/7 and 2009/10, the overall waterbird assemblage numbers have been relatively stable since 1998.

Breeding populations

Table 8.4 summarises the most recent counts of breeding gulls and terns at the two key breeding sites within Poole Harbour. Common Terns and Sandwich Terns breed at Brownsea Island Lagoon on specially created gravel islands. The main breeding site for the nationally important Mediterranean Gull, as well as Black-headed Gull, is in the Wareham Channel adjacent to Holton Heath on three islands of *Spartina anglica* saltmarsh, known locally as 'Gull Islands'. These islands are shrinking due to erosion, and instances of illegal egg harvesting have been documented (Chown 2015). Between 2015 and 2018, the number of Black-headed Gull nests recorded fluctuated between ~2,600 and ~6,400. Mediterranean Gulls are present in lower numbers than Black-headed Gulls but represent

Table 8.4 Breeding seabird data from two key sites (Brownsea Island and *Spartina* Gull Islands) within Poole Harbour

Common name	Latin name	Average count (pairs)		% of GB breeding population (for combined site estimates)*
		Brownsea Island	Spartina Gull Islands	
Mediterranean Gull	*Ichthyaetus melanocephalus*	3[1]	51[3]	4.5
Black-headed Gull	*Chroicocephalus ridibundus*	177[1]	4710[4]	3.5
Sandwich Tern	*Thalasseus sandvicensis*	173[2]	–	1.4
Common Tern	*Sterna hirundo*	129[2]	–	1.3
Great black-Backed Gull	*Larus marinus*	11[1]	2[5]	0.1

Data for Brownsea Island collected by Dorset Wildlife Trust volunteers. Data for Spartina Islands recorded in the Seabird Monitoring Programme (SMP) database and in Hopper (2016) for Mediterranean Gull. Based on average counts of apparently occupied nests (AONs) in the following years:

[1] 2015–19.

[2] 2015, 2017–19.

[3] 2015, 2016, reported in Hopper (2016).

[4] 2015–18.

[5] 2016.

* Based on breeding season population estimates from Woodward *et al.* (2020).

a greater proportion of the GB breeding population estimate (4.5%). Gulls breeding in the urbanised sites around the harbour were surveyed in 2016. Numbers of AONs were 811 Herring Gull, 76 Lesser Black-backed Gull *Larus fuscus* and 11 Great Black-backed Gull *Larus marinus* (Hopper 2016).

A 2014 survey of breeding Redshank in Poole Harbour estimated between 74 and 147 pairs, depending on the methodology used (Archer and Branston 2014). The maximum estimate suggests that Poole Harbour could support up to 1.2% of the UK saltmarsh breeding Redshank (Archer and Branston 2014). The figures also suggest an increase on the previous estimate of 69 pairs in 2004, which could be linked to saltmarsh recovery resulting from the Sika Deer control programme implemented at Arne in 2006 (Archer and Branston 2014). As the national breeding Redshank population is rapidly declining – it has been placed on the Amber List of Birds of Conservation Concern (Eaton *et al.* 2015) and in the 'vulnerable' category of the International Union for Conservation of Nature Red List (Birdlife International 2015) – Poole Harbour could be considered an important breeding site for this species.

Comprehensive surveys of Poole Harbour's breeding Ardeidae (Grey Heron *Ardea cinerea* and Little Egret) were conducted in 2017 and 2018 (Hopper 2018). Arne Heath Heronry is the key site in the harbour for both species. Nine pairs of Grey Heron nested at the Arne Heath Heronry in 2017 and 11 pairs in 2018 (Hopper 2018). Brownsea Island was the first documented UK breeding site for Little Egrets in 1996, but heavy predation by Ravens in 2005 led the colony to relocate to the Arne Heath Heronry (Hopper 2018). In 2017, 32 Little Egret pairs nested at Arne Heath, but in 2018, only 11 pairs nested (Hopper 2018).

A breeding season survey of Water Rail in 2013 recorded 177 pairs and 122 single birds in reedbed areas surrounding the harbour, mainly on the western side (Hopper 2013).

Recent waterbird research in Poole Harbour

Poole Harbour's abundant bird life, the various anthropogenic activities that occur on the site and its proximity to an urban centre have provided ideal conditions for research into human-wildlife conflicts and other conservation issues. In recent years, a number of research projects have been undertaken on the birds within Poole Harbour. Here we provide a brief summary of work that has been completed between 2010 and 2020.

Several studies have assessed the impacts of disturbance on Poole Harbour's waterbirds. Liley and Fearnley (2012) assessed the effects of recreational disturbance on Poole Harbour's wintering waterbirds. They showed that disturbance led to a redistribution of birds around the harbour and peak disturbance coincided with the holiday period in December. Off-lead dog walking disturbed the birds more than any other activity. While the severity of disturbance was higher for water-based activities (e.g. canoeing, pump-scoop dredging), these events occurred infrequently. However, the extent to which these disturbances were affecting the fitness of the birds was uncertain. Collop (2017) found the levels of human disturbance were not currently reducing the carrying capacity of Poole Harbour for wintering waders and wildfowl. However, simulation modelling using an individual-based model (IBM) indicated that current levels of disturbance in combination with predicted future stresses such as reduction in prey availability through overexploitation (bait-digging) and loss of habitat through sea level rise could lead to a reduction in birds' ability to meet their energetic requirements (Collop 2017).

The effects of bait-digging on bird distribution and foraging behaviour in Arne Bay and Holes Bay were examined by Fearnley *et al.* (2013). Clarke (2018) also looked at the effects of intertidal harvesting on benthic communities and the implications for wader populations. IBM modelling work indicated a compensatory shift in feeding behaviour of Oystercatchers from feeding on preferred bivalve prey to feeding on worms (Clarke 2018).

However, field surveys showed no significant effect of shellfish dredging on food intake rates or bird distribution within the harbour – though continued monitoring was recommended (Clarke 2018).

The invertebrate prey availability within the intertidal zone was surveyed by Herbert *et al.* (2010). This study included a comparison of the invertebrate biomass availability and energetic requirements of birds present in different sectors of the harbour (those used for WeBS Low Tide Counts). Other modelling work by Bowgen (2015, 2017) using an IBM of Poole Harbour showed the effects of various changes in invertebrate availability on wader species (Dunlin, Redshank, Black-tailed Godwit, Oystercatcher and Curlew). This work showed that larger birds with specific feeding strategies, like Curlew, were more susceptible to changes in invertebrate abundance than more generalist feeders like Oystercatcher. Invertebrate prey availability for Avocet within Poole Harbour was examined by Ross (2013). The Avocet population was shown to be vulnerable to predicted sea level rise, particularly if Brownsea Island Lagoon became inundated. The importance of nektonic prey, such as gobies (*Pomatoschistus* sp.), was also shown to be a critical food resource for Avocet (Ross 2013). The impact of algal mats on wading birds and their invertebrate prey was assessed by Thornton (2016). The presence of residual algal mats led to changes in foraging behaviour of wintering birds; furthermore, an effect was also observed in areas where mats had been present the previous summer. The algal mats generally led to a decrease in energy availability for birds due to an increase in small, low-quality prey such as small worms and bivalves. Some species were observed to be adapting to these changes by feeding on smaller prey items than they would under 'normal' conditions, whereas other species were found to feed on prey items from the surface of the mats (Thornton 2016).

Conservation management issues and future research

Issues of coastal erosion, habitat loss due to sea level rise and recreational disturbance have all been identified as potential causes of concern to Poole Harbour's bird populations (Chown 2015; Ross 2013; Collop 2017).

As the long-term prospects of tern colonies are under threat from sea level rise and breaching of the sea wall that separates the Brownsea Lagoon from the harbour, an alternative site has been created at Little Sea, Studland. In addition, new lagoon habitat has been created at Arne Bay and Poole Park to offset future habitat loss due to sea level rise. Improvement works to the two freshwater lakes and lagoon at Poole Park were completed in April 2019. These works were designed to enhance the wildlife and amenity function of the site. Poole Park Lagoon was dredged to increase its depth and the dredged material was used to create islands for birds to roost.

While much research effort to date has been focused on the overwintering bird population, more studies on the important tern and gull breeding populations would be useful to determine the relative impacts of disturbance and predation. The risks to the breeding gull colonies of habitat loss on the *Spartina* 'Gull Islands' due to erosion and sea level rise should be assessed. More intensive monitoring to ensure egg collecting is not occurring would also be beneficial.

Foraging locations of terns originating in Poole Harbour and other south coast SPAs were recorded as part of a wider study of tern foraging ranges (Perrow *et al.* 2010; 2016). The study showed that Common Terns had a fairly restricted range, foraging mainly in the harbour itself, whereas Sandwich Tern ventured up to 30 km west along the Purbeck Coast, as far as Osmington Mills. This research informed the boundary selection for the Solent and Dorset Coast SPA, which was classified in January 2020.

Reintroduction programmes for two avian apex predators are currently underway in the vicinity of Poole Harbour, supported by the Birds of Poole Harbour, the Roy Dennis

Wildlife Foundation and Forestry England. Between 2017 and 2021, 60 Osprey *Pandion haliaetus* chicks were being translocated from Scotland and released in the Poole Harbour area. Natural England has also issued licenses for the translocation of up to 60 White-tailed Eagles *Haliaeetus albicilla* from Scotland to the Isle of Wight between 2019 and 2023, with the hopes that the reintroduced population will disperse along the south coast to sites including Poole Harbour. Restoring breeding populations of these top predators to the south coast could provide a boost to tourism in Poole Harbour, in addition to the ecological benefits.

Acknowledgements

This chapter contains Wetland Bird Survey (WeBS) data from Waterbirds in the UK 2018/19© copyright and database right 2020. WeBS is a partnership jointly funded by the BTO, RSPB and JNCC, in association with WWT, with fieldwork conducted by volunteers. Breeding seabird data were extracted from the Seabird Monitoring Programme (SMP) database (https://app.bto.org/seabirds). Data have been provided to the SMP by the generous contributions of nature conservation and research organisations, and many volunteers throughout Britain and Ireland. Data on breeding gulls and terns were collected by Dorset Wildlife Trust volunteers and kindly provided by Luke Johns. The authors would also like to acknowledge the Birds of Poole Harbour website (https://www.birdsofpooleharbour.co.uk/) as an invaluable source of up-to-date information from the Poole Harbour birding community.

References

Archer, R. and Branston, T. 2014. Poole Harbour breeding redshank survey 2014. Royal Society for the Protection of Birds. Unpublished report. Available from: https://www.birdsofpooleharbour.co.uk/surveys/ (accessed 28 September 2021).

BirdLife International. 2015. European Red list of birds. Office for Official Publications of the European Communities, Luxembourg. https://doi.org/10.2779/975810

Bowgen, K.M., Stillman, R.A., and Herbert, R.J.H. 2015. Predicting the effect of invertebrate regime shifts on wading birds: Insights from Poole Harbour, UK. *Biological Conservation* 186: 60–8. https://doi.org/10.1016/j.biocon.2015.02.032

Bowgen, K.M. 2017. Predicting the effect of environmental change on wading birds: Insights from individual-based models. PhD Thesis. Bournemouth University & HR Wallingford. Available from: http://eprints.bournemouth.ac.uk/27010/ (accessed 28 September 2021).

Chown, D. 2015. A report to birds of Poole Harbour: Census of breeding gulls on Spartina Islands in Wareham Channel, Poole Harbour, May 2015. Birds of Poole Harbour, Poole, UK. Available from: https://www.birdsofpooleharbour.co.uk/surveys/ (accessed 28 September 2021).

Clarke, L. 2018. Ecosystem impacts of intertidal invertebrate harvesting: From benthic habitats to bird predators. PhD Thesis. Bournemouth University.

Available from: http://eprints.bournemouth.ac.uk/31136/ (accessed 28 September 2021).

Collop, C. 2017. Impact of human disturbance on coastal birds: Population consequences derived from behavioural responses. PhD thesis. Bournemouth University. Available from: http://eprints.bournemouth.ac.uk/27019/ (accessed 28 September 2021).

Eaton, M.A., Aebischer, N.J., Brown, A.F., Hearn, R., Lock, L., Musgrove, A.J., Noble, D.G., Stroud, D.A., and Gregory, R.D. 2015. Birds of conservation concern 4: The population status of birds in the United Kingdom, Channel Islands and the Isle of Man. *British Birds* 108: 708–46. Available from: https://britishbirds.co.uk/ (accessed 28 September 2021).

Fearnley, H., Cruickshanks, K., Lake, S., and Liley, D. 2013. The effect of bait harvesting on bird distribution and foraging behaviour in Poole Harbour SPA. Unpublished report by Footprint Ecology for Natural England. Available from: https://www.footprint-ecology.co.uk/work/reports-and-publications (accessed 28 September 2021).

Frost, T.M., Calbrade, N.A., Birtles, G.A., Mellan, H.J., Hall, C., Robinson, A.E., Wotton, S.R., Balmer, D.E., and Austin, G.E. 2020. Waterbirds in the UK 2018/19: The Wetland bird survey. BTO/RSPB/JNCC, Thetford, UK. Available from: https://www.bto.org/our-science/publications/

waterbirds-uk/waterbirds-uk-201819 (accessed 28 September 2021).

Green, A.J. and Elmberg, J. 2014. Ecosystem services provided by waterbirds. *Biological Reviews* 89: 105–22. https://doi.org/10.1111/brv.12045

Herbert, R.J.H., Ross, K., Hübner, R., and Stillman, R.A. 2010. Intertidal invertebrates and biotopes of Poole Harbour SSSI and survey of Brownsea Island Lagoon. Report to Natural England. Available from: http://eprints.bournemouth.ac.uk/16360/ (accessed 28 September 2021).

Holt, C.A., Austin, G.E., Calbrade, N.A., Mellan, H.J., Hearn, R.D., Stroud, D.A., Wotton, S.R., and Musgrove, A.J. 2015. Waterbirds in the UK 2013/14: The Wetland bird survey. BTO/RSPB/JNCC, Thetford, UK. Available from: https://www.bto.org/our-science/publications/waterbirds-uk/waterbirds-uk-201314-wetland-bird-survey (accessed 28 September 2021).

Hopper, N. 2013. Breeding season survey of Water Rails in Poole Harbour, Spring 2013. Birds of Poole Harbour, Poole, UK. Available from: https://www.birdsofpooleharbour.co.uk/surveys/ (accessed 28 September 2021).

Hopper, N. 2014. Population and distribution of wintering Woodcock in Poole Harbour: Winter 2013/14. Birds of Poole Harbour, Poole, UK. Available from: https://www.birdsofpooleharbour.co.uk/surveys/ (accessed 28 September 2021).

Hopper, N. 2016. Breeding gulls of Poole Harbour, Summer 2016. Birds of Poole Harbour, Poole, UK. Available from: https://www.birdsofpooleharbour.co.uk/surveys/ (accessed 28 September 2021).

Hopper, N. 2018. The Ciconiiformes of Poole Harbour – Herons and their allies, November 2016–April 2018. Birds of Poole Harbour, Poole, UK. Available from: https://www.birdsofpooleharbour.co.uk/surveys/ (accessed 28 September 2021).

JNCC. 2017. Natura 2000 standard data form: Poole Harbour UK9010111. Joint Nature Conservation Committee, Peterborough, UK. Available from: https://jncc.gov.uk/jncc-assets/SPA-N2K/uk9010111.pdf (accessed 28 September 2021).

Liley, D. and Fearnley, H. 2012. Poole Harbour disturbance study. Report for Natural England. Footprint Ecology Ltd., Wareham, UK. Available from: https://www.footprint-ecology.co.uk/work/reports-and-publications (accessed 28 September 2021).

Musgrove, A., Aebischer, N., Eaton, M., Hearn, R., Newson, S., Noble, D., Parsons, M., Risely, K., and Stroud, D. 2013. Population estimates of birds in Great Britain and the United Kingdom. *British Birds* 106: 64–100. Available from: https://britishbirds.co.uk/ (accessed 28 September 2021).

Natural England. 2016. Poole Harbour pSPA Departmental Brief Final. Unpublished report. Available from: https://assets.publishing.service.gov.uk/government/uploads/system/uploads/attachment_data/file/492838/poole-harbour-departmental-brief.pdf (accessed 28 September 2021).

Ogden, J.C., Baldwin, J.D., Bass, O.L., Browder, J.A., Cook, M.I., Frederick, P.C., Frezza, P.E., Galvez, R.A., Hodgson, A.B., Meyer, K.D., Oberhofer, L.D., Paul, A.F., Fletcher, P.J., Davis, S.M., and Lorenz, J.J. 2014. Waterbirds as indicators of ecosystem health in the coastal marine habitats of southern Florida: 1. Selection and justification for a suite of indicator species. *Ecological Indicators* 44: 148–63. https://doi.org/10.1016/j.ecolind.2014.03.007

Perrow, M.R., Gilroy, J.J, Skeate, E.R., and Mackenzie, A. 2010. Quantifying the relative use of coastal waters by breeding terns: Towards effective tools for planning and assessing the ornithological impacts of offshore wind farms. ECON Ecological Consultancy Ltd. Report to COWRIE Ltd. Available from: https://tethys.pnnl.gov/sites/default/files/publications/Perrow-et-al-2010.pdf (accessed 28 September 2021).

Perrow, M.R., Harwood, A.J.P., and Caldow, R.W.G. 2016. Tern verification surveys for marine sites. Natural England Commissioned Reports, Number 212. Available from: http://publications.naturalengland.org.uk/publication/6688364374786048 (accessed 28 September 2021).

Pickess, B. 2007. Important birds of Poole Harbour and their status (1988/99–2004/05). Poole Harbour study group, publication no. 10. Wareham, UK. Available from: https://www.birdsofpooleharbour.co.uk/surveys/ (accessed 28 September 2021).

Pickess, B.P. and Underhill-Day, J.C. 2002. Important birds of Poole Harbour. Poole Harbour study group, publication no. 2. Wareham, UK. Available from: https://www.birdsofpooleharbour.co.uk/surveys/ (accessed 28 September 2021).

Ross, K.E. 2013. Investigating the physical and ecological drivers of change in a coastal ecosystem: From individual- to population-scale impacts. PhD Thesis. Bournemouth University & HR Wallingford. Available from: https://eprints.bournemouth.ac.uk/21351/ (accessed 28 September 2021).

Thornton, A. 2016. The impact of green macroalgal mats on benthic invertebrates and overwintering wading birds. PhD Thesis. Bournemouth University. Available from: https://eprints.bournemouth.ac.uk/24874/ (accessed 28 September 2021).

Walsh, P.M., Halley, D.J., Harris, M.P., del Nevo, A., Sim, I.M., and Tasker, M.L. 1995. *Seabird monitoring handbook for Britain and Ireland.* JNCC/RSPB/ITE/Seabird Group, Peterborough, UK. Available from: https://hub.jncc.gov.uk/assets/bf4516ad-ecde-4831-a2cb-d10d89128497 (accessed 28 September 2021).

Woodward, I., Aebischer, N., Burnell, D., Eaton, M., Frost, T., Hall, C., Stroud, D., and Noble, D. 2020. Population estimates of birds in Great Britain and the United Kingdom. *British Birds* 113: 69–104. Available from: https://britishbirds.co.uk/ (accessed 28 September 2021).

Woodward, I.D., Frost, T.M., Hammond, M.J., and Austin, G.E. 2019. Wetland bird survey alerts 2016/2017: Changes in numbers of wintering waterbirds in the constituent countries of the United Kingdom, special protection areas (SPAs), sites of special scientific interest (SSSIs) and areas of special scientific interest (ASSIs). BTO Research Report 721. BTO, Thetford, UK. Available from: www.bto.org/webs-reporting -alerts (accessed 28 September 2021).

CHAPTER 9

An Overview of Seals in Poole Harbour

SARAH HODGSON and JULIE HATCHER

Abstract

Anecdotal evidence suggests that seals have long been present in Poole Harbour, although prior to Dorset Wildlife Trust's Dorset Seal Project, which began in 2014, there were no systematic or scientific studies of them. Since then, remote cameras have been used to monitor seal behaviour at a known haul-out in the harbour. This, along with casual sightings data and the establishment of a seal photo identification catalogue, has provided an insight into the local population dynamics of Common Seal (also called Harbour Seal) *Phoca vitulina* and Grey Seal *Halichoerus grypus*, the threats they face and their role within the harbour ecosystem.

Keywords: Common Seal, Grey Seal, haul-out, photo identification

Correspondence: SHodgson@dorsetwildlifetrust.org.uk

Introduction

Speak to people who have known and used Poole Harbour since childhood, and many remember seeing seals as children and young adults. This anecdotal evidence of seals in the harbour dates back at least to the 1940s and includes every decade since. There are photographs from the 1990s showing seals hauled out on a tethered human-made platform. This platform was enlarged in 1998 and remains in regular use.

With Common Seals known to be present, Poole Harbour has been the chosen site for a number of releases of this species over the years. The first documented example was one rescued in Southampton's Town Quay and rehabilitated at Gweek Seal Sanctuary in Cornwall, which was released in 1997 (J. Mallinson, pers. comm.). Between 2016 and 2017 another six seal pups were released by RSPCA West Hatch. Further rationale for releasing rehabilitated seals into Poole Harbour includes the proximity to the original rescue sites, ecological suitability and opportunity for post-release monitoring. In January 1997, a juvenile Harp Seal *Phoca groenlandica* was rescued from Poole Harbour and taken to the Cornish Seal Sanctuary with the aim of relocating it north of Scotland once recovered (Cornish Seal Sanctuary 2018). The Harp Seal is native to Arctic waters and is regarded as a vagrant in the UK.

The most westerly known breeding colony of Common Seal *Phoca vitulina* on the Channel coast is in the Solent, including Chichester and Langstone harbours. There are occasional sightings of Grey Seals in Dorset, especially along the open coast, although the nearest known breeding colonies are further west in Cornwall. This makes Dorset an interesting place to study seals as changes in their distribution could occur as a result of expanding populations and may be more noticeable here. Other causes of changes in distribution could be linked to climate change, for example changes in the distribution of prey species, rising sea levels swamping haul-outs and increased storminess affecting pup survival.

In 2014, as a result of the anecdotal stories and in the absence of any systematic or scientific records, the Dorset Seal Project was initiated by Dorset Wildlife Trust (DWT). The aims were to obtain information on the number and species of seal in Dorset, learn about their movements and whether they were transient or if any were resident or semi-resident in the county, and identify any key areas such as breeding and haul-out sites. In Poole Harbour, remote cameras have been used to monitor seal behaviour. Along with casual sightings data and the establishment of a seal photo identification catalogue, this has provided an insight into the local population dynamics of both Common and Grey Seals, the threats they face and their role within the harbour ecosystem.

Species

There are two species of seal native to UK waters, the Grey Seal *Halichoerus grypus* and the Common Seal, both of which have been recorded in Dorset. The majority, 86%, of Common Seal sightings in Dorset are from Poole Harbour. Research conducted by DWT has discovered that a small number of Common Seals are present year-round in Poole Harbour.

Both Common and Grey Seals feature on the IUCN Red List of threatened species as least concern. Seals have legal protection under the Conservation of Seals Act 1970 (England and Wales), which prohibits the killing, injuring or taking of seals.

Population status and distribution

In 2018, the estimated UK population of Common Seals was 45,800 (SCOS 2019), which represents approximately 5% of the worldwide population of the species (JNCC 2021) and

Figure 9.1 Common Seals in Poole Harbour (Sarah Hodgson. Photo taken from distance using long zoom).

30% of the European subspecies *P. vitulina vitulina* (SCOS 2019). Scotland remains a strong-hold for Common Seals, with 79% of the UK population found in this region. Around 16% of the UK population of Common Seals are found in England, predominantly along the east coast in the Greater Thames Estuary and the Wash (SCOS 2019). More locally, along the south-east Channel coast, there is a small breeding colony of Common Seals in the Solent. The Solent Seal Tagging Project in 2010 estimated this population to number between 22 and 25 individuals (Chesworth *et al.* 2010), expanding to between 43 and 50 in 2015 (Bird Aware Solent blog 2020).

A major reduction in the UK Common Seal population was recorded following phocine distemper virus epidemics in 1988 and 2002. There has since been a recovery with numbers returning to levels previously seen in the 1990s. Recent results indicate that while there have been population increases in some areas, other areas are seeing significant declines (Thompson *et al.* 2019).

Of the two resident species, Grey Seals are more abundant in the UK with population estimates of 152,800 in 2018: approximately 38% of the global population (SCOS 2019). There are Grey Seal colonies all around the country including the south-west and east of England, Wales, Northern Ireland and Scotland, which is home to around 88% of the UK breeding population (SCOS 2019).

Species descriptions

Common Seals can be identified by their smaller size, concave head profile, mottled grey-brown fur and V-shaped nostrils. Adults can reach weights of up to 130 kg and live for 20 years (males) and up to 30 years for females (Mammal Society – Harbour 2021). Common Seals prefer to haul out in sheltered estuaries, on sandbanks and mudflats, although they may be found on rocky shores in some areas (JNCC 2021). During their annual moult, which occurs in August, Common Seals will spend longer periods of time hauled out, sometimes adopting a banana-shaped pose with head and flippers raised to reduce heat-loss from their extremities which have less insulating blubber. The breeding season for Common Seals is between June and July with females giving birth to a single pup which will be able to swim almost straight away. While Common Seal pups have been recorded within Poole Harbour in 2015 and 2019, it has not been concluded whether they were born within the harbour or have migrated from elsewhere. Seals are highly mobile marine mammals, capable of covering hundreds of kilometres and remaining at sea for days at a time. Indeed, a seal regularly observed in Poole Harbour is known to have travelled from northern France to the Dorset coast, a straight-line distance of over 200 km, at only a few months old. Typically, Common Seals would forage within a range of up to 50 km from a haul-out (SCOS 2019). They can dive to depths of 50 m and remain submerged for up to 10 minutes at a time (Mammal Society – Harbour 2021).

Grey Seals are the larger species, weighing up to 300 kg for males. They can be distinguished by their characteristically long and broad muzzles giving them a flat head profile with parallel nostrils. Like the Common Seal, their pelage is of a grey or brown colouration but with a blotchier pattern. Grey Seals are long-lived marine mammals – up to 35 years for females and 25 years for males (Mammal Society – Grey 2021).

The estuarine habitats of Poole Harbour are unlikely to be utilised by Grey Seals as a pupping site. Grey Seals are more likely to seek out remote, inaccessible locations to give birth to their young. Beaches surrounded by sheer cliffs, uninhabited off-shore islands and secluded sea caves are generally preferred with a notable exception for the population in the east of England which haul out in substantial numbers on beaches during the breeding season. Grey Seals spend around two-thirds of their life at sea (Mammal Society – Grey 2021). Outside of the breeding (August–December) and moulting seasons (December–April), Grey Seals range over vast distances, regularly travelling over 100 km in open seas on foraging excursions which can last several days (SCOS 2019).

Grey Seals account for 24% of the sightings in Poole Harbour, where the sighting can be identified to species. Through photo identification eight individuals have been catalogued. With the exception of one juvenile Grey Seal which has been recorded visiting the harbour over a four-year period, the remainder have only been recorded on between one and three occasions, suggesting they are much more transient.

Dorset Seal Project

Sightings data

The Dorset Seal Project was set up to obtain information on the location, frequency and species of seal present along the Dorset coastline including Poole Harbour by collecting casual sightings data.

This ad hoc data has been invaluable in gaining a better overall understanding about seals in Dorset. Prior to the project only a handful of seal sightings were recorded each year, but data collected shows an average of 125 sightings per year between 2014 and 2020. As these figures are not obtained through systematic survey methods, and there are inconsistencies in surveyor effort between sites and times of year, they do not enable conclusions to be drawn about variances in abundance and distribution of seals.

Tracking seal movements using photo identification

Seals have unique markings and pelage patterning which are retained for life. By photographing and cataloguing these recognisable features, along with any scars or tags if present, individuals can be identified and their movements tracked. Photo identification is an inexpensive, non-invasive survey tool which can be used to study seals with minimal impact. Other study techniques are available which may be more efficient or provide more detailed data, such as aerial surveys and satellite telemetry; however, these are costly and more intrusive.

While photo identification is a useful seal study method, it presents some challenges. Good quality, clear photos need to be obtained, taken from a distance so as not to disturb the seals while still retaining enough detail to capture identifiable markings. As the seals frequently haul out on the mudflats, their fur can often be coated with a layer of mud hiding their unique natural markings. The time of year that photographs are taken can affect the quality of the image for photo identification purposes. Before seals undergo their annual moult, their old fur changes colour and distinctive features are temporarily less visible.

Between 2014 and 2020 a total of 73 seals were added to the Dorset Seal photo identification catalogue including eight Grey Seals and nine Common Seals from Poole Harbour. Around 40% of all the seals from Poole Harbour in the catalogue have only been recorded once. Four of the catalogued Common Seals have been recorded multiple times over several years and include an adult female, two juvenile females and a juvenile male. Other adult Common Seals have been recorded on several occasions; however, images were unsuitable for photo identification analysis. Of the six seals released between 2016 and 2017, two seals are known to have since died, and another was relocated for welfare reasons.

Sharing the photo identification catalogue with other seal recorders can help inform how far and how often seals travel between counties and even countries. This research has shown that seals recorded in Poole Harbour have been linked to other sites in Dorset, around the Isle of Wight and France.

In 2016 a large bull Grey Seal was recorded hauled out near Brownsea Island and Sandbanks. Through photo identification it was discovered that the same individual had been recorded at Yarmouth and Bembridge on the Isle of Wight several times in the previous weeks. The seal left the harbour after a few days and was last seen off Swanage, but its subsequent whereabouts are not known.

In December 2017, a deceased Common Seal was found at Thorness Bay on the Isle of Wight with a tag on its rear flipper. These tags carry unique reference numbers and are used to monitor the post-release progress and movements of rehabilitated seals. The tag revealed that the seal had been released in Poole Harbour by the RSPCA one month before.

An adult Common Seal added to the catalogue in 2016 was re-identified a number of times all within the harbour boundaries. On one occasion, a flipper tag was observed, and the number traced to a rescue centre in France. The female seal had been rescued as a pup in 2007 and released the same year in the Baie du Mont Saint-Michel fitted with a satellite telemetry tag. The young seal made the Channel crossing soon after release and travelled along the Dorset coast from Lyme Bay to Poole Harbour before the satellite tag stopped transmitting (Association Chene 2021).

Studying haul-out behaviour

Seals are semi-aquatic marine mammals and as such need to come onto land, or haul out, to rest, aid digestion, breed and moult. The expansive mudflats of Poole Harbour uncovered at low tide provide numerous opportunities for seals to haul out.

To study the haul-out behaviour of seals, remote cameras were installed at a site in the harbour to provide data on the number of seals using the haul-out, the frequency, daily and seasonal variations. The cameras, installed in 2016, capture a still image at five-minute intervals during daylight hours. The time-lapse function was found to be more reliable and preserved battery life, reducing the number of site visits needed therefore minimising disturbance. While this effort-based sampling technique provides better opportunities for comparison and analysis of data over time, there are still limitations of this method:

- It only shows the behaviour at one site, so may not be indicative of all seals in the Poole Harbour area.
- Difficult to ascertain the exact size of the local population. Seals may use other haul-outs which are not being monitored.
- Does not account for nocturnal behaviour.
- The artificial haul-out can only accommodate a limited number of seals at any one time.
- Unlike the mudflats, which are influenced by the tide, the artificial haul-out is accessible at any tidal state.
- The positioning of the cameras can vary and the entire haul-out may not always be visible.
- Issues with camera functionality can lead to gaps in data.
- Resolution is not high enough to identify individual seals for photo ID.

Various technical and operational issues have meant that it has not been possible to obtain a complete set of data for each year that the cameras have been installed, making it difficult to identify trends and compare behaviour between years. However, initial analysis of the footage that has been obtained has revealed that seals are using the haul-out throughout the year. Figure 9.2 (a)–(d) shows how many haul-out events were recorded each year

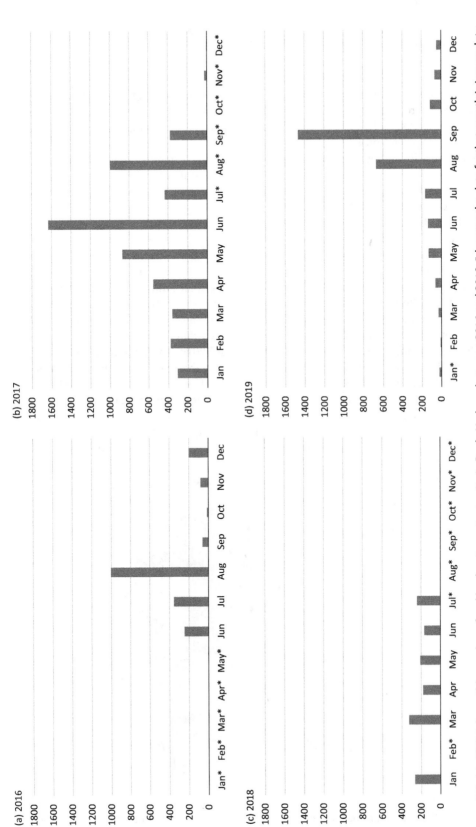

Figure 9.2 (a)–(d) Total monthly number of seal haul-out events at a site in Poole Harbour between 2016 and 2019. *denotes the data for that month is incomplete.

with a monthly breakdown. In this context a haul-out event is defined as a single image that shows any number of seals on the haul-out.

There does appear to be more usage of the haul-out by seals during the summer period, between May and September, but the month that this activity peaks varies from year to year. It should be noted that with longer daylight hours offering more opportunity for images to be recorded, there will inevitably be more haul-out events in the summer; however, this does coincide with the Common Seals' annual moult, when it is anticipated that seals will spend more time ashore.

Variations in the levels of haul-out activity during different times of day were also analysed and show that seals were present at the haul-out most between dusk and 10 am, with 53% of all haul-out events recorded in this timeframe (Figure 9.3).

Data collected by the remote cameras shows the highest number of Common Seals recorded at the haul-out was four. It is difficult to utilise this data to determine the size of the local Common Seal population, as the physical space on the haul-out is restricted and seals are known to haul out at other locations, and therefore this only provides a minimum estimate. To get a more accurate population count, it would be beneficial to conduct more widespread surveys around the harbour simultaneously during low tide.

In addition to the Common Seals, a juvenile female Grey Seal is regularly seen hauling out alongside them (Figure 9.4). Through a combination of photo identification and the remote camera monitoring, it has been discovered that this seal is a frequent summer visitor, with confirmed sightings between 2016 and 2019. There are no sightings of this individual in Poole Harbour during the winter, and the seal has not been identified at any other location in Dorset or further afield. Interactions between this Grey Seal and the Common Seals have been observed. Most of these interactions appear passive, with both species seemingly content to tolerate each other at the haul-out. Occasionally, more energetic and potentially aggressive behaviour has been witnessed; however, the reasons for this are not understood. In other areas where there is spatial overlap between Grey Seals

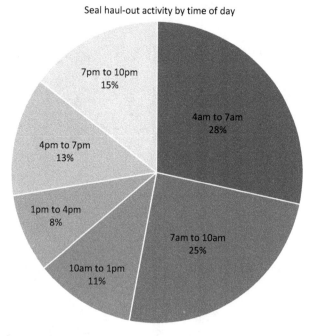

Seal haul-out activity by time of day

Figure 9.3 The time of day that haul-out activity was recorded based on data collected between 2016 and 2019.

Figure 9.4 Mixed-species haul-out in Poole Harbour (Sarah Hodgson).

and Common Seals, interspecies aggression has been observed at haul-outs where space is limited, possibly as a form of interference competition (Wilson and Hammond 2019).

Feeding

Diet

Seals are apex predators and as such are indicators of a healthy marine ecosystem. They maintain the balance between populations of their prey species and the rest of the ecosystem, removing any weak or injured individuals and driving the population to a more robust, healthy level. They also prevent any one of their prey species becoming dominant and impacting on populations of animals lower down in the food chain. Both Grey and Common Seals appear to have a similar diet, mainly fish, making use of a wide variety of species in a rather opportunistic way, taking fish from the seabed as well as from midwater. Common Seals also forage up estuaries into rivers and may take some freshwater species such as grayling and trout. The prey species of choice varies widely from location to location as well as seasonally and most likely depends on what is available at any one time in their foraging area. This includes both commercial and non-commercial species of fish. As there is the potential for conflicts between seals and the fishing industry over fish availability, a number of studies on seal diet have been carried out in the UK and Ireland. In the UK these have been mostly in Scotland, Northern Ireland and the North Sea (Brown *et al*. 2012). On the south coast, a study of resident Common Seals in the Solent included some aspects of foraging behaviour but did not investigate prey species, although it did look at the size of foraging areas and the variety of seabed habitats within them (Chesworth *et al*. 2010). As top predators, seals also bioaccumulate toxins obtained through their food. A healthy breeding population of seals is therefore an indicator of good water quality in an estuary or harbour.

Seal diet studies have used fish otoliths recovered from seal scat to identify the common prey species. When combined, the results from these studies indicate that Sand Eel and Whiting make up around 45% of the Common Seal's diet while Sand Eel and Cod make up around 33% of the Grey Seal's diet (Brown *et al.* 2012). However, it was noted that a wide variety of other species are also taken, for example Dragonets, which, it is believed, can make up a substantial proportion of a Grey Seal's diet (Cornwall Seal Group 2021). Attempts have been made to estimate the biomass of food seals require. In order to maintain a healthy insulating layer of blubber, it is estimated that seals need to ingest up to 5% of their body weight each day (Chesworth *et al.* 2010), averaging at between 3 and 5 kg per day for an adult (SCOS 2019), but it has to be noted that seals do not feed every day, and males do not eat anything at all in the breeding season.

As seals are opportunistic feeders, catching a wide variety of prey depending on location and season, it is difficult to draw conclusions from these diet studies as to the specific dietary composition of the Poole Harbour seals. Most prey items are ingested underwater and are only brought to the surface if large or difficult to manoeuvre. Identifying prey species from photographs is therefore very limited. There is photographic evidence of them feeding on species of wrasse, mullet, bass and Plaice *Pleuronectes platessa*, although how much of their overall diet these species contribute is unknown. However, the fact that a small number of seals are present in the harbour year-round and have been for a number of years, including at least one individual for over 12 years, would suggest that the ecosystem is healthy and diverse enough to support them.

Foraging

In 2010 the Solent Seal Tagging Project successfully tracked five Common Seals to investigate their foraging behaviour among other things (Chesworth *et al.* 2010). It found that each individual seal had a unique pattern of foraging, some ranging quite widely within the Solent area and others keeping to a much more restricted area. Although no foraging behaviour studies have been carried out on the Common Seals in Poole Harbour, there is no evidence of them outside the harbour entrance although there are regular sightings at the mouths of and along the rivers running into the harbour. The furthest upriver a seal has been recorded is North Bridge in Wareham along the river Piddle and Holmebridge between Wareham and Wool along the river Frome.

Of the Grey Seals recorded in the harbour, nearly all have been transient, only staying for a few days or months before disappearing. Only one Grey Seal has returned to the harbour after having left for a significant period.

Threats

As apex predators here in Dorset, seals do not have any natural predators, although they face a number of anthropogenic threats. As well as being one of the world's largest natural harbours and an internationally important area for nature conservation, Poole Harbour is home to a busy commercial shipping port, a fishing fleet, a military presence and is a very popular recreational area for a wide variety of water sports.

Boat-strike

Although there are quiet areas within the harbour, inevitably there is the potential for conflict between the seals and motorised vessels. While there are speed limits in areas of the harbour there are also designated zones for waterskiing and personal watercraft (PWC), although a large part of the southern half of the harbour is out-of-bounds to PWC (Poole

Harbour Commissioners 2021). However, seals do not restrict themselves to these quieter areas and may sometimes be seen by the harbour entrance, around Brownsea Island, along Poole Quay and close to the commercial port area where they are at risk of boat or ship-strike. In 1998 a dead male Common Seal was recovered from Round Island with injuries consistent with a boat-strike (J. Mallinson, pers. comm.).

Disturbance

Disturbance is defined as a change in natural behaviour caused by human activity (Bellman *et al.* 2019). Unmotorised watercraft, for example sailing dinghies, kayaks and paddleboards, can pose a threat to seals as their quiet approach means they can often get very close before the seal realises they are there. Seals resting at a haul-out are vulnerable to disturbance which often results in them returning to the water before they have replenished their oxygen supplies, heat and energy (Bellman *et al.* 2019). This can have an adverse impact, causing them to become chilled (especially while moulting), preventing digestion and separating mothers and pups and can ultimately reduce life expectancy.

There are three recognised levels of disturbance of hauled-out seals:

1. a change from relaxed/resting to being alert with a raised head;
2. movement towards the sea;
3. entering the water.

Seals resting in the water can also be disturbed resulting in a sudden 'crash dive', which creates a noticeable splash. This happens when a seal is forced to dive before it is ready (Bellman *et al.* 2019).

Other causes of disturbance include land-based activities, for example walkers, anglers, wildlife photographers and dogs off leashes. Disturbance happens when people get too close to a seal and trigger its predator response. A report commissioned by the Seal Alliance noted a high level of disturbance in the UK at sensitive haul-out sites close to popular recreational and tourist areas (Bellman *et al.* 2019).

Our own observations have shown that disturbance has been caused in Poole Harbour with boats and personal watercraft approaching the haul-out when seals were present resulting in a change in their behaviour from resting to alert and even temporary abandonment of the haul-out until the perceived danger moved away.

Human habituation

Habituation is defined as the reduction of an instinctive or natural response to a frequently repeated activity (Bellman *et al.* 2019). Habituation to human contact occurs when seals routinely interact with people. Inquisitive young seals sometimes approach people to investigate them and if no immediate harm occurs, they may come to view this as normal behaviour. Feeding seals rewards them for this close interaction and may result in less time spent foraging for their own food and more time in close proximity to boats and people. Unfortunately, the result of this usually ends in premature death for the seal often from close contact with boats and propellers. Feeding seals therefore has the same ultimate impact as disturbance.

Pollution/litter

Pollution and litter pose a threat to the health of seals. As apex predators, seals bioaccumulate toxins that have worked their way up through the food chain. Historically, toxic pollutants from industry and human activity have accumulated in sediments in parts of Poole Harbour, especially where water circulation is limited such as in Holes Bay (Wardlaw 2005). As an example of the effects industrial chemicals can have on wildlife, it has been found that female Common Seals feeding on fish with high levels of polychlorinated biphenyls (PCBs) may fail to breed (Mammal Society – Harbour 2021).

Marine litter poses a threat to all marine wildlife, not least seals, through the potential for ingestion and entanglement which can cause injury and death. As seals are intelligent, inquisitive animals they are particularly vulnerable to entanglement and lost rope and fishing nets are common hazards. Research by the Cornwall Seal Group Research Trust, carried out at a Grey Seal haul-out in north Cornwall, found that entanglement caused constrictions and open wounds, affecting movement and feeding and increasing the risk of infections (Allen *et al.* 2012). Research has found a correlation between entanglement and reduced survival. Another potential threat is the ingestion of fishing hooks, which can become lodged in the mouth or internally. One seal was observed in Poole Harbour with a fishing hook embedded in its mouth. This type of injury can result in infection and impair feeding.

Conclusion

Seals have long been associated with Poole Harbour and as top predators their presence can be viewed as indicative of a healthy marine ecosystem. Our research indicates that Poole Harbour is an important site for Common Seals, with 86% of sightings in Dorset recorded in the harbour and adjacent rivers. Grey Seals are regular visitors but much more transient accounting for less than one-quarter of seal sightings in Poole Harbour where the species can be identified.

Monitoring the seals by deploying remote cameras at a haul-out has shown that Common Seals are utilising the harbour all year round. Despite the challenges we have encountered leading to inconsistencies in data, we have gained a valuable insight into the seals' behaviour and identified that haul-out activity increases during the summer months. However, with water-based recreational activities increasing at the same time of year, potential for disturbance is elevated.

With at least four Common Seals regularly re-sighted in Poole Harbour, this small population is nevertheless regionally important, with the only known breeding colony of Common Seals in the southeastern Channel in the Solent. Being a small group potentially makes the seals more sensitive to anthropogenic pressures and less adaptable to environmental changes. In Poole Harbour the seals may not have any natural predators but disturbance, boat-strikes, pollution and entanglement have all been identified as significant threats.

The Dorset Seal Project is ongoing, and DWT will continue to collect casual sightings data, conduct photo identification work and monitor haul-out behaviour; however, to enhance our knowledge of seals in Dorset, further research is required. Systematic surveys which can provide more accurate population figures and help assess changes in abundance in the future are being developed. A dedicated Poole Harbour seal diet survey would be beneficial as previous research into seal diets in other areas has identified variances in prey seasonally and regionally. With seals facing increasing pressures from human activities on or near the water, a study to monitor the extent of disturbance particularly around haul-outs would be useful and could help to inform future recommendations on how to maintain the health of the seals in Poole Harbour for years to come.

Acknowledgements

The Dorset Seal Project is extremely grateful to the following people: Steve Sheppard for giving up his time and providing boat transport and advice/support with monitoring of the seals. The Palmer family for allowing us to work from Round Island and for supporting our project. Sue Sayer of the Cornwall Seal Group Research Trust for always being at the end of a phone offering her experience and expert knowledge. Bournemouth Oceanarium's donation enabled the purchase of cameras, and a grant from Sea-Changers made it possible to create our online sightings form.

References

Allen, R., Jarvis, D., Sayer, S., and Mills, C. 2012. Entanglement of grey seals *Halichoerus grypus* at a haul out site in Cornwall, UK. *Marine Pollution Bulletin* 64. https://doi.org/10.1016/j.marpolbul.2012.09.005

Association Chene. Available from: https://www.associationchene.com/centre-de-sauvegarde/les-phoques/ (accessed 3 April 2021).

Bellman, K., Bennett, S., James-Hussey, A., Watson, L., Ottaway, A., and Sayer, S. 2019. PLEASE DO NOT DISTURB! The growing threat of seal disturbance in the United Kingdom: Case studies from around the British Coast. Commissioned by the Seal Alliance; funded by the Seal Protection Action Group and Cornwall Seal Group Research Trust.

Bird Aware Solent blog – 23rd June 2020 – Seals of the Solent. Available from: https://solent.birdaware.org/article/34101/23rd-June-2020---Seals-of-the-Solent (accessed 23 June 2020).

Brown, S., Bearhop, S., Harrod, C., and Mcdonald, R. 2012. A review of spatial and temporal variation in grey and common seal diet in the United Kingdom and Ireland. *Journal of the Marine Biological Association of the United Kingdom* 92. https://doi.org/10.1017/S0025315411002050

Chesworth, J.C., Leggett, V.L., and Rowsell, E.S. 2010. *Solent Seal Tagging Project Summary Report.* Wildlife Trusts' South East Marine Programme, Hampshire and Isle of Wight Wildlife Trust, Hampshire.

Cornish Seal Sanctuary. 2018. S.O.S. News Update Issue 150. Available from: https://www.seal-sanctuary.co.uk/html/gweekaugust2018css.html (accessed 25 March 2021).

Cornwall Seal Group and Research Trust. Available from: https://www.cornwallsealgroup.co.uk/about-seals/typical-behaviour/ (accessed 15 March 2021).

JNCC. 1365 Harbour seal *Phoca vitulina*. Available from: https://sac.jncc.gov.uk/species/S1365/ (accessed 21 March 2021).

Poole Harbour Commissioners. Available from: https://www.phc.co.uk/ (accessed 26 March 2021).

Special Committee on Seals (SCOS) Scientific Advice on Matters Related to the Management of Seal Populations. 2019. Available from: http://www.smru.st-andrews.ac.uk/files/2020/08/SCOS-2019.pdf (accessed 21 March 2021).

The Mammal Society. Species – Grey seal. Available from: https://www.mammal.org.uk/species-hub/full-species-hub/discover-mammals/species-grey-seal/ (accessed 21 March 2021).

The Mammal Society. Species – Harbour seal. Available from: https://www.mammal.org.uk/species-hub/full-species-hub/discover-mammals/species-harbour-seal/ (accessed 21 March 2021).

Thompson, D., Duck, C.D., Morris, C.D., and Russell, D.J.F. 2019. The status of harbour seals (Phoca vitulina) in the UK. *Aquatic Conservation: Marine and Freshwater Ecosystems* 29(S1): 40–60. https://doi.org/10.1002/aqc.3110

Wardlaw, J. 2005. Water quality and pollution monitoring in Poole Harbour. In: Humphreys, J. and May, V. (eds) *The Ecology of Poole Harbour.* Elsevier, Amsterdam, pp. 219–22. https://doi.org/10.1016/S1568-2692(05)80023-3

Wilson, L.J. and Hammond, P.S. 2019. The diet of harbour and grey seals around Britain. Examining the role of prey as a potential cause of harbour seal declines. *Aquatic Conservation: Marine and Freshwater Ecosystems* 29(S1): 71–85. https://doi.org/10.1002/aqc.3131

CHAPTER 10

Holes Bay Nature Park:
Ecology and Human Activity

EMMA RANCE

Abstract

The Holes Bay Nature Park covers 286 ha and lies within the northern aspect of Poole Harbour. The Bird Sensitive Area is a haven for some 26 species of international important wildfowl and waders attracted to the safe shelter of the prized salt meadows, reedbeds, saltmarsh and mudflats. The busy Backwater Channel attracts important subtidal species – native seahorse species, flat oyster and a nationally rare sea sponge.

The area has suffered great impact. Since 1924, 80 ha of intertidal area have been lost through land reclamation after the building of the railway embankment, power station and A305. Past industries have discharged untreated sewage and heavy metals which still reside in the sediments of these sheltered and unflushed waters. Seagrass-dominated mudflats have been replaced with algal mats largely through nutrient-rich agricultural run-off.

Despite these challenges, Holes Bay remains of high commercial, recreational and conservational importance.

Keywords: Holes Bay, Poole Harbour, marine, human activity, conservation

Correspondence: emma@noctiluca-marine-consulting.co.uk

Introduction

In 2019, a desktop report was written by Dorset Wildlife Trust (Rance *et al.* 2019) to gather published information on the ecological status and human uses of Holes Bay. The review of available data, historical anecdotes and research provided far greater understanding, thus enabling best practice in the conservation management of the area. The report forms the basis of this chapter and includes work from the contributing authors.

In 2014, 286 ha of saltmarsh and intertidal mudflats in Holes Bay were purchased by Poole Harbour Commissioners (PHC) and Dorset Wildlife Trust (DWT), under the Great Heath Living Landscape Initiative (Dorset Wildlife Trust 2020). In 2015, the Holes Bay Nature Park was established (see Figure 10.1) and is now managed by a consortium of landowners and stakeholders including PHC and DWT, Marina Developments Limited, Borough of Poole, Davis' Boatyard and Royal National Lifeboat Institution (RNLI).

Emma Rance, 'Holes Bay Nature Park: Ecology and Human Activity' in: *Harbour Ecology*. Pelagic Publishing (2022).
© Emma Rance. DOI: 10.53061/XUFY6761

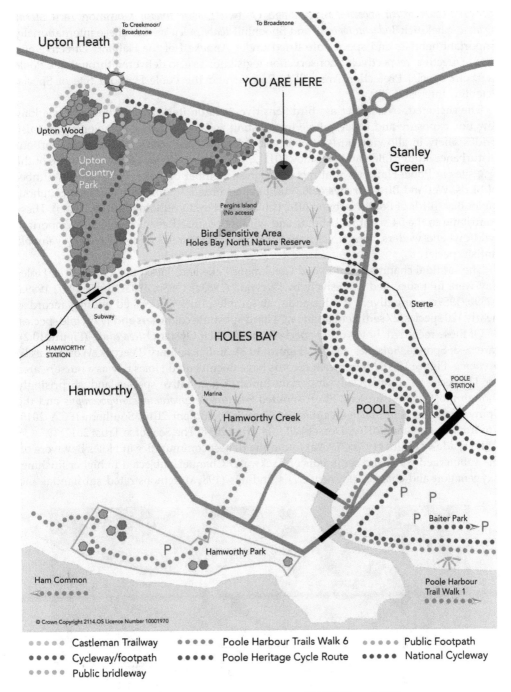

Figure 10.1 Holes Bay public interpretation map (© Dorset Wildlife Trust, 2015).

Important habitats and species

Holes Bay is situated in the northern aspect of Poole Harbour and together with its surroundings is an incredibly special area providing habitats for a huge diversity of wildlife. The area boasts prized salt meadows, reedbeds, saltmarsh and intertidal mudflats, which provide a rich food source and haven for overwintering waterfowl and waders, including

Avocet *Recurvirostra avosetta*, Black-tailed Godwit *Limosa limosa*, Common Tern *Sterna hirundo*, Shelduck *Tadorna tadorna* and Spoonbill *Platalea leucorodia*. These internationally important habitats and species are listed under Annexe I of the Habitats Directive and Birds Directive, respectively. Conservation legislation is also delivered through the Poole Harbour Special Protection Area (SPA) Ramsar and the Poole Harbour Site of Special Scientific Interest (SSSI).

The sheltered, overwintering, Bird Sensitive Area on its northern shore offers a long low tide exposure and a safe refuge for roosting and feeding. This large open area also prides safety in allowing birds a good visual to anticipate danger. In a Poole Harbour disturbance study (Liley and Fearnley 2012), Holes Bay north was found to be one of the few sites in Poole Harbour to attract the highest number of species and largest number of birds. Wetland Bird Surveys (2003–16) have recorded 26 different species throughout Holes Bay (Birds of Poole Harbour 2016; British Trust for Ornithology 2016a, 2016b). These contribute to the 54 separate habitats and wildlife recorded as internationally important wildfowl and waders, coastal vegetation, marine invertebrates or commercially valuable finfish species.

The subtidal channels (Upton and Creekmoor Lake and Backwater Channel) of Holes Bay were first surveyed extensively by Dyrynda (1983a, 1983b, 1983c, 1985, 1987, 1989a, 1989b, 1989c) using divers, dredge and grab samples. In 1985, detailed findings recorded nearly 120 species: 27 sediment infauna, 74 hard substrate colonisers and 17 mobile species.

Of those recorded, noteworthy species were the Flat Oyster *Ostrea edulis* (Figure 10.2), rare Sea Sponge *Suberites massa* (Figure 10.3) and a solitary Peacock Worm *Sabella pavonina* (Figure 10.4). More recent records have documented Holes Bay as a nursery area for 11 species of commercially important finfish, 8 non-native species, and, surprisingly though most likely transient, Short-snouted Seahorse *Hippocampus hippocampus* and the Spiny Seahorse *Hippocampus guttulatus* (Garrick-Maidment 2010; Southern IFCA 2015; Environment Agency 2016a; Dorset Wildlife Trust 2017; The Seahorse Trust 2017).

Dyrynda's studies (1983a, 1989a) found the richest communities in Holes Bay were on the submerged walls and sediments of Backwater Channel, subjected to higher flushing, oxygenation and coarser sediments. His findings (1989a) demonstrated substantial and

Figure 10.2 Flat Oyster *Ostrea edulis* (© Tinsley, P., 2009).

Figure 10.3 Rare Sea Sponge *Suberites massa* (© Tinsley, P., 2009).

ecologically significant communities of marine species. Nearly forty years on, Dyrynda's baseline data still forms the most detailed subtidal collection available to Holes Bay.

While additional studies in Holes Bay have shown low biodiversity, attributed to the poor flushing and eutrophication, it was found to inhabit a higher abundance of individual, opportunistic and short-lived species (Howard and Moore 1988; Langston *et al.* 2003).

Historical changes

Holes Bay has suffered great changes over the last 150 years: used as a dumping ground for vessels, industrial chemicals and raw sewage. The area has faced many challenges in addition to heavy urbanisation and infilling of the natural environment through the building of the railway embankment, in 1847 and the coal-fired power station in 1946 (Hubner 2009; Le Pard 2010). This was followed by the creation of the Holes Bay Relief Road (A305) in 1988. The impact upon loss of the natural environment becomes more apparent when calculated spatially. Between 1924 and 1985, Holes Bay intertidal area reduced from 330 ha to 250 ha (Gray and Pearson 1983; Scopac 2004).

Another surprising change to note is the difference we see in the coastal vegetation of Holes Bay compared to that of the mid-nineteenth century. Acres of intertidal mudflats were dominated by seagrass *Zostera* spp. (Haigh 1976). Today, the mudflats are draped with algal mats *Ulva* spp.

Pollution

The building of the railway embankment effectively created tertiary (north) and secondary (south) lagoons, dramatically changing and reducing the flushing capabilities and therefore larval dispersal, oxygenation, sediment dispersal and food availability (Dyrynda 1985). This 'lagoonization' increased the vulnerability of Holes Bay, details of which have been documented by Dyrynda's baseline studies (1983a), some ten years after the effects of pollution and eutrophication became known. Further insight on the biological pollution

Figure 10.4 Peacock Worm *Sabella pavonina* (© Tinsley, P., 2009).

in Holes Bay can be found at Langston *et al.* (2003), Hübner (2009), Hubner *et al.* (2010), Underhill-Day *et al.* (2010) and Kite *et al.* (2012).

Historically, Holes Bay was regarded as the most polluted embayment of Poole Harbour. Consented raw sewage and industrial waste were previously discharged into the north-eastern aspect until legislation tightened in the 1960s (Drake and Bennett 2011; Wardlaw 2005; Falconer,1993). Research in the 1970s showed copper, nickel and zinc to exceed typical levels in the English Channel (Langston *et al.* 2003). This contamination was exacerbated by the extremely low fluvial and tidal flushing rate, locking heavy metals away into the sheltered sediments. Tributyltin is still above probable effect levels in the water column of Backwater Channel, contributing to the failure for Poole Harbour to meet 'Good Chemical Status' for tributyltin under the current classification of the Water Framework Directive (Environment Agency 2014, 2016b, 2018).

The Poole Harbour nitrogen load has doubled in 50 years, and the levels of macroalgae, dissolved oxygen and dissolved inorganic nitrogen do not meet 'Good' status under the Water Framework Directive (Kite *et al.* 2012). It is understood that nitrate leaching from historical agriculture is the main source of nitrogen causing eutrophication within the Poole Harbour. A review conducted for the Dorset Local Nature Partnership (Taylor 2015) concluded that even if the standing biomass of macroalgae was removed from Poole Harbour annually, this would only account for 6% of the proposed reduction in annual nitrogen input to Poole Harbour, and it is uncertain whether macroalgal removal would itself lead to net negative conservation impacts (Dorset Wildlife Trust 2014). This along with other pollutants has contributed to the bright green algal mats covering 75% of Holes Bay (Thornton 2016), which are clearly visible at low tide.

Fishing

Permitted commercial fishing activity in Holes Bay includes bait-dragging, angling, netting and fyke netting. However, Holes Bay is 'unclassified' for the safe gathering or production of bivalves (clams, cockles, oysters, mussels, etc.) due to chemical contamination (Borough of Poole 2020). These historical contaminants prohibit legal shellfish classification, and any such bivalve fishing is prohibited, creating de facto protection. The area is now a shell-fish refuge and will receive indirect benefits from the nearby Marine Stewardship Council certified Poole Harbour cockle and clam fishery (MSC 2018).

Human uses

The Poole Bay and its environs provide a popular open space for people to enjoy a variety of activities. Poole Bay also provides a good visual overview to passers-by, with its eastern boundary being adjacent to the A305 – one of Poole's busiest roads. Despite being in the heart of a busy commercial area, Holes Bay majorly contributes to the high-quality natural environment enjoyed by residents, visitors and local businesses.

The area also remains of high commercial importance; it is home to the harbour's largest marina at Cobbs Quay, RNLI headquarters, slipways, yacht club and the iconic Twin Sails and Poole lifting bridges.

The Holes Bay Nature Park reflects current access that exists, the challenges this brings to balance and achieve priorities for wildlife and habitats, increasing recreational pressure and the public health role it provides. It is a highly popular area for a range of recreational and commercial users and is accessed widely by all transport. Holes Bay Nature Park provides an ideal opportunity for engagement with nature, green travel and public health initiatives. Green travel options meet a hub in Holes Bay with links to the national cycle network, Poole Harbour trail walks, local bus and train routes.

The primary impact of land-based human activities is attributable to direct access and recreation and is likely to be highly seasonal and weather-dependent. Visitor numbers associated with Upton Country Park (Pearce 2015) show steady increase year on year, and associated land-based activities such as the exercising of dogs, associated waste, reedbed trampling and general litter deposition appear to be the strongest influences. Longer-term changes driven by local development plans are likely to lead to further increased visitor pressure.

Conclusions

Given the informal status of Holes Bay Nature Park, the primary challenge is to achieve management coordination and clear designation of focal points within responsible partner agencies. Management will build upon the framework matrix completed for the Poole Harbour Aquatic Management Plan and respond to the identified changes in anthropogenic pressures.

It is anticipated that the findings made in the report (Rance et al. 2019), along with various recommendations made for further survey work and management of the area, will continue to inform planning applications and develop the Holes Bay Nature Park

Memorandum of Understanding and Management Plan. The resulting public engagement will highlight potential human impacts and the need for sensitive access promoting Holes Bay as a commercially important, highly prized and ecologically sensitive nature park.

References

Borough of Poole. 2020. *Shellfish*. Available from: https://www.poole.gov.uk/environmental-problems/food-safety-and-standards/shellfish/ (accessed 2 November 2020).

British Trust for Ornithology. 2016a. Data from: The Wetland Bird Survey (WeBS): Holes Bay, Poole Harbour 2003–2016, electronic data set.

British Trust for Ornithology. 2016b. The Wetland Bird Survey (WeBS). Available from: http://www.bto.org/volunteer-surveys/webs (accessed 17 March 2016).

Dorset Wildlife Trust. 2014. *Dorset State of the Environment Report*. Published report for Local Nature Partnership.

Dorset Wildlife Trust. 2017. Data from: Living *Seas Sightings Database* 2014–2017, electronic data set.

Dorset Wildlife Trust. 2020. *The Great Heath Living Landscape*. Available from: https://www.dorsetwildlifetrust.org.uk/TGH (accessed 4 November 2020).

Drake, W. and Bennett, L. 2011. *Poole Harbour Aquatic Management Plan 2006 Amended 2011*. Available from: https://www.phc.co.uk/environment/management/aquatic-management-plan/ (accessed 8 November 2020).

Dyrynda, P.E.J. 1983a. Investigation of the sublittoral ecology of Holes Bay, Poole Harbour. Report to the Nature Conservancy Council.

Dyrynda, P.E.J. 1983b. *Lower Backwater Channel & Industry, Holes Bay [map]. Investigation of the Sublittoral Ecology of Holes Bay, Poole Harbour*. Report to the Nature Conservancy Council.

Dyrynda, P.E.J. 1983c. *Changes to Holes Bay Before and After Creation of the Railway Embankment 'Curve' and Other Development –1849 & 1982*. Investigation of the sublittoral ecology of Holes Bay, Poole Harbour. Report to the Nature Conservancy Council.

Dyrynda, P.E.J. 1985. *Poole Harbour Subtidal Survey: General Assessment & Northern Sector Survey (1985) Report*. Report to the Nature Conservancy Council.

Dyrynda, P.E.J. 1987. *Poole Harbour Sublittoral Survey – IV: Baseline Assessment*. Report to the Nature Conservancy Council.

Dyrynda, P.E.J. 1989a. Holes Bay [map]. *Marine Biological Survey of the Bed and Waters of Holes Bay, Poole Harbour, Dorset –1988*. Report to Dorset County Council.

Dyrynda, P.E.J. 1989b. *Marine Biological Survey of the Bed and Waters of Holes Bay, Poole Harbour, Dorset –1988*. Report to Dorset County Council.

Dyrynda, P.E.J. 1989c. *Holes Bay Bridge Replacement Environmental Impact Assessment: Detailed Marine Biological Survey of the Proposed Marina Relocation Site*. Report to Dorset County Council.

Environment Agency. 2014. *Water Framework Directive Implementation in England and Wales: New and Updated Standards to Protect the Water Environment*. Available from: https://assets.publishing.service.gov.uk/government/uploads/system/uploads/attachment_data/file/307788/river-basin-planning-standards.pdf (accessed 18 October 2020).

Environment Agency. 2016a. *TraC Fish Counts for All Species for All Estuaries and All Years*. Available from: https://www.gov.uk/guidance/environmental-data (accessed 18 October 2020).

Environment Agency. 2016b. *WFD Cycle 2 TraC Macroalgae Classification*. Available from: https://www.gov.uk/guidance/environmental-data (accessed 18 October 2020).

Environment Agency. 2018. *Catchment Data Explorer: Poole Harbour*. Available from: https://environment.data.gov.uk/catchment-planning/WaterBody/GB520804415800 (accessed 4 November 2020).

Falconer, R. 1993. *Application of Numerical Models for Water Quality Studies. Proceedings of the Institution of Civil Engineers-civil Engineering*. Available from: https://www.researchgate.net/publication/245411865_Application_of_numerical_models_for_water_quality_studies (accessed 27 March 2021).

Garrick-Maidment, N. 2010. *Seahorses in Poole Harbour*. The Seahorse Trust, 2010. Available from: https://www.theseahorsetrust.org/userfiles/PDF/Seahorses%20in%20Poole%20Harbour%20in%20Dorset.pdf (accessed 9 October 2020).

Gray, A.J. and Pearson, J.M. 1983. *Holes Bay: Saltmarsh Vegetation Survey*. Natural Environment Research Council, Dorset.

Haigh, M.J. 1976. *A Biogeographical Reconnaissance of the Coastal Marshlands of Poole Harbour, Dorset (1975)*. Keele University Library, Oklahoma.

Howard, S. and Moore, J. 1988. *Surveys of Harbours, Rias and Estuaries in Southern Britain: Poole Harbour*, 35 pp. Nature Conservancy Council, Peterborough. Report 896.

Hübner, R. 2009. *Sediment Geochemistry – A Case Study Report*. DPhil Thesis. Bournemouth University.

Hübner, R., Herbert, R.J.H., and Astin, K.B. 2010. Cadmium release caused by the die-back of the saltmarsh cord grass Spartina anglica in Poole Harbour (UK). *Estuarine, Coastal and Shelf Science* 87(4): 553–60. https://doi.org/10.1016/j.ecss.2010.02.010

Kite, D.J., Bryan, G., and Jonas, P. 2012. *Nitrogen in the Poole Harbour Catchment*. Published technical report for Environment Agency and Natural England.

Langston, W.J., Chesman, B.S., Burt, G.R., Hawkins, S.J., Readman, J., and Worsfold, P. 2003. *Characterisation of the South West European Marine Sites: Poole Harbour SPA*. Available from: https://core.ac.uk/reader/78755770 (accessed 28 October 2020).

Le Pard, G. 2010. Boundaries, quays, jetties, slipways and marinas. In: Dyer, B. and Darvill, T. (eds) *The Book of Poole Harbour*. The Dovecote Press, Dorset.

Liley, D. and Fearnley, H. 2012. *Poole Harbour Disturbance Study*. Report for Natural England. Footprint Ecology Ltd, Wareham, Dorset.

Marine Stewardship Council. 2018. *The Poole Harbour Clam and Cockle Fishery*. Available from: https://fisheries.msc.org/en/fisheries/the-poole-harbour-clam-cockle-fishery/ (accessed 9 November 2020).

Pearce, R. 2015. *Summary of Visitor Monitoring at Upton Country Park SANG*. Report for the Urban Heaths Partnership.

Rance, E.L., Taylor, D., Lagden, B., and Bleese, B. 2019. *Holes Bay Nature Park: Ecology and Human Activity*. Dorset Wildlife Trust, Dorset.

Scopac. 2004. Poole Harbour. Available from: http://www.scopac.org.uk/scopac_sedimentdb/pharb/pharb.htm (accessed 27 March 2021).

Southern IFCA. 2015. *Small Fish Survey Lytchett and Holes Bay Proposal 2015*. Unpublished.

Taylor, D. 2015. *Review of Potential to Remove Harmful Algae and Reduce Nitrogen Load in Poole Harbour*. A study commissioned by the Dorset Local Nature Partnership.

The Seahorse Trust. 2017. Data from: *National Seahorse Records 2017*, electronic data set.

Thornton, A. 2016. *The Impact of Green Macroalgal Mats on Benthic Invertebrates and Overwintering Wading Birds*. DPhil Thesis. Bournemouth University. Available from: http://eprints.bournemouth.ac.uk/24874/1/Ann%20Thornton%20BU%20Thesis.pdf (accessed 26 October 2020).

Underhill-Day, J., Underhill-Day, N., White, J., and Gartshore, N. 2010. *Poole Harbour SSSI Condition Assessment*. A report to Natural England. Wareham.

Wardlaw, J. 2005. Water quality and pollution monitoring in Poole Harbour. In: Humphreys, J. and May, V. (eds) *The Ecology of Poole Harbour: Proceedings in Marine Science 7*. Elsevier, Amsterdam. https://doi.org/10.1016/S1568-2692(05)80023-3

PART III

Fisheries

CHAPTER 11

Fisheries of Poole Harbour

ROBERT W.E. CLARK

Abstract

The fishers of Poole target a large variety of fish and shellfish, and the ability to adapt to the mixed fishing opportunities emerges as a key attribute of the fishing fleet. The Poole fishery is broadly occupied with either fishing in the harbour or into Poole Bay, utilising a wide variety of methods which, with the exception of the shellfish dredge fisheries, predominantly employ static gear. The importance and strength of the local governance and management arrangements are detailed, as are the legislation and other regulatory measures in place to manage the fishery. Many fisheries are showing signs of stock decline (with the exception of the clam fishery). The reasons for this are varied but the impacts on the fishery, as a consequence of climate change and pollution, will require further management attention; meanwhile, the fisheries will need to be able to continually adapt and the fisheries management regime must be sufficiently flexible (within the environmental designations constraints) to enable this to happen. Poole Harbour is a busy place, and development pressures coupled with declines in certain fishing opportunities will mean that addressing the broader livelihood issues of the fishery participants will continue to require co-management between fishers and managers so as to facilitate institutional innovation and adaption.

Keywords: fisheries, commercial, shellfish, static fishing, pump-scoop dredging, management, Poole Harbour

Correspondence: robert.clark@association-ifca.org.uk

Introduction

Poole Harbour is the home port of one of the largest inshore fishing fleets in England. With around one-hundred commercial fishing vessels either based in Poole or regularly visiting the harbour, the relative significance of the fleet is a consequence of a number of factors, including the size of the harbour itself and the sheltered conditions provided therein, the relative abundance of the fishing opportunities in Poole Harbour and its surrounds, plus the well-established facilities and governance arrangements. This relative importance has increased due to the general trend towards a reduction in the English inshore fishing fleet elsewhere.

At 6 m, the average overall length of the commercial registered and licensed vessels based in Poole Harbour is one of the smallest in England. This average size masks two broadly distinct vessel class categories: vessels that fish predominantly within the harbour

Robert W.E. Clark, 'Fisheries of Poole Harbour' in: *Harbour Ecology*. Pelagic Publishing (2022). © Robert W.E. Clark. DOI: 10.53061/KZRS7935 (typically, as one might expect, smaller and with shallower draft) and those that fish in Poole Bay and beyond (generally 8–12 m in length). The upper limit of 12 m is a consequence of a local vessel length byelaw restricting fishing to this size when inside the British 6-mile limit.

Almost all of the commercial fishing vessels in Poole are rigged and geared to fish with a variety of static fishing methods (pots, traps, nets, etc.). In addition to this static gear arrangement, some 40 or so craft are permitted to use mobile pump-scoop dredges, where and when they are engaged in the clam fishery (as defined by the conditions associated with those permits).

A large proportion of the shellfish catch is exported directly to the Continent (where better prices than in Britain are often offered) via the local ferry ports, while the finfish supply local markets or are transported to Billingsgate or to the markets in Brixham and Plymouth. The wide variety of fish along this coast has encouraged local fishermen to be versatile, and many of the inshore boats are equipped to work a number of fishing methods corresponding to seasonal fisheries. This flexibility has also allowed them to exploit new species and to cope with frequently changing market conditions. Typical examples in recent years have been the collapse of the Bass stocks, the expansion of the Asian whelk markets, increased demand for Cuttlefish *Sepia officinalis*, the closure of the Native Oyster *Ostrea edulis* beds and the emergence of the Manila Clam *Ruditapes philippinarum* fishery (Table 11.1).

The majority of the fleet is based at Fisherman's Dock, which is near to the heart of Poole Quay, although there is a further cluster of vessels moored in Lytchett Bay and accessed shore side from Turlin Moor. Wareham provides a base for a few further craft. The size of the main fleet and its proximity to Poole Quay mean that fishing remains important to the identity of the town. The location of the majority of the fleet in the centre of Poole is significant as it means that the historic association of fishing with Poole is maintained as a characteristic of the town, albeit today this primary industry is juxtaposed with the development of accommodation, leisure, tourism and retail in common with many seaside towns on the south coast of England (Figure 11.1).

Recreational fishing in Poole Harbour and Poole Bay, as well as the wider coastal area, is very significant. The harbour is home to one of the largest charter fishing fleets in England (see Williams, this volume). These vessels are also predominantly based in Fisherman's Dock and they fish mostly outside of the harbour. The high concentration of moorings and marinas in the harbour more generally provides for many privately owned leisure craft which, along with large numbers of shore anglers, target a variety of fish.

There is a long tradition of fishing in and from Poole. John Sydenham, writing in 1839, observed that 'The Harbour and the adjoining coasts, are tolerably productive of fish'. Indeed, in the nineteenth century, 'The fish principally caught' were noticeably similar as one might see landed into Poole today: 'mackerel, herring, whiting, cod, turbot, brills, soles, place, skate, gray [sic] mullet, red mullet, barce [sic], lobsters, crabs, oysters, cockles, muscles [sic], and periwinkles'. The principal fishery of the sixteenth to the early twenty-first century Native Oysters, however, are today commercially extinct. Today this fishery

Table 11.1 Top ten species by landed weight in Poole, 2017

1. Whelks	6. Lobsters
2. Crabs	7. Cockles
3. Scallops	8. Sole
4. Mullet	9. Plaice
5. Cuttlefish	10. Manila Clam

Figure 11.1 Aerial view of the Fisherman's Dock.

has been replaced by the commercial cultivation of the non-native Pacific Oysters (see Humphreys, this volume) and the naturalisation of another newer arrival, the Manila Clam (see Birchenough, this volume).

In this chapter the governance of the fishery is described and an overview of the main fisheries is provided. Where particularly relevant to the characteristics that define the fishery, socio-economic and environmental aspects are discussed.

Governance

Inshore fisheries (0–6 m from shore) in England are managed by Inshore Fisheries and Conservation Authorities (IFCAs). The inshore fleet is also managed and administered by the Marine Management Organisation (MMO). The MMO, alongside other duties such as marine planning, administers inshore fishing vessels licensing and quotas in accordance with government policies set by the Department for Environment, Food and Rural Affairs (Defra). Until the UK's decision to leave the European Union, much of this national policy emanated from origins in the Common Fisheries Policy (CFP).

Southern IFCA, one of the ten IFCAs in England, is responsible for the management of Poole fisheries. Southern IFCA has duties and powers stemming from the Marine and Coastal Access Act 2009. This act requires the IFCA to seek to achieve sustainable inshore fisheries and to protect the marine environment. Southern IFCA is a joint committee of local government. Its district extends to 6 nm from territorial baselines and covers the coastal waters of Hampshire, Dorset and the Isle of Wight. The membership of the IFCA comprises locally elected councillors (including two from Bournemouth, Christchurch and Poole Council), two from Dorset Council and also 'general members' appointed for their knowledge and skills relevant to the work of the IFCAs. While members do not represent particular sectors, their skills in these sectors ensure that when the IFCA is discharging its duties, it is informed by local and specialist knowledge. Management is principally delivered through the implementation of byelaws and codes of practice. The IFCA develops locally tailored policy, implements aspects of national legislation, and to do so, employs scientists and enforcement officers to undertake its work.

Poole fishermen are represented by the Poole and District Fishermen's Association (PDFA). PDFA exists to support its members in their commercial endeavours and to represent their views at local and national levels. PDFA has good contacts within the government departments and NGOs that oversee the fishing industry, and PDFA always partakes in the consultation process when legislative changes are proposed.

The PDFA has over 90 members and can trace its origins back to a 1903 agreement when the fishermen of Poole were granted the right to use an area to the east of Town Quay in Poole. The PDFA are responsible for the management of the Fish Landing Area, the Fishermen's Slipway and the 120 berth Fishermen's Boat Haven. They act as agents for Poole Harbour Commissioners in the collection of dues and mooring fees.

Poole Harbour Commissioners (PHC) is a trust, that is an independent statutory body, governed by legislation, the latest of which is the Poole Harbour Revision Order 2015. Trust ports hold a unique place in the UK ports industry; there are no shareholders or owners and any surplus is invested back into the harbour and port operations for the benefit of the stakeholders of the trust. PHC seek to ensure that all the varied interests operate in harmony, both for the common good and for the long-term sustainability of the whole harbour.

Fisheries interests from Poole and the PDFA are represented by way of appointment on merit to the Southern IFCA and the PHC. Through these mechanisms Poole fishers are actively engaged in the management of the Poole fisheries and the management of the harbour and port more generally.

Regulation

The MMO administers inshore fishing vessels licensing and quotas. A powered boat for the purpose of commercial fishing must be registered and licensed with the MMO. Depending on the size of the vessel and the species being targeted the license will dictate, for certain 'pressure stocks', that is, those for which a total allowable catch (TAC) is set, the quantity of fish which may be taken within a certain time period. The TAC and quota system originated through the EU's Common Fisheries Policy and is a way of distributing fishing opportunities across the EU fleet, where that fleet fishes on 'shared stocks'. The distribution of fishing opportunities, by country and by region, has been set at a point in time according to agreements reached at the accession to the EU (which attempted to reflect historical access and activity). Whereas the historical access rationale is becoming diminished by Brexit, the overall TAC is still informed by the overall assessment of stock biomass and is set annually according to annual agreements.

Fishing boats under 10 m in length (the majority of the Poole fleet) fish a fixed quota allocation known as the 'under 10 m pool'. This collective 'pool' consists of the total amount of fish which may be taken from a sea area by the under 10 m fleet in a period of time. The administration of the pool is (usually) achieved by setting monthly limits (by way of license variation) applicable to the relevant sea area while also monitoring that uptake and adjusting the monthly limits accordingly so as to not exceed the annual TAC.

The inshore fleet in England, despite supporting the majority of jobs in the fishing industry and representing 77% of UK vessels, receives only 1.5% of the share of fishing rights for quota species. In part the distribution of these fishing opportunities can explain the reliance of the Poole fleet on 'non-quota' species; however, the distribution and abundance of crustacea and molluscan shellfish, the catching capacity of the fleet, as well traditional reliance on a variety of mixed species are all additional factors.

More recently, as well as central government administration of quotas, restrictions on fleet capacity and entrance into the fishery for certain key species have emerged. This is in response to wider concerns for the sustainability of stocks in coastal seas. To cap and to manage fishing effort, inshore vessels are now required to hold a shellfish entitlement and Bass entitlement in addition to their licence. Furthermore, the fishing license can be either capped or un-capped and restrictions on the consolidation of licenses have been introduced. This creates increasing barriers to entry to the Poole fishery, reduces the opportunity for the fishery to diversify and has the potential to concentrate effort into other fisheries.

Local management

While national management regimes are preoccupied with ensuring the administration of fishing opportunities across national jurisdictions, the local regulations attempt to ensure that the fisheries (often outside of the national and EU quotas control) are managed in a sustainable manner, the needs of different interests in fishing are accommodated and the fisheries do not have a significant long-term damaging effect on the marine environment.

Southern IFCA manages fishing in Poole Harbour through codes of conduct and byelaws. Section 156 of the Marine and Coastal Access Act 2009 sets out a non-exhaustive list of the types of activities for which IFCAs may make byelaws (including emergency byelaws) to manage sea fisheries resources in their district.

The provisions that may be made by a byelaw under Section 156 include prohibiting or restricting the exploitation of sea fisheries:

(a) In specified areas or during specified periods
(b) Limiting the amount of sea fisheries resources a person or vessel may take in a specified period.

The provisions cover:

- permits (including conditions for the issue, cost and use of permits)
- vessels
- methods and gear (including the possession, use, retention on board, storage or transportation of specified items).

protection of fisheries for shellfish, including monitoring by:

(a) requiring vessels to be fitted with specified equipment;
(b) requiring vessels to carry on board specified persons for the purpose of observing activities carried out on those vessels:

- marking of gear
- identification of items
- information that those involved in the exploitation of sea fisheries resources in an IFCA district must submit to the IFCA.

More specifically byelaws may:

- prohibit or restrict the exploitation of sea fisheries resources in specified areas or periods or limiting the amount of resources that may be exploited or the amount of time a person or vessel may spend exploiting fisheries resources in a specified period;
- prohibit or restrict the exploitation of sea fisheries resources in an IFCA district without a permit. IFCAs will be able to recover the costs of administering and enforcing a permit scheme, attach conditions to permits and limit the number of permits they issue under a particular scheme;

- prohibit or restrict the use of vessels of specified descriptions and any method of exploiting sea fisheries resources. The possession, use and transportation of specified items or types of items used in the exploitation of sea fisheries resources may also be prohibited or restricted. This would enable an IFCA to require the use of a particular method of sea fishing or an item used in sea fishing (e.g. to reduce by-catch) by means of a prohibition on the use of other method and items;
- protect and regulate shellfisheries including, but not limited to, requirements for shellfish to be re-deposited in specified places and for the protection of shellfish laid down for breeding purposes and culch, which is the substrate/material on which the spat or young of shellfish may attach and grow;
- establish a district of oyster cultivation, allowing an IFCA to prohibit the sale of oysters between certain dates, and allows IFCA authorities to disapply the defence concerning the taking and sale of certain crabs and lobsters as set out in Section 17(2) of the Sea Fisheries (Shellfish) Act 1967;
- make provision for monitoring the exploitation of sea fisheries resources. This includes requirements as to the fitting of particular equipment, the carriage of on-board observers and the marking or tagging of items used in the exploitation of sea fisheries resources;
- require people involved in the exploitation of sea fisheries resources in their district to provide them with specified information so that it is an offence if certain information is not provided.

Environmental protection

Poole Harbour is a large natural harbour comprising of extensive tidal mudflats and salt-marshes together with associated reedbeds, freshwater marshes and wet grasslands. It also includes seagrass beds located towards the north-east of the harbour and subtidal channels in which 68 seaweed species, 159 invertebrate species and 32 fish species have been recorded. Because of the importance of Poole Harbour and its surrounds for species and habitats of conservation importance, various nature conservation designations afford protections which require appropriate fisheries management.

The origins of these protections are both domestic and international. Poole Harbour is a Site of Special Scientific Interest (SSSI) designated to protect the very high proportion of the harbour comprising intertidal marshes and mudflats. These, together with the permanent channels, support large numbers of non-breeding waterbirds, for which Poole Harbour is of international significance. Fringing habitats of heathland, grassland and the islands provide additional interests, in turn supporting further scarce and restricted flora and fauna. Several rare marine invertebrates also occur within the harbour. In 2019 the SSSI boundaries were extended to protect additional subtidal features and 'it is the first SSSI specifically to include subtidal areas' (Defra 2019)

The harbour is also a Special Protection Area (SPA). The site qualifies under Article 4 of the Birds Directive (2009/147/EC) for the following reasons:

- The site regularly supports more than 1% of the Great Britain populations of five species listed in Annex I of the EC Birds Directive.
- The site regularly supports more than 1% of the biogeographic population of two regularly occurring migratory species not listed in Annex I of the EC Birds Directive.

Outside the harbour itself Poole Bay and the surrounds are similarly designated under both domestic and international legislation.

The Studland to Portland Special Area of Conservation (SAC) designations protect the reef and Mearl communities around Handfast Point west, while the Poole Rocks, Studland and Southbourne Rough Marine Conservation Zones (MCZs) protect a variety of habitats and features – notably, from a fisheries management perspective, Black Seabream *Spondyliosoma cantharus*, Native Oysters and habitats such as seagrass and subtidal mixed sediments.

Management implications

Accordingly, local fisheries managers must exercise duties to further the conservation objectives of the MCZs and in the case of an SAC or SPA (collectively termed a European Marine Site or EMS) to determine whether a plan or project may have a significant effect on an EMS (see Clark *et al.* 2016). Consequently, the management of fisheries in Poole Harbour and the surrounds is significantly influenced by nature conservation policy.

Mollusc fisheries

Native Oysters

Until the 1970s Native Oyster fisheries were important in Poole; however, the fishery is commercially extinct today. Sydenham (1839) noted that 'the oyster [Poole] fishery for many years constituted a lucrative field for the exertions of the fishermen and its preservation has been and object of great attention to the authorities'. The native oyster is characterised by slow growth rate and sporadic recruitment success, and the number of oysters surviving to maturity can vary hugely from year to year, making it a particularly vulnerable species.

Evidence suggests that the fishery was not self-sustaining to the extent that it could withstand significant fishing pressure, relying instead on recruitment from the shellfish caught elsewhere in the English Channel and relayed in or near the harbour. 'The earliest beds existed outside the harbour; and it is believed that they were originally formed from oysters brought from the Channel Islands for sale here [in Poole], but thrown overboard at times when the market was over stocked.'

> Forty sloops were employed in this branch of the fishery, for two months every spring, which season was the fisherman's harvest and during which time they were said to receive upwards of £3,000. The catch of the last day of the season was, by prescriptive custom, thrown into the channels within the harbour, and thus were formed what are now termed the channel beds, which at present are the most productive, the off-ground [in Poole Bay] fishery being indeed almost extinct. (*Op. Cit.*)

At that time, reflecting upon the future of the oyster fishery Sydenham (1839) noted that 'it is apprehended that the beds will, in a few years, dwindle to a state of almost exhaustion'. He considered this as a consequence of a lack then of 'wholesome regulation' following the cessation of the admiralty jurisdiction in the harbour. We might now attribute the collapse of the oyster stocks (which has been repeated throughout Europe) to be as a consequence of not just overfishing (and the lack of regulation thereof) but to a lack of recruitment associated with the absence of relaying from elsewhere (enhancing the spawning stock biomass) coupled with disease and other environmental stresses including habitat degradation, reduced water quality, as well as the prevalence of invasive species (notably *Crepidula fornicata*).

In the early 2000s as today, Jensen *et al.* (2005a) noted that native 'oysters are naturally occurring within the harbour. They were laid down as seen on some of the Several shellfish beds some years ago [the 1970s and 1980s (Wordsworth pers. comm.)] and oysters from the Solent fishery were re-laid to grow on for a year before harvest,[1] but some stocks were reduced significantly by an outbreak of Bonamia in the 1980s.'

Today some natural resettlement still occurs in the harbour and there is some evidence of the shellfish in the bay. Occasionally, fishermen will report (small) catches of oysters caught in nets, sparking interest in a return of the fishery. The subsequent investigations (both organised and 'informal'), however, demonstrate the relative scarcity of the species.

Renewed interest in the biological conservation value of oyster beds, being associated as they are with species diversity and richness (as well as other ecosystem functions such as water filtration), has prompted conservation efforts. Notably, the inclusion of an objective of 'recover' is attributed to the Poole Rocks MCZ. In the absence or restoration, how fisheries and conservation managers achieve this objective is unclear, and this is an area of emerging policy in the UK as well as elsewhere in Europe.

Pacific Oyster

Pacific Oysters *Magallana gigas* were introduced into British waters in 1890 and were intended to support an industry suffering from the decline of the native oyster. Pacific Oysters were introduced into Poole as part of an aquaculture regime originally associated with native oyster mariculture. Pacific Oyster farms are situated in Poole Harbour around Brownsea Island, and Othniel Oysters Ltd is one of the aquaculture businesses, which operated under leases issued under the Poole Harbour Fishery Order, 2015. The farm, at peak production, produces 300–400 tonnes of Pacific Oysters a year, making it one of the largest Pacific Oyster aquaculture production sites in the UK. Juvenile oysters are brought into the harbour from France and on-grown in bags before being laid out on beds for on-growing prior to harvest.

There is a tension between the economic importance of the fishery and the management of biosecurity/conservation risks associated with Pacific Oysters (Herbert *et al.* 2016), with fisheries regulators (Southern IFCA) required to undertake assessments of the fishery and its compatibility with the conservation objectives of the SPA designation. Wild settlement in the harbour, however, appears sparse and irregular (Deane *et al.* 2013). The shellfish, as a consequence, does not appear in significant quantities in the wild capture fisheries.

Manilla Clam and Cockle fisheries

The Manila Clam *Ruditapes philippinarum* was introduced to Poole Harbour in 1989 as a novel species for aquaculture. Contrary to expectations this species has become naturalised in the harbour. The fishery is now significant and more recently (2015) has been subject to regulation in the form of the Poole Harbour Dredge Fishery Byelaw.

This byelaw enables the Authority to regulate the use of dredges within Poole Harbour. Under this byelaw, a restricted number of permits are issued each year by the Authority with accompanying conditions relating to catch restrictions and reporting, gear types, gear construction and restrictions, spatial and temporal restrictions and the fitting of specified equipment to vessels.

These pump-scoop dredge fisheries use a small towed dredge which pumps seawater through a small metal basket. This method is employed in shallow water areas. The

---

[1] In the 1980s, the Solent oyster fishery was the largest 'self sustaining [*sic*]' fishery in Europe. The fishery has collapsed and has been virtually closed since 2014.

Figure 11.2 A 'pump-scoop' shellfish dredge being prepared for redeployment following clearing (© Emma Rance Noctiluca marine with permission).

injection of water from the pump is sometimes directed into the seabed when fishing for cockle due to the sandy sediment habitats, but in the clam fishery the water is directed into the basket to clean the sediments from the catch (Figure 11.2).

Under the terms of the byelaw strict regulation ensures the fishery is managed in accordance with the duties under Regulation 63 of the Conservation of Habitats and Species Regulations 2017. Southern IFCA undertakes an annual appropriate assessment for the issue of permits; the purpose of this assessment is to determine whether or not in the view of Southern IFCA, the issue of permits will hinder the achievement of the conservation objectives of the Poole Harbour SPA and lead to an adverse effect on site integrity. The stock appears stable and self-sustaining.

Mussel fishery

The Blue Mussel or Common Mussel *Mytilus edulis* is the most commonly produced species in aquaculture in Europe. The species is naturally occurring in intertidal and subtidal areas in the whole of the UK coastline. The species has a relatively complex lifecycle, but in summary it consists of a number of larval stages during which they float on prevailing currents and commence filter feeding before settling on a primary settlement location. They will remain there for a number of weeks usually doubling in size before detaching and finding a permanent settlement point (often attaching to each other to form large mussel beds). This settlement can be on the seabed or on suitable vertical or horizontal structures.

There is limited natural settlement of mussels in Poole Harbour. Certainly, extensive natural settlement does not occur to the extent that it forms the basis of a commercial fishery, so instead and until more recently (2014) juvenile mussels were fished from a designated area off Portland, Dorset, and re-laid into Poole Harbour onto leased bed grounds (again under the provisions of the Poole Harbour Fishery Order, 2015). Since the severe storms of 2014, the available mussel seed has been absent from the previously stable beds off Portland and consequently the activity has been severely interrupted.

Whelks

Static, baited pots designed specifically to catch *Buccinum undatum* are generally used by Poole fishers operating in Poole Bay and beyond. The type of pots comprises a weighted cylindrical structure with a mesh-covered opening. These can take two common forms: pots made from discarded, typically approximately 25 L, plastic containers weighted down with concrete or purpose-built (and therefore more expensive) pots. Escape gaps are included to allow undersized whelks to exit. This type of trap is generally associated with a low by-catch. Pots are laid down in strings, set between 15 m and 20 m apart, with the number placed by each vessel varying with vessel size and local regulations. Strings of pots can number hundreds at a time, and each is baited with dead brown crab and fish (of a variety of species), attractive to whelks. Catches are collected daily, with pots re-laid every 24 hours for two main reasons: bait quality declines rapidly, and once *B. undatum* enters the pot the bait is quickly consumed. Once collected, a riddle of parallel bars is often used to sort the catch, to retain whelks above the EU minimum size of 45 mm (Council Regulation (EC) No. 850/98), allowing for the return of any undersized whelks caught.

Cuttlefish

Cuttlefish *Sepia officinalis* have a short lifespan of between 18 and 24 months which is terminated by a reproduction event followed by mass mortality of the adult stock (Bloor *et al.* 2013). Reproduction occurs all along the south coast and in inshore areas along Poole Bay. Nets and traps are set for the mollusc in spring close inshore, where the cuttlefish come to congregate. The fishery has expanded in the last two decades as markets have emerged in continental Europe and more recently in Asia (in part as a consequence of the collapse of Indian Ocean fisheries). The exploitation of the English Channel cuttlefish population has nearly doubled during the last decade.

The intermittent terminal spawning employed by *S. officinalis* means that the fishers target fully grown cuttlefish at or near the end of their lives and as such, the fishing mortality of these inshore fisheries is close to natural mortality (Bloor 2013. Notwithstanding the inherent sustainable nature of the inshore fishery in the recent decade, catches have declined as a consequence of the emergence of offshore fisheries elsewhere in the English Channel, which target premature congregations.

Scallops

Scallops are caught by dredging, using Newhaven-type scallop dredges, or by hand with SCUBA. There are no significant scallop beds in Poole Bay, and their appearance in the landing statistics for Poole is largely associated with catches made elsewhere in West Dorset by hand divers or visiting scallop vessels.

Crustacea

As many as ten full-time boats of 10–12 m set crab *Cancer pagarus* and lobster *Homarus gammarus* pots year round with the majority of the effort in spring and summer associated with the lobster fishery. The fishery extends from close inshore on reefs off Purbeck and

Christchurch and out to 10 m offshore south-east off Swanage. Perhaps as much as 150,000 pot hauls per month in the peak season return some 1,000 to 2,000 kg of lobster per month for the main fishery participants.

There has been a slight increase in effort over the past ten years, and the catches have declined slightly. More recently, an expansion in the offshore crab fishery associated with access to the Asian markets has increased fishing pressure on the offshore crab stocks, and the crab fisheries have seen a more marked decline in landings into Poole.

Finfish fisheries

Finfish are targeted by commercial vessels both within Poole Harbour and into Poole Bay and beyond. There is extensive netting along the coast except when weed becomes a problem in summer.

Fish targeted commercially in the harbour are mullet, bass, flounder, sole and plaice, which are caught using fixed, drift, ring nets, beach 'D' nets and hand lines. There are minimum sizes on all the commercial species which are contained within both European legislation and specific Southern IFCA byelaws for different species. These byelaws are 'Minimum Fish Sizes', 'Grey Mullet – Minimum Size' and 'Skates and Rays – Minimum Size', copies of which are available from Southern IFCA.

The use of fixed nets in the harbour is restricted by the 'Fixed Engines' byelaw, which prohibits the use of fixed nets in the harbour between 1 April and 30 September each year.

Bass

The popularity of Bass *Dicentrachus labrax* for domestic and export markets has seen increased effort directed towards this fishery. The commercial rod and line fisheries are notable in Poole and the premium quality of the product demands a premium in the market.

Following a general collapse in the stock in around 2015, which was associated with colder winters reducing recruitment and a lack of corresponding reduction in fishing pressure (particularly associated with the French mid-channel pair trawl fishery), European emergency measures were introduced to protect the spawning stock biomass. This included prohibition of pair trawling, a by-catch only allowance for the trawl fishery, closed periods for both commercial and recreational fisheries as well as a monthly limit on the allowance for commercial rod and line and net fisheries. In 2017, the minimum size for Bass was increased from 36 cm to 42 cm.

The Bass fishery is important to the local fleet, which uses mainly hand lines and some longlines and nets sometimes seaward of the Isle of Wight. During the bass season up to 30 boats may fish the harbour entrance and Christchurch Ledge, some taking out angling parties and catching sand eels off Hook Bank for bait. Recognising the importance of the harbour to juvenile Bass, most of Poole Harbour is a designated Bass nursery area.

Pelagics

Mackerel *Scomber scombrus*, Herring *Clupea harengus* and Sprats *Sprattus sprattus* are caught using drift nets in the harbour, but the activity is limited for the latter two species because the market demand for these species is limited in the quantities that are produced in the relatively small-scale fishery. While the market for all these species exists, it predominantly serves local demand and retail with the price being depressed in the event of large quantities being landed. This is because large-scale offshore fisheries provide the bulk of supply, and processing costs and transportation costs are barriers to wider exploitation by small-scale vessels generally.

Sand Eels *Ammodytes tobianus* are caught in fisheries to supply bait for commercial and recreational rod and line fisheries for Bass, but the activity is limited with pre-processed supply servicing the larger of the two markets the recreational demand.

Diadromous and catadromous fish

Atlantic Salmon *Salmo salar* are associated with Poole Harbour's main rivers. These fish have a relatively complicated lifecycle that involves several stages which occur in different habitats. Eggs are laid in freshwater rivers during winter, and over the course of three months or more these develop into very small salmon called alevin, which then emerge from the eggs. In the spring, the alevin swim out of the gravel and develop into fry, at which point they begin to eat small, water-living insects. Fry grow larger and become camouflaged to suit their river environment, when they are called parr. These parr live and grow in the river for between one and four years depending on the river, its temperature and the amount of prey available, until they are ready to 'smoltify', at around 13 cm in size, when they develop saltwater tolerance. Smolts migrate to the ocean, where they go to feed. Finally, they return to their home river after between one and four years to spawn the next generation.

Numbers of adult Atlantic Salmon and Trout in Poole (as elsewhere in England and Wales) have declined over the last two decades. A decline is also seen elsewhere around the Atlantic Ocean. Salmon numbers in Dorset rivers have been falling for a number of years and the decline shows no sign of slowing. On the river Frome in Dorset the total number of adult Salmon returning to spawn has been monitored since 1973 by a fish counter at East Stoke. These long-term records represent the most comprehensive account of Salmon movement in England and Wales.

The Atlantic Salmon and the Brown Sea Trout *Salmo trutta* are referenced within the citations of the River Frome SSSI and therefore receive consideration as a faunal component of the 'Rivers and Streams' feature. To reduce the risk of net fishery interaction with Salmon and Sea Trout, Poole Harbour is subject to a seasonal (April–September) fixed engine closure under the Southern IFCA Fixed Engines byelaw.

Historically, Salmon and Sea Trout fisheries were important in Poole, but their declines have resulted in the commercial fisheries being prohibited. To reduce incidental capture in net fisheries the use of fixed engines is prohibited above a line drawn from Bower Point to Bucks Cove this being the boundary of the Environment Agency owned and controlled fishery of the rivers Frome and Piddle. Furthermore the Poole Harbour Net Limitation Order for catching of Salmon and Sea Trout (regulated by the Environment Agency) is now used for research purposes, with all fish returned.

European Eel

The European Eel *Anguilla Anguilla* once a prolific fishery in Poole is now 'Critically Endangered' by the IUCN Red List of Threatened Species. The severe decline in European Eel recruitment prompted the International Council for the Exploration of the Seas (ICES) to declare the stock as 'outside safe biological limits' (ICES 2011). Authorisations for commercial yellow and silver eel fishing are restricted to those already licensed to fish. Elver fishing is restricted to certain locations with fishers only authorised to use hand-held 'dip nets'. There are silver eel racks at East Burton and on the Piddle at Trigon.

Fyke nets were and still are used in Poole, now under strict regulation by the Environment Agency. A fyke net is a type of fish trap. It consists of a long cylindrical netting bag usually with several netting cones fitted inside the netting cylinder to make entry easy and exit difficult. This net is then mounted on rigid rings or other rigid framework and fixed on the seabed by anchors, ballast or stakes. It also has wings or leaders to help guide the fish towards the entrance of the bag.

Knights and White (1997) offer a useful overview of the modern history of the fishery in the harbour. Fyke nets were introduced in the 1950s, and catches in the 1960 to 1970s were about 30 t per year (Morrice 1989). At the end of the 1960s, five 'punts' were laying up to 650 fyke-net ends per night. Increasing effort (as much as a tenfold increase in numbers of ends deployed per day) was needed to maintain catches in the early 1980s. Today one fyke-net license holder fishes in Poole and the traps are set in the harbour including in Holes Bay.

Flounder

Flounder *Platichthys flesus* is a migratory fish, which for most of the year is found in Poole Harbour. Adults occur in the mud and sand bottom in shallow water, at sea and in brackish waters, often entering freshwaters. During winter, adults retreat to deeper, warmer waters, where they spawn in spring. Juvenile Flounder live in Poole Harbour and the surrounding shallow coastal waters, which are also the summer feeding grounds for the adults.

Flounder was once an important fishery in Poole Harbour, but populations, in common with the situation in estuaries in Europe, have declined considerably in recent decades. There was once a trawl fishery in the harbour as well as a fixed-net fishery, but the species is not targeted routinely now due to the limited number of fish as well as the lack of value. Changes in environmental factors, including salinity, hypoxia, temperature and eutrophication, are identified as probable drivers; this is, however, not well studied in Poole or England generally.

Grey Mullet

Both ring and beach D nets are predominantly used in the harbour and target Thin-lipped *Chelon labrosus* and Thick-lipped Grey Mullet *Chelon ramada*. They typically have a soak time of a maximum of 15 mins or less. This is the time it takes to enter into the net circle to 'scare' the fish, exit the circle and then begin hauling. Ring and beach nets do not drift. Netters fish mostly on high, low or slack to avoid the drifting motion of the net.

Grey Mullet species exhibit slow growth, late maturity and high site fidelity. Grey Mullet spawn in the English Channel and Irish Sea and potentially in estuaries in Eastern England. Thick-lipped Mullet are thought to spawn on alternate years. They spawn in open water and after around two to six weeks, the then juvenile fish move into inshore waters, especially estuaries. Their occurrence is impacted by temperature, and they are often observed near the surface (0–10 m depth). Because of this and their large size, they can be popular with anglers.

Recreational fishery

There is a large recreational finfish fishery in the harbour including club competitions. There has been an increase in charter (and casual) angling vessels, which is now a year-round industry of considerable importance to the local economy. Catch data are not recorded for anglers and hobby fishermen, but it is likely that they contribute a considerable proportion of some species' landings.

Hand gathering and bait collection

Bait-digging in the harbour includes both commercial and recreational operations. To manage bait collection in Poole a memorandum of agreement for bait-digging was produced from working group discussions with interested parties, and includes agreement to spatial and temporal avoidance of bird sensitive areas with respect to Poole Harbour's European Marine Site status and seagrass beds and prohibited bait-digging areas.

Manila Clam and cockles are fished commercially by handpicking. Handpicking of these species is typically accessed by walking on the mudflats at low water from the shore because hand fishing from vessels requires licensed and registered vessels.

The principal driver behind commercial handpicking is market demand outside of the dredge fishing season. No licensing is required for commercial handpickers, and the practice is regulated through the Poole Harbour Shellfish Hand Gathering byelaw. This byelaw limits temporal access (closure 1 November to 31 March) and spatial access. In addition to commercial handpicking for cockle and clams, recreational catches of these species as well as American Hardshell *Mercenaria mercenaria* and Razor Clams *Ensis ensis* are permitted within the harbour. Monitoring of recreational handpicking is undertaken by Southern IFCA, but there are no estimates on total annual take. All species are subject to minimum size limits: razor clam 10 cm, cockle not to fit through a square opening measuring 2.38 cm on each side, Manila Clam 3.5 cm, American Hardshell Clam 6.3 cm. A closed season (1 February to 30 April) for handpicking and hand raking of cockle is controlled through the Fishing for Cockle byelaw. Spatial closure of seagrass beds is applied through the Prohibition of Gathering (Sea Fisheries Resources) in Seagrass Beds byelaw.

Conclusions

This description of the Poole fisheries highlights a number of important challenges facing the fishing industry. The review of fisheries highlights that many of the key stocks have experienced declines and many of these are out with the control of local fisheries management systems and are instead associated with wider ecological or harvest strategies across the range of the stocks being targeted.

The resilience of the fleet can be in part ascribed to the culture that shapes the interactions of fisheries participants with the management structures. This emerges as one of the key themes that has enabled the continued evolution and development of the fishery, particularly the co-management structures which have allowed the emergence of the fisheries for novel fisheries to develop when traditional fisheries have declined.

It is too early to say whether the perceived unfairness of the CFP, which has led to failed implementation of aspects of the policy (i.e. the distribution of fishing opportunities) will deliver dividends for the inshore fishers of Poole. As the UK develops its ambition to move English fisheries management from the paradigm of labyrinthine institutions of the CFP, the need to recognise the importance of socially embedding decision-making, which is geographically contextualised and historically sensitive, requires continued consideration in Poole as it does elsewhere in England.

The impacts on the fishery of climate change and pollution will require further management attention and the fisheries will need to be able to continually adapt and the fisheries management regime will need to be sufficiently flexible (within the environmental designations constraints) to enable this to happen. Poole Harbour is a busy place and development pressures coupled with declines in certain fishing opportunities will mean that addressing the broader livelihood issues of the fishery participants will continue to require co-management between fishers and managers so as to enable institutional innovation and adaption.

References

Bloor, I., Attrill, M., and Jackson, E. 2013. A review of the factors influencing spawning, early life stage survival and recruitment variability in the common cuttlefish (*Sepia officinalis*). *Advances in Marine Biology* 65: 1–65.

Clark, R.W.E., Humphreys, J., Solandt, J.-L., and Weller, C. 2016. Dialectics of nature: The emergence of policy on the management of commercial fisheries in English European marine sites. *Marine Policy* 78: 11–17. https://doi.org/10.1016/j.marpol.2016.12.021

Deane, S., Jensen, A., and Collins, K. 2013. Distribution, abundance and temporal variation of the Pacific oyster, *Crassostrea gigas* in Poole Harbour. Available from: https://assets.publishing.service.gov.uk/government/uploads/system/uploads/attachment_data/file/313003/fcf-oyster.pdf (accessed 12th November 2020).

Defra. 2019. Strengthened protection for Poole Harbour's unique range of wildlife. Press release. Available from: https://www.gov.uk/government/news/strengthened-protection-for-poole-harbours-unique-range-of-wildlife#:~:text=Natural%20England%20has%20confirmed%20the,increase%20of%2040%20per%20cent (accessed 12th November 2020).

Herbert, R.J.H., Humphreys, J., Davies, C.J., *et al.* 2016. Ecological impacts of non-native Pacific oysters (*Crassostrea gigas*) and management measures for protected areas in Europe. *Biodiversity and Conservation* 25: 2835–65. https://doi.org/10.1007/s10531-016-1209-4

ICES 2011. Report of the 2011 session of the Joint EIFAAC/ICES Working Group on Eels. Lisbon, Portugal, from 5 to 9 September 2011. EIFAAC Occasional Paper. No. 48. ICES CM 2011/ACOM: 18. Rome, FAO/Copenhagen, ICES. 2011. p. 246.

Jensen, A., Carrier, I., and Richardson, N. 2005a. Marine fisheries of Poole Harbour. In: Humphreys, J. and May, V. (eds) *Proceedings in Marine Science, Volume 7*. Elsevier, Oxford, pp. 195–204.

Jensen, A., Humphreys, J., Caldow, R., and Cesar, C. 2005b. 13. The Manila clam in Poole Harbour. In: Humphreys, J. and May, V. (eds) *Proceedings in Marine Science*, Volume 7. Elsevier, Oxford, pp. 163–74.

Knights, B. and White, E.M. 1997. An appraisal of stocking strategies for the European eel, *Anguilla anguilla* L. In: Cowx, I.G. (ed.) *Stocking and Introduction of Fish*. Fishing News Books, Oxford, pp. 121–40.

Morrice, C.P. 1989. Eel fisheries in the United Kingdom. MAFF Internal Report No. 18. MAFF, Lowestoft, England.

Sydenham, J. 1839. The history of the town and county of Poole. Poole Historical Trust; facsimile of 1839 edition. June 1986.

CHAPTER 12

A Tale of Three Fisheries: The Value of the Small-scale Commercial Fishing Fleet, Aquaculture and the Recreational Charter Boat Fleet to the Local Economy of Poole

CHRIS WILLIAMS and WILLIAM DAVIES

Abstract

This chapter focuses on the total economic contribution of small-scale coastal fishing, aquaculture and charter boat fishing to the economy of Poole in Dorset. Considering aquaculture, coastal fisheries and recreational charter boat fishing, these sectors together generate a gross output of £6,719,958, an indirect output of £5,802,973 and support £12,522,931 of total economic activity. This total economic activity equates to 17.5% of the 'coastal visits spend' for Poole (£71,573,000).

Keywords: aquaculture, fisheries, charter boat fishing, tourism, local economies

Correspondence: Williams_Chris@ITF.org.uk

Tourism in Dorset

Balancing the economic and social needs of coastal communities with the sustainability of the marine environment is a key challenge. The goal to ensure that both return to health and prosperity while building their resilience is a shared aspiration for local authorities, residents and businesses alike. Improving the marine environment and the coastal economy is not mutually exclusive; indeed, the opportunity exists for them to enhance each other (Balata and Vardakoulias 2016).

Dorset is home to numerous fishing towns and villages, attracting visitors from around the world. Estimates of the total visitor-related spend in Dorset was £1,832,771,000 for 2017 and the total estimated employment related to tourism was 46,254 (34,628 full-time equivalent jobs (FTEs)), representing 12% of all employment in the county. In terms of coastal visits, 11,768,000 were undertaken for 2017 and £402,805,000 was spent on coastal visits in the same year (The South West Research Company Ltd 2018).

Dorset harbours remain as hubs of activity within communities, and the fishing indus-try has had to adapt over the years to the available fish quotas, the state of key stocks

Chris Williams and William Davies, 'A Tale of Three Fisheries: The Value of the Small-scale Commercial Fishing Fleet, Aquaculture and the Recreational Charter Boat Fleet to the Local Economy of Poole' in: *Harbour Ecology*. Pelagic Publishing (2022). © Chris Williams and William Davies. DOI: 10.53061/ATWH9910 and changes in national and export demand (Dorset and East Devon FLAG 2016). The Dorset coastline is rich in environmental designations (terrestrial and marine-protected areas) and the local, small-scale, inshore commercial fishing fleet prides itself on using low-impact fishing techniques (Williams 2019). As the fleet uses mainly pots and nets, or rod and line fishing, the associated impact on non-target species and the wider environment is considerably lower than that caused by large-scale, heavy, towed gear such as beam trawls or scallop dredges. Although uncoordinated in their approach, numerous fishing associations exist in the area. Fishing boats in Dorset are predominately under 10 m in length, although there are also some larger vessels in the area (Williams and Davies 2018).

Poole Harbour and tourism

Poole Harbour is a highly protected marine area with a diversity of nationally important wildlife and fish and shellfish species, with a national and international reputation. It has been designated as an area of environmental importance (classified as a Site of Specific Scientific Interest (SSSI), Area of Outstanding Natural Beauty and EU Special Protection Area) (Dorset Coastal Community Team 2016) and is also recognised by having engaged stakeholders (e.g. in the Marine Conservation Zone planning process). The harbour is of ecological, recreational and commercial importance to residents and visitors alike, with Ramsar (wetland), SSSI, Special Protection Area and Special Area of Conservation (SAC) designations. Nearby is the Studland to Portland SAC (designated under the Habitats Directive) and the Poole Rocks Marine Conservation Zone designated under the Marine and Coastal Access Act, 2009. The town of Poole contains a number of historic maritime listed buildings, and it is located within a few miles of the Jurassic Coast World Heritage Site, the New Forest and the Purbeck Hills (Dorset Coastal Community Team 2016).

In terms of the value of tourism to Poole, in 2017, £82,455,000 was spent by staying visitors. £216,097,000 of that can be considered direct visitor spend, supporting 3,506 FTE jobs. This represents 7% of all employment in Poole. Around 2,091,000 visits to the coast were calculated and the coastal visits spend for Poole was £71,573,000 (The South West Research Company Ltd 2018).

Sectors considered in this study

As competition for space intensifies in the region, so commercial fishing, recreational angling and aquaculture production are 'squeezed' by coastal development needs. Property developers, in particular, prize waterside access for recreational amenities (marinas) and the immediate hinterland for housing. Poole Harbour and Poole Bay are considered attractive places to live and tourism is significant – consequently they are busy places, and local decision-making needs to consider the social, economic and environmental value generated by key marine sectors (Williams and Davies 2018). Coastal areas are becoming cold-spots for social mobility. Poole is among the worst performing 20% of local authorities in England for social mobility and one of the worst towns on the south coast of England in the same regard (Social Mobility and Child Poverty Commission 2016).

There are three main extractive marine industries of note in the Poole region when it comes to the local marine environment: aquaculture (the growing of bivalve shellfish), small-scale commercial fishing (targeting both finfish and shellfish) and recreational (charter boat) fishing, targeting finfish such as bass, plaice or cod using rod and line (Southern IFCA). A description and valuation of each is presented further.

Aquaculture in Poole Harbour

Aquaculture is a major activity in Poole Harbour, focused on bivalve shellfish. The main species grown are oysters (both Pacific and native), clams (both *Mercenaria* and Manila), mussels and cockles. Today, Poole Harbour is the largest Pacific Oyster production area in England. Pacific Oyster *Magallana* (formerly *Crassostrea) gigas* production in Poole ranges between 300 and 400 tonnes per year; Mussel *Mytilus edulis* production for the financial year 2014/15 was 262,536 kg. Total aquaculture (including clam and cockle) production for Poole Harbour for 2014/15 was 700,000 kg, where oyster landings contributed around 270 tonnes of the total.

The Poole Harbour Fishery Order 2015 (a Several Order) covers an area of approximately 838 ha. Presently, less than one-quarter of this area is under aquaculture production. With careful management, this can be increased to improve the economic output (Southern IFCA). The Poole Harbour Fishery Order 2015 grants Southern IFCA the opportunity to lease ground for aquaculture. The ability to do so has been in place since 1915. As of 2015, Orders are granted for a period of 20 years, enabling long-term planning for the industry. There are currently 31 beds leased within the extent of the Order (Southern IFCA).

Commercial aquaculture species cultivated in Poole Harbour

The Pacific Oyster was introduced into British waters in 1890 to support an industry suffering from the decline of the native oyster (Britain was thought to contain Europe's richest natural oyster beds, generating revenues and employment including hundreds of fishers and vessels) (Humphreys *et al.* 2014). The dramatic decline in the native oysters provided a need for importing oyster seeds to grow-on in the UK, and these included the non-indigenous American Oyster *Crassostrea virginica* and the Portuguese Oyster *C. angulata* (Humphreys *et al.* 2014). The Pacific Oyster was introduced into Poole Harbour in 1890, leading to the establishment of self-sustaining populations, meaning the species is effectively naturalised (this is part of an ongoing debate about what constitutes a 'native' species). Given the global market for oysters, especially in China it is clear that British Pacific Oyster production could be significantly increased, delivering socio-economic benefits for Poole and its local economy. The growth of bivalve shellfish aquaculture is important for marine planning and is a key area promoted in the Blue New Deal action plan (Balata and Vardakoulias 2016).

The Common or Blue Mussel is the common name for a number of species of the family Mytilidae, of which *M. edulis* is the species most common in UK waters. In Poole Harbour, all mussels are produced under the Poole Harbour Fishery Order 2015, and stocks are managed by Southern Inshore Fisheries and Conservation Authority. Private operators manage 'lays', which are leased areas for growing on seed mussels that have been collected from elsewhere (usually from open water along the Dorset coast). Management aims for compliance with the conservation objectives for European Marine Sites (Seafish Risk Assessment for Sourcing Seafood 2018). Native oyster *Ostrea edulis*, the European Flat Oyster, is classified as highly endangered throughout the EU, and stocks are currently severely depleted compared to its historic abundance (University of Portsmouth 2017). The decline is attributed to a combination of habitat destruction, disease mortality, water quality, fishing and the interplay between these factors.

The Hardshell Clam *Mercenaria mercenaria* originates from the east coast of North America and was purposefully introduced into Southampton waters in 1925. Dumping and deliberate introduction attempts, as well as natural larval dispersal, are all thought to be the reasons for the species' current occurrence in Poole Harbour. American Hardshell Clams were first targeted as a commercial prospect in the 1970s, whereas historically, the species had been a 'boom and bust' fishery. The clams are sought in their own right and are also retained as a by-catch species when fishing for Manila Clams (described later). The species is harvested in the Pacific Oyster aquaculture beds within Poole Harbour, using oyster barges (Southern IFCA).

The Manila Clam *Ruditapes philippinarum* is one of the top five, most commercially valuable bivalve species globally, but it is also found within the protected site of Poole Harbour. The Manila Clam was first introduced into Britain in 1986 by the UK government's Ministry of Agriculture, Fisheries and Food (MAFF) (Humphreys *et al.* 2007). It was assumed that the species would not naturalise, owing to water temperature restricting their reproduction; however, this proved to be incorrect. Extensive naturalised populations are now found in Poole Harbour, the Solent and other English estuaries. Manila Clams live buried in coastal sediments and are well adapted to estuarine habitats, such as the mudflats of Poole Harbour. Although they are naturalised and play a significant role in the food chain, they do not appear to be aggressively invasive.

The Common Cockle *Cerastoderma edule* is indigenous to UK waters and is widely distributed in estuaries and sandy bays. The species is retained predominantly as a by-catch when fishing for other bivalve species, namely Manila Clam, although it can be targeted in its own right. The species is harvested from Pacific Oyster and clam aquaculture beds. In Poole Harbour the pump-scoop dredge method is used for harvesting cockles. The injection of water from the pump may also be directed into the seabed when fishing for cockle, due to the sandy sediment habitat.

Through interviews with producers and buyers and using official harvest and price data, various economic aspects of the aquaculture sector were examined for the species produced.

In terms of the economic contribution of aquaculture, gross output was estimated as £1,590,000, the indirect output at £1,025,250 and the total economic activity supported by aquaculture in Poole Harbour at £2,615,250.

The small-scale commercial fishery

Although known for its fisheries heritage (in particular the sprat and native oyster fishery), fisheries in Poole have been declining since 2010. This has a number of causes, including limited access to quotas (for the inshore fleet who do not hold quota) and licences, which both had a significant effect on the fishermen in Poole from the 1990s onwards. As a consequence, a number of fishermen sold up and moved out of the industry, while others diversified into aquaculture or took up charter angling trips on a full-time basis. The local fishing industry has declined in recent years, due to multiple factors, including the lack of young people joining the sector; competition from larger vessels; the sector not being seen as economically viable due to high start-up costs; and increased regulatory processes surrounding access to quota and discarding. Poole is the third largest port for landings in Southern IFCA district (Southern IFCA) (landing 583 tonnes in 2017, worth £1.6 million). Landings from the Poole Harbour fishing fleet have halved since 2010, a decreasing trend both in terms of volume and value during recent years.

In terms of the main species, plaice, sole and bass are the main catch for finfish. In terms of shellfish: whelks, cockles and clams comprise the main species landed. The Manila Clam fishery, whelks and the brown crab potting fishery now comprise the major wild fisheries by landed value.

There are currently around 100 active fishermen in the Poole Harbour area. Poole and District Fishermen's Association (PDFA) membership includes full-time commercial fishermen, aquaculture owners and charter skippers (Southern IFCA 2016). There are 76 commercial fishing vessels operating from Poole Harbour (registered in Poole although a small number of these are currently unlicensed) (MMO 2014). An average length of 6 m for the fleet means the fleet is considered very small scale.

The Manila Clam fishery forms the basis of the largest production area of this wild fishery in the UK. Live, landed weight has decreased by nearly 50% during the past six years, while the value of landings has decreased by 25% since 2010.

Over half of the vessels engage in the clam and cockle fishery, while the majority of the fleet uses static nets and pots to catch crab, lobster, sole, plaice, bass, mullet and whelks. The main species harvested (whelks and farmed Pacific Oysters and clams) are exported. Whelks represent the major landing in terms of weight, followed by brown crab and scallops. Whelks are exported to South Korea, and the other main species are consumed locally, nationally and exported; the exact percentage of each varies according to the quantity that each individual vessel lands (and therefore whether they choose to transport their catch to another port and market, such as Plymouth or Brixham).

Since 2012, the Sustainable Food City Partnership, for Bournemouth and Poole, has been campaigning to raise understanding and awareness of the need for local organisations, businesses and consumers to source fish and shellfish only from sustainable fisheries. The success of the campaign has led to the area becoming the first Sustainable Fish City in the world (Sustainable fish cities 2015). The partnership wishes to build on this initial success by working with key geographical locations to ensure that all relevant businesses have joined the campaign and that they pledge to serve only sustainable fish. Although there are clear opportunities to promote fisheries products on the back of small-scale sustainable fishery and the high-quality marine environment, the branding and identity of fisheries products are not as well known as for some other foods. For shellfish especially, the opportunities are clear (e.g. the Marine Stewardship Council accreditation of the clam and cockle fishery).

To calculate the local economic impact of commercial fishing, an average price for each species caught and the average landing size and prices were used in combination with output multiplier values, created by Seafish (a Non Departmental Public Body set up by the Fisheries Act 1981). Indirect and induced GVA value represented approximations of indirect output. An output multiplier calculates the amount of economic change that occurs as a result of changes in an industrial sector. In order to provide a useful indicator for the wider value created by the commercial and charter boat fishing fleets, we need to look beyond income or first sale value of the landed catch. The economic impacts of these activities do not cease once the fish is landed or when a charter boat customer has returned from sea. Therefore, we conducted interviews with local stakeholders and industry experts to establish some assumptions, values and proxies for use in this research and to develop the local economic impact calculations.

In terms of economic contribution, the top 11 species were considered for the purposes of this research and the gross output was estimated at £2,000,271. The indirect output was estimated at £2,177,620 and in terms of the total economic activity supported £4,177,891 was estimated for the small-scale coastal fishery (top 11 species).

The recreational charter boat fishery

Recreational sea angling by individuals (fishing with a hook and a line for non-commercial purposes) does not require a licence in the UK. Most recreational fisheries operate under a 'regulated open access management' regime, whereby individuals follow fishing gear and catch restrictions (such as the EU Minimum Conservation Reference Sizes) as well as national and local regulations. Sea angling takes place all around the UK but the South West and South coasts of England have the largest concentrations of angling charter vessels (Tinch et al. 2015). The largest charter boat ports in the UK are Weymouth and

Poole. According to research conducted, 184,400 households in the South West participate in sea angling, with an estimated 240,900 individual sea anglers who live in the South West. Visiting anglers contribute significantly to the region's economy, with an estimated 750,000 days spent sea angling. This is worth £165 million annually in terms of expenditure within the region (£110 million from resident anglers and £55 million from visiting anglers). An estimated 3,000 jobs are linked to sea angling in the region through charter boats, mackerel-fishing trips and tackle shops (Cappell and Lawrence 2005).

Poole has the second largest charter boat fleet in the UK, with 33 registered charter boats operating from Poole and a further 9 from Swanage that moor in Poole. A total of 42 charter boat skippers are employed in the Poole area (33 from Poole and 9 from Swanage); all are represented by the PDFA (Poole and District Fishermen's Association). Poole is one the most popular angling destinations in the country due to the ease of access and diversity of species (Higgins 2016). The most popular species for anglers to target is bass, followed by cod, mackerel, rays and sharks, conger eels and pollack. The recreational fishery in Poole also has a strong reliance on bass. In recent years, bass fishing has been subject to considerable reductions in fishing mortality, in response to a decline in stocks at EU level, where Emergency Measures were introduced in 2015. The area is very important as a destination for bass angling, and there is a need to support the marketing and diversification (where appropriate) of this fishery. In 2016 the fishery was subject to a 'catch and release only' policy for the first part of the year (MMO 2018) and during 2018 the fishery has been subject to a year-round 'catch and release only' policy.

Charter boats and tackle shops do not provide the only turnover related to recreational angling. Owners of hotels, B&Bs, food and drink, fuel and transport businesses, as well as those involved in boat manufacture and maintenance, will also benefit from sea angling tourism; even parking fees collected by the borough council can be significant. Charter skippers' expenditure includes moorings, insurance, maintenance, licences and so on.

The open-book accounts for two charter boats were used as the basis on which to estimate the gross and indirect output for each charter boat (an average was taken). The owners requested that the information was presented anonymously.

These sentinel vessels were used to calculate ranges for the contribution of the charter boat fleet, and for 33 vessels the gross output was estimated at £3,129,687, the indirect output at £2,600,103 and the total economic contribution at £5,729,790.

Economic contribution to Poole

Considering the three fisheries/sectors together, a gross output of £6,719,958 was calculated, an indirect output of £5,802,973 and £12,522,931 of total economic activity supported. This total economic activity equates to 15% of the staying visitor spend for Poole (£82,455,000) or 17.5% of the 'coastal visits spend' for Poole (£71,573,000). These are shown in Table 12.1.

This research suggests that these three sectors are valuable to Poole's local economy and, although not generally part of local economic or tourism regeneration strategies, they

Table 12.1 Gross output, indirect output and total economic activity supported by the three fisheries sectors in Poole, 2017

Poole Harbour	Gross output	Indirect output	Total economic activity
Aquaculture	£1,590,000	£1,025,250	£2,615,250
Commercial fisheries (top 11 species)	£2,000,271	£2,177,620	£4,177,891
Charter vessels (33 vessel estimate)	£3,129,687	£2,600,103	£5,729,790
Total economic activity for three sectors combined	**£6,719,958**	**£5,802,973**	**£12,522,931**

should be considered in developing plans for local food production to sustainable tourism. Making sectors more visible and presenting their socio-economic contribution and environmental credentials must be part of local development plans.

References

Balata, F. and Vardakoulias, O. 2016. *Turning Back to the Sea: A Blue New Deal to Revitalise Coastal Communities*. New Economics Foundation, London. Available from: https://neweconomics.org/2016/11/turning-back-to-the-sea/ (accessed 22 February 2022).

Cappell, R. and Lawrence, K. 2005. Invest in fish South West: The motivation, demographics and views of South West recreational sea anglers and their socio-economic impact on the region. Report on recreational sea angling in the South West.

Dorset and East Devon FLAG. 2016. Dorset and East Devon Fisheries Local Action Group community-led local development strategy. Available from: https://www.dorsetcoast.com/wpcontent/uploads/2017/10/Dorset-and-East-Devon-Fisheries-Local-Action-GroupCommunity-led-Local-Development-Strategy-Aug-2016.pdf (accessed 22 February 2022).

Dorset Coastal Community Team. 2016. Dorset Coastal Community Team connective economic plan: Poole coastal area plan. Available from: https://www.dorsetcoast.com/wpcontent/uploads/2017/09/Poole-Coastal-Area-Plan-V3.pdf (accessed 22 February 2022).

Higgins, P. 2016. 2 Annexe 9 – P. Higgins, Director Professional Boatman's Association (pers. comm.) email to Gary Wordsworth, Chairman of the Poole and District Fisherman's Association 23 Feb 2016 17:34.

Humphreys, J., Caldow, R.W.G., McGrorty, S., West, A.D., and Jensen, A.C. 2007. Population dynamics of naturalised Manila clams *Ruditapes philippinarum* in British coastal waters. *Marine Biology* 151: 2255–70. https://doi.org/10.1007/s00227-007-0660-x

Humphreys, J., *et al.* 2014. A reappraisal of the history and economics of the Pacific oyster in Britain. *Aquaculture* 428–9. http://dx.doi.org/10.1016/j.aquaculture.2014.02.034

Marine Management Organisation. 2014. UK fishing vessel lists. Details of registered and licensed fishing vessels over 10 metres and 10 metres and under. Marine Management Organisation. Crown copyright. Available from: https://www.gov.uk/government/collections/uk-vessel-lists (accessed 22 February 2022).

Marine Management Organisation. 2018. Bass fishing guidance 2017. Crown copyright. Available from: https://www.gov.uk/government/publications/bass-fishing-guidance/bass-fishing-guidance (accessed 22 February 2022).

Seafish Risk Assessment for Sourcing Seafood (website). Available from: http://www.seafish.org/rass/index.php/profiles/mussel-in-the-pool-harbour-regulatingorder-mussel-dredge/ (accessed 22 February 2022).

Social Mobility and Child Poverty Commission. 2016. The social mobility index, January 2016. Crown copyright. Available from: https://www.gov.uk/government/publications/social-mobility-index (accessed 22 February 2022).

Southern IFCA. Poole Harbour Fisheries. Available from: https://www.southern-ifca.gov.uk/poole-harbour-fisheries (accessed 13 July 2022).

Southern IFCA. 2016. Southern IFCA quarterly report of the chief officer, 1 November 2015–31 January 2016, p. 219. Available from: https://www.southern-ifca.gov.uk/authority-reports (accessed 27 July 2022).

Sustainable fish cities. Available from: http://www.sustainweb.org/sustainablefishcity/bournemouthandpoole/ (accessed 22 February 2022).

Tinch, R., *et al.* 2015. Comparing industry sector values, with a case study of commercial fishing and recreational sea angling. Eftec for the UK Fisheries Economists Network, supported by Seafish, Defra, Marine Scotland. Available from: http://www.seafish.org/media/publications/eftec_comparing_industry_sector_values_FINAL_Aug_2015.pdf (accessed 22 February 2022).

The South West Research Company Ltd 2018. The economic impact of Dorset's visitor economy 2017. Dorset and Districts. Produced on behalf of the Dorset Tourism Partnership. Available from: https://www.visit-dorset.com/dbimgs/the-economic-impact-of-visitor-economy-Dorset-and-Districts-2017.pdf (accessed 22 February 2022).

University of Portsmouth. 2017. New European alliance to save the flat oyster (News release). EurekAlert. Available from: https://www.eurekalert.org/pub_releases/2017-12/uop-nea120117.php (accessed 22 February 2022).

Williams, C. 2019. Defining criteria for low-impact fisheries in the UK. Technical report. New Economics Foundation. Available from: https://www.researchgate.net/publication/334964793_Defining_criteria_for_low-_impact_fisheries_in_the_UK (accessed 22 February 2022).

Williams, C. and Davies, W. 2018. A tale of three fisheries: The value of the small-scale commercial fishing fleet, aquaculture and the recreational charter boat fleet, to the local economy of Poole. NEF Consulting. Available from: https://bit.ly/2vv8sxR (accessed 22 February 2022).

CHAPTER 13

The Manila Clam in Poole Harbour: A Journey to Sustainability

SARAH ELIZABETH BIRCHENOUGH

Abstract

The fishery for the non-native Manila Clam *Ruditapes philippinarum* in Poole has faced a long and, at times, challenging road to sustainability. From the first introduction in 1988 for the purposes of aquaculture, to the emergence of a commercial fishery for wild stock in 1994, to the development and implementation of the Poole Harbour Dredge Permit Byelaw in 2015, the fishery has been associated with innovation in fishing techniques and fisheries management. The challenges posed by the fishery, namely achieving environmental protection for the Poole Harbour Marine Protected Area (MPA) and removing the culture of illegal fishing, set the targets which management needed to meet. Through a collaborative approach, encompassing extensive community engagement and participation and utilising the tools made available by the Marine and Coastal Access Act (2009), a robust management scheme was delivered which achieved the required balance of environmental and socio-economic protections. Achieving sustainability has led to economic and environmental successes, and independent verification of these through certification under the MSC Standard has allowed the achievements of the fishery to be outward facing and has set the tone for continued development.

Keywords: Manila Clam, inshore fisheries, fisheries management, Marine Protected Areas, sustainable fisheries, sustainability

Correspondence: sarah.birchenough@southern-ifca.gov.uk

Introduction

The Manila Clam *Ruditapes philippinarum* is a marine bivalve mollusc of the family Veneridae, native to the Indo-Pacific region between latitudes of 25° and 45° north (Scarlato 1981). The species was first described by Adams and Reeve in 1850 and since that time has been associated with 28 scientific and 17 common names (Goulletquer 1997), having been located at different times within seven different genera including *Protothaca*, *Venerupis* and *Tapes* (Howson and Picton 1997). The current taxonomic name is *R. philippinarum*, although *Venerupis philippinarum* is an accepted alternative.

The Manila Clam is similar in appearance to the European native Grooved Carpet Shell Clam *Ruditapes decussatus* (Hurtado *et al.* 2011). The species is distinguishable by a more angulated shell and more pronounced decussate sculpture; however, the best identifying

Sarah Elizabeth Birchenough, 'The Manila Clam in Poole Harbour: A Journey to Sustainability' in: *Harbour Ecology*. Pelagic Publishing (2022). © Sarah Elizabeth Birchenough. DOI: 10.53061/QMDP6661 characteristic is the almost fully fused, short exhalent siphons compared to the longer separate siphons of *R. decussatus* (Hurtado *et al.* 2011). The Manila Clam is found in fine sediments from the intertidal and upper sub-littoral zones at a depth of 3–5 cm below the surface (Jensen *et al.* 2005a). The species is well suited to estuarine environments having been documented to be euryhaline, surviving in salinities from 7 g/L to 42 g/L (Elston *et al.* 2003; Carregosa *et al.* 2014a; Carregosa *et al.* 2014b) and tolerant to a wide temperature range, growing between 10 and 30°C with the optimum temperature for fastest growth at 20–25°C (Laing and Child 1996; Numaguchi 1998). The Manila Clam is a suspension feeder, ingesting food particles from the water column including algae, cyanobacteria, diatoms, bacterioplankton and microzooplankton (Sorokin and Giovanardi 1995; Watanabe *et al.* 2009). It plays a significant role in nutrient and particulate organic matter recycling through the processes of feeding, growth and nutrient regeneration (Nakamura 2004), with the species' long lifespan and high biomass allowing it to function as an organic matter reservoir and therefore minimising nutrient loading in the water column (Ozbay *et al.* 2014; Koo and Seo 2020).

The Manila Clam is a broadcast spawner, the onset of gametogenesis being dependent on environmental cues with temperature seen to be the most important (Drummond *et al.* 2006). The species is capable of multiple spawning events within a single season with peak spawning activity occurring in the summer period up to September/October (Beninger and Lucas 1984; Rodriquez-Moscoso *et al.* 1992; Laruelle *et al.* 1994; Drummond *et al.* 2006; Moura *et al.* 2017). The pelagic larval stage is key to the ability of the Manila Clam to spread and settle on new habitats (Laruelle *et al.* 1994; De Montaudouin 1997); however, the extent to which this can occur is seen to be dependent on suitable physiological conditions (Herbert *et al.* 2012). The species has the ability to achieve high densities in suitable habitats with reports ranging from 259 to 5744 m^{-2}; however, densities in a wild Manila Clam population towards the species' northern limit in Europe were seen to be considerably lower at 156 m^{-2} (Humphreys *et al.* 2007).

The Manila Clam as an invasive species

The progression of the Manila Clam from its native distribution to numerous other locations worldwide is well documented (Flassch and Leborgne 1992; Jensen *et al.* 2004; Ruesink *et al.* 2006; Pranovi *et al.* 2006; Bidegain and Juanes 2013; Humphreys *et al.* 2015; Cordero *et al.* 2017). The introduction to new locations has occurred both intentionally and accidentally, with intentional introductions primarily driven by the species' high economic value (Jensen *et al.* 2005a), although other benefits of large stock biomass, fast growth, adaptability, ease of capture and rapid depuration of toxins also highlighted this species as particularly suited for aquaculture (Bidegain and Juanes 2013; Bidegain *et al.* 2015).

Accidental introduction to the Pacific coast of North America occurred in 1936 as a result of Pacific Oyster *Magallana gigas* seed imports (Chew 1989) and resulted in a spread of the species down the Pacific coast from California to British Colombia (Magoon and Vining 1981). Purposeful introduction into France from the North American population occurred between 1972 and 1974 (Flassch and Leborgne 1992) and into the Venice Lagoon, Italy, in 1983 (Pranovi *et al.* 2006). Both of these introductions aimed to support aquaculture during an unproductive period for native species. The introduction into the Venice Lagoon resulted in a rapid expansion into the wider coastal area and other areas of the Adriatic (Pranovi *et al.* 2006). Export of stock from European aquaculture led to further aquaculture enterprises in Spain, Norway, Ireland and Germany (Drummond *et al.* 2006; Sladonja *et al.* 2011; Cordero *et al.* 2017) and more recently in Portugal, where introductions into the Tagus Estuary (2000) and Sado Estuary (2010) have resulted in the Manila Clam becoming the dominant bivalve species (Cabral *et al.* 2020).

Introduction into the UK occurred in 1980 through the import of stock from the Pacific coast of North America by the then UK Ministry of Agriculture, Fisheries and Food (MAFF) (Humphreys *et al.* 2015). Imported for the purposes of aquaculture, initial trials were carried out in Wales with further study sites in the river Exe, Devon and Poole Harbour, Dorset (Jensen *et al.* 2005a). The introduction of this non-native species into UK waters was met with outrange by some conservation groups, dubbing it the 'alien monster' amid concern over the potential for displacement of native species (Utting 1995; Humphreys 2010; Humphreys *et al.* 2015) and other ecological impacts as seen elsewhere in Europe (Pranovi *et al.* 2006). However, MAFF maintained the position that the conditions in the UK would not allow for successful reproduction (Laing and Utting 1994) and trials continued. The introduction of Manila Clam into Poole Harbour resulted in the first documented wild population for the UK and kick-started a fishery which continues to develop to this day.

The Manila Clam in Poole Harbour

The Manila Clam was introduced into Poole by a local aquaculture business for the purposes of farming in 1988, and by 1994 it was determined that reproduction had occurred and the species was being exploited commercially by local fishers (Jensen *et al.* 2004). Work by Jensen *et al.* (2004) proved that the Manila Clam had successfully formed wild populations in the intertidal areas of Poole Harbour and that regular recruitment was taking place.

Poole Harbour has a number of characteristics which make it well suited to the establishment and development of the Manila Clam. Located on the south coast on the UK in the county of Dorset (Figure 13.1), it is one of the largest lowland estuaries in Europe, fed by several freshwater inputs, with a total area of water around 3,600 ha on a high-water spring tide (Humphreys and May 2005). Formed during post-glacial sea level rise, the harbour consists of a main basin with two smaller embayments and five islands, the largest of which is Brownsea Island (Humphreys and May 2005). There are extensive intertidal mudflats (~1,325 ha) caused by a shallow average water depth (~0.5 m) and a narrow entrance, which results in a poor flushing capacity (Natural England 1990). These factors, combined with a small tidal range, result in temperatures not being extensively moderated by water exchange between the harbour and the open sea, leading to summer temperatures for the water and intertidal mud peaking in the optimum growth range of the Manila Clam between 20 and 25°C (Jensen *et al.* 2004; Jensen *et al.* 2005a). The harbour also exhibits a surface water salinity gradient from the mouth to the upper reaches of the harbour with salinity values varying between an average of 29.8 at the harbour entrance to 15.7 in the Wareham Channel in the west of the harbour (Humphreys 2005), well within the tolerance range for the Manila Clam. Herbert *et al.* (2012) found that the differences in salinity in the harbour can lead to retention of Manila Clam larvae and, in areas such as the Wareham Channel where salinity is lowest, can result in self-sustaining populations. The distribution pattern of Manila Clam within Poole Harbour is thought to be primarily driven by a combination of salinity and flushing capacity (Herbert *et al.* 2012), with the low flushing rates helping to retain larvae within the boundary of the harbour (Jensen *et al.* 2004).

A variety of sediment types are also found across the harbour, including muddy sands and sand and gravel, caused by different levels of wave and tide exposure. The variety of different sediments and the habitats they support, including seagrass, saltmarsh and reedbeds, give rise to a diverse infaunal community and a highly productive environment for grazing, deposit- and suspension-feeding species (Dyrynda 2005). The double high-water results in the tidal level being held at or above mean level for approximately 16 out of 24 hours each day (Humphreys 2005). This greatly increases the time in which

Figure 13.1 Poole Harbour, Dorset, located on the south coast of the UK.

the Manila Clam is able to feed (Humphreys 2005) and, at the same time, in combination with the poor flushing capacity, increases the potential for food particles to be retained within the harbour (Jensen *et al.* 2004). The freshwater inputs to the harbour along with other sewage discharges also increase the nutrient load further increasing the available food resource (Jensen *et al.* 2004). This rare combination of circumstances may indicate why commercially viable wild populations of Manila Clam developed so extensively in Poole Harbour. Although there have been wild settlements of Manila Clam in other areas along the south coast including Southampton Water, Portsmouth and Langstone harbours to the east (Humphreys *et al.* 2015), in which commercial fisheries have also developed, populations are not as consistent or well developed as that in Poole Harbour.

The initial wild settlement and the subsequent establishment of a targeted fishery in the early 1990s led to a requirement for management intervention by the then Southern Sea Fisheries Committee (now the Southern Inshore Fisheries and Conservation Authority) and in 1997, a licensing system was introduced under the hybrid Poole Fishery Order 1985 creating a fishing season and a regulation of fishing practices for the first time.

The Poole Harbour dredge fishery

Poole Harbour has been a key fishing port on the south coast of England since the earliest records, and it became one of the main oyster ports of Britain in the mid-1800s (Humphreys *et al.* 2014). Despite fisheries for a number of shellfish species including Lobster *Homarus gammarus*, Edible Crab *Cancer pagurus* and cockles, it was the development of the wild population of Manila Clam which drove forward development and innovation in fishing methods and a large increase in intensity of fishing effort (Birchenough *et al.* 2019).

The modern-day fishery uses the pump-scoop dredge system to harvest Manila Clam along with other shellfish species, including Common Cockle, the Native Clam and the American Hard-shelled Clam *Mercenaria mercenaria* (Birchenough *et al.* 2019). The American Hard-shelled Clam is also an introduced species; however, it has been present in British waters for much longer than the Manila Clam with the first record in the Humber in 1884 (Mitchell 1974). The species is now found to occur at lower levels than when it was first introduced and, similar to the Manila Clam, has been seen to provide a benefit as a commercially fished species (Eno and Sanderson 1997).

The fishing methods used in Poole have evolved over time, moving from hand-raking to hand scoops to mechanised dredges over the 1990s to mid-2000s. The evolution of the pump-scoop dredge came from a combination of the need for improved efficiency and the safety of the fishers, and is a method almost unique to Poole Harbour (Jensen *et al.* 2005a). The current dredge design consists of a toothed dredge with water jets (powered by a separate generator) (Figure 13.2), where water is directed towards the back of the dredge basket resulting in a flow of sediment through the dredge (Jensen *et al.*

Figure 13.2 The pump-scoop dredge system used to harvest Manila Clam and other shellfish species in Poole Harbour.

2005a; Birchenough *et al.* 2019). A regulation on the dimensions of the dredge (460 × 460 × 300 mm) was introduced by the Southern Sea Fisheries Committee under the licensing system for fishing for Manila Clam.

In the early period of the targeted Manila Clam fishery, yields were seen to be more than double those of areas outside Poole Harbour with approximately 300 tonnes of Manila Clam landed in 2001 and 2002 (Jensen *et al.* 2005b). The fishery also proved to be valuable to the local industry with first sale price reaching up to £5 per kg (Jensen *et al.* 2005b). The value of the fishery has maintained with the price in 2015–19 ranging from £4 to £6 per kg (price fluctuates during the fishing season with an average of £5 per kg) which results in the fishery being worth over £1 million each year (data from Southern IFCA). The short timeframe over which this fishery emerged and developed did, however, result in several issues, mainly relating to the fact that a novel fishery involving a non-native species and a newly developed type of bottom towed fishing gear was operating within a Marine Protected Area (MPA) (Birchenough *et al.* 2019). In addition, the high value of the target species and complex management, split across a number of different regulations, led to issues with enforcement and the emergence of illegal, unreported and unregulated (IUU) fishing activity.

In 2011, under the Marine and Coastal Access Act 2009 (MaCAA) the management authority for the Manila Clam fishery in Poole, the Southern Sea Fisheries Committee became the Southern Inshore Fisheries and Conservation Authority (IFCA). The vision for IFCAs set out under the act emphasised the need for balance between sustainable fisheries and safeguards for the marine ecosystem (Phillipson and Symes 2010). To accomplish this, the act increased the range of regulatory options available to IFCAs as well as improving methods of enforcement (Birchenough *et al.* 2019). This was accompanied by a drive for more inclusive methods of governance that promoted stakeholder participation and community involvement. The resulting new regulatory framework introduced the potential for inshore fisheries management to be more effective, more efficient and ultimately more inclusive (Rodwell *et al.* 2014). Four years after the Southern IFCA was created, the main regulatory mechanism for the Manila Clam fishery in Poole, the Poole Fishery Order 1985, came up for its 30-year review. At this point, the issues associated with the fishery were clearly apparent and the perception of the Manila Clam in Poole had become inextricably linked with illegality and unsustainability. In order to move towards a sustainable fishery which achieved the objective of balance between industry and the environment, the Southern IFCA took the opportunity to overhaul the management measures and address the issues that persisted in holding the fishery back from meeting its potential.

Identifying the problem

A project run by Seafish (a public body set up to support the UK's seafood industry) in 2012 titled 'Project Inshore' set out to map inshore fisheries in England and assess them for sustainability using criteria based on the Marine Stewardship Council (MSC) Standard for sustainable fishing. One of these fisheries was the dredge fishery in Poole Harbour. The results of the assessment indicated that clam and cockle species in Poole Harbour were priority species for the Southern IFCA District (Project Inshore Stage 3 Strategic Sustainability Review Report). The assessment also highlighted many of the known issues faced by the fishery but, importantly, indicated that should these issues be addressed, the fishery had the potential for full certification under the MSC Standard as a sustainable fishery. The framing of these issues in the context of scoring against the MSC Standard under Project Inshore helped to indicate priority areas for consideration under any new management scheme for the fishery.

The main issue, and the one which ultimately would prevent the fishery from being seen as sustainable, was that it was operating within the boundary of a MPA. Poole Harbour is designated as a Special Protection Area (SPA) under the EU Birds Directive (2009), a Site of Special Scientific Interest under the Wildlife and Countryside Act (1981) and a Ramsar site under the Convention on Wetlands of International Importance. The SPA designation also qualifies the harbour as a European Marine Site as defined under the Conservation of Habitats and Species Regulations (2010). Each of the three designations comes with a set of features and supporting habitats which must be protected, the SPA designation providing protection for nationally and internationally important bird species as well as an internationally important waterbird assemblage. The Southern IFCA, as a competent authority under the legislations underpinning these designations, is required to appropriately assess fishing activity against the features and supporting habitats for an MPA and ensure that those activities are not having an adverse impact. The potential for adverse impact in this case was through interactions between fishing gear and habitats which support the bird features of the SPA and through direct impacts to those bird features through visual and noise disturbance.

Connected to the status of Poole Harbour as an MPA, the status of the Manila Clam being a non-native species also had the potential to create a barrier to the fishery being viewed as sustainable. Poole Harbour faced the common concerns regarding the Manila Clam as a non-native species, namely, outcompeting native species via consumption of the same food resource, displacement of species through occupation of the same ecological niche and settlement changing seabed topography and habitat complexity (Moura *et al.* 2017). The primary impact of these concerns was a decline in important prey species for the bird features of the harbour.

Finally, and potentially, the most difficult issue to overcome was that of the illegal, unreported and unregulated (IUU) fishing activity. IUU fishing had developed in Poole from a combination of a readily available and high value species in the Manila Clam, a complex management system comprising several different regulations and a lack of movement in licence holders which had prevented other fishers from legitimately engaging with the fishery. The different requirements created by different regulations were a result of a single-species approach to fisheries management, harking back to a time when provision of food resources was the primary factor in determining the management required (Clark *et al.* 2017). While once appropriate, the emergence of the Manila Clam fishery had exposed several regulatory loopholes in existing management measures which had become readily exploited (Birchenough *et al.* 2019). The compounding factor was that the increase in illegal fishing had led to a breakdown in the relationship between fishers and fisheries managers where fishers felt that the then Sea Fisheries Committee was not tackling the problem of IUU fishing (Birchenough *et al.* 2019). The lack of communication between different parties and the absence of processes to facilitate stakeholder involvement further promoted this sense of distrust (Birchenough *et al.* 2019). In considering the issues facing the fishery, it was clear that simply defining a new management method would not be enough, there was a need to revisit the process through which management is developed and ensure that the key regulatory aspects, that is, protection of the marine environment and sustainability of target species stocks, could be met while also changing the perception of the fishery and engaging with those who participate in it.

The path to sustainability

In 2015, the Southern IFCA introduced the 'Poole Harbour Dredge Permit Byelaw', a restricted-entry permit scheme which introduced flexible permit conditions to manage dredge fishing in Poole Harbour. Management of all dredge fishing in Poole Harbour under a single regulation immediately removed the complexities associated with multiple management measures for different species using the same fishing gear. This was the most important step in addressing the issues of managing a dredge fishery within an MPA and IUU fishing as it provided a clear regulatory framework which encompassed all fishers: those licensed under the Poole Fishery Order 1985 to fish for Manila Clam as well as those engaged in a long-standing cockle fishery.

The ability to introduce flexible management, through a permit byelaw, was a result of the new provisions for IFCAs set out in the Marine and Coastal Access Act (2009). This form of regulation allows for management to be reactive to changes in the fishery or the marine environment through a review of conditions introduced under the permit, which can be done at the level of the local IFCA, rather than a need to change the overarching byelaw which requires approval at a national level. There are a range of conditions that can accompany a permit. For the Poole Harbour Dredge Permit Byelaw the permit sets out regulations in relation to catch restrictions and reporting, gear type, gear construction and restrictions on use, spatial and temporal restrictions and the potential for the fitting of specified equipment to vessels.

The introduction of spatial and temporal restrictions on gear usage enabled the fishery to demonstrate that it could exist with no adverse effect on the features and supporting habitats of the Poole Harbour MPA (Figure 13.3). Measures were developed in close consultation with nature conservation advisors and non-governmental organisations to ensure that protections were adequate and robust to the potential impacts of the fishery.

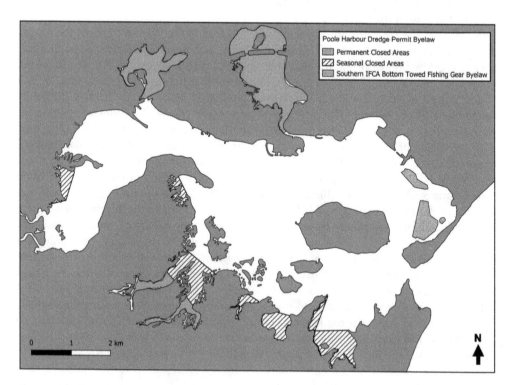

Figure 13.3 Spatial and temporal restrictions on dredge fishing under the Poole Harbour Dredge Permit Byelaw permit conditions. Areas are defined as permanent closed areas (dark hashed) and seasonal closed areas (diagonal lined). Also shown are permanent closed areas under the Southern IFCA Bottom Towed Fishing Gear 2016 byelaw (light grey hashed), these areas protect intertidal seagrass beds.

Key sensitive areas for roosting, feeding and breeding of the bird features of Poole Harbour were protected seasonally, during periods of increased sensitivity to disturbance, and an overarching dredge season ensured that no dredge activity would take place during the most sensitive period for the bird features, from January to March. Further provisions included effort limitation through limiting the number of permits issued and regulations on the construction of fishing gear, requiring a size limit on the dredge and specifications on the water system used with the dredge to avoid fluidisation of the sediment. Further protections are provided through a code of conduct for saltmarsh habitats and a prohibition on fishing in areas of intertidal seagrass beds under the Southern IFCA Bottom Towed Fishing Gear 2016 byelaw.

In addressing the status of the Manila Clam as a non-native species, a series of studies that had been emerging since the introduction of the Manila Clam to Poole Harbour gave confidence in the fact that the harbour was not seeing negative environmental effects as a result of its introduction. Although there is a greater population of the Manila Clam than the Native Clam R. decussatus in Poole Harbour, it is not thought that the former species is the driver for this. Rather than outcompete food sources, the Manila Clam was found to exist in relatively low densities compared to its native population and that of other introduced populations (60 m^{-2} to 156 m^{-2}) (Jensen et al. 2004; Humphreys et al. 2007). In addition, other factors such as a shallow burial depth made them more susceptible to predators indicating that predation rather than competition may provide a limiting factor on their density (Bidegain and Juanes 2013). The Manila Clam has also been demonstrated to provide a new and valuable food source for several shorebird populations, including the Oystercatcher Haematopus ostralegus where winter mortality was reduced by the presence of the Manila Clam (Caldow et al. 2007). It is deemed to be the eighth most important macro-invertebrate species on the intertidal in terms of biomass density (Jensen et al. 2004), with the species' shallow burial depth making it a readily available food source and therefore supporting the requirements of the bird features for which Poole Harbour is designated.

The issue of IUU fishing arguably posed a more complex problem. The first step in addressing this was through the implementation of more clearly defined regulation which removed the historic loopholes that provided for illegality. This came as a result of the byelaw creating a single management scheme, regulating fishing gear rather than species and having spatial and temporal measures that restricted the activity of fishing and the presence of dredge gear on a vessel rather than the action of taking a particular species. However, in order for other improvements, such as the ability for the fishery to operate sustainably within an MPA, to be successful it was necessary that compliance levels within the fishery remained high. This would ultimately require the fishery to address the distrust between fishers and managers and foster a culture of compliance (Birchenough et al. 2019).

The main pathway by which this issue was addressed was through creating a value associated with the permit which outweighed the perceived value associated with illegal fishing. The limited-entry system adopted for the permit scheme was based on historic track record and therefore allowed legitimate fishers access to the new fishery. This created a sense of ownership and investment in the fishery by those who held a permit (Birchenough et al. 2019). The risk of losing a permit as a result of illegal activity and therefore losing access not only to the Manila Clam fishery but also to any other species historically fished using the pump-scoop dredge method incentivised fishers to comply with the regulations (Birchenough et al. 2019). The ability to enforce all of the regulations associated with the permit scheme was also a defining factor in improving the levels of IUU fishing. The extended range of powers available to IFCAs under the Marine and Coastal Access Act (2009), coupled with the development of processes to assess risk and appropriately assign assets, allowed for an improved enforcement regime which further emphasised the value of compliance and increased the risk for fishers of non-compliance being detected. Reconciling the issues around IUU fishing are perhaps the best marker of success for the Manila Clam fishery in Poole. Within a year of the introduction of the Poole Harbour Dredge Permit Byelaw the recorded level of IUU fishing had decreased by 95%. The number of compliance inspections relating to IUU vessels declined from between 34% and 23% of all inspections in 2013 and 2014 respectively, prior to the introduction of the byelaw, to 1% by 2016. Since the drop in IUU activity following the introduction of the byelaw, the percentage of inspections relating to IUU vessels has stayed at 1% (most recent data available is for 2019) (Figure 13.4).

The development of the permit scheme, and all of the provisions contained within the permit conditions, took place over a two-year period to ensure that the appropriate level of engagement could be undertaken encompassing government advisors, non-governmental organisations and other authorities as well as the fishing industry and the wider stakeholder community. The process for engagement developed through the design and implementation of the Poole Harbour Dredge Permit Byelaw was the key component in ensuring the success of the byelaw and has been mirrored for other fisheries across the Southern IFCA District. A combination of engagement methods including drop-in sessions/workshops, one-to-one meetings and both written and verbal communications ensured that all interested stakeholder parties could participate. By implementing this programme of communication at the earliest stages in the development of the byelaw and maintaining the momentum of this engagement throughout the process, relationships between fishers and fisheries managers in particular improved. Fishers felt

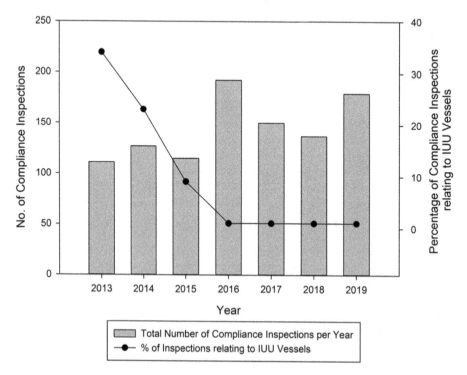

Figure 13.4 The total number of compliance inspections carried out for the Poole dredge fishery each year (grey bars) and the percentage of those inspections that related to illegal, unreported and unregulated (IUU) fishing vessels (black line), for the period 2013 to 2019. The Poole Harbour Dredge Permit Byelaw was introduced in 2015.

a sense of involvement and that their ideas and comments were being taken on board (local Poole Harbour fishers, per comm.) and in this way had a greater desire to actively participate in the development of their fishery. This improved relationship also contributed to the successes around reducing IUU fishing adding to fishers' sense of involvement and pride in their fishery.

In order for the success of the fishery to be recognised on a wider scale and for the achievements in sustainability to be independently verified, an ambitious project was started to build on the initial work of Project Inshore and to see if the fishery had improved enough to reach the standard of sustainability set down by the MSC.

An outward-facing success story

Once the Poole Harbour Dredge Permit Byelaw had been created and introduced into the fishery, it was determined that as a joint effort between the Poole and District Fishermen's Association and the Southern IFCA, the potential for the fishery to achieve certification under the MSC Standard should be re-explored. A project was set up, funded by Seafish and the Resources Legacy Fund in the United States to work towards a joint aim of certification under MSC as well as under the Seafish Responsible Fishing Scheme (RFS), a certification that recognises the use of industry best practice and high standards of social responsibility by individual fishers. This ambitious aim was the first of its kind globally as simultaneous certification under both schemes had never been achieved before. At the time of the assessment, improvements in the fishery and practice of the fishers were noted across the board and the scores awarded to the fishery were raised in all areas, including those identified as problematic by Project Inshore. In March 2018, the Poole Harbour clam and cockle fishery achieved a global first double certification, attaining the MSC Standard and having many of the fishers certified under the RFS.

This achievement provided independent and international recognition of the fishery as sustainable and introduced opportunities to promote the success of the fishery beyond those immediately involved with it out to the wider community. Historic perceptions of the fishery as unsustainable and driven by IUU fishing could finally be addressed and backed up by independent verification. This introduced new market opportunities for the fishers and gave them a sense of pride in their fishery which further encouraged continued active participation in the development of the fishery going forward and engagement with fisheries managers. The MSC certification, in particular, also provided a framework under which the fishery could continue to develop. During the process of certification, conditions were defined for two areas of the fishery, where it was determined that additional measures could be introduced to further sustainable practices over a three-year period following the initial certification. This led to the introduction of a reporting process for any interactions between the fishery and any Endangered, Threatened and Protected species and improved spatial resolution of catch data, defining fishing zones within the harbour so that catch data could be better related to data from the annual stock survey. In this way, the MSC certification provided a forward direction for the development of the fishery, and the annual audit for the certification ensures that high standards continue to be met in all areas. At the second annual surveillance audit in April 2020, one of the conditions was signed off as fully completed a year ahead of schedule showing the continued work put in by the fishers and fisheries managers to ensure that the fishery continues to strive for ever higher standards in sustainability. The outward-facing nature of the MSC certification and the process of annual feedback ensure that this progress and the ongoing achievements of the fishery can be made known to all those who participate in the fishery as well as the wider community.

Continued growth and potential for the future

The development of the Poole Harbour Dredge Permit Byelaw has shown that it is possible for a dredge fishery with a non-native target species to operate sustainably within a MPA, balancing the needs of the marine environment with those of the fishing industry. Since the byelaw was introduced the fishery has thrived, with the average catch per unit effort, measured as quantity of shellfish (kg) harvested per hour, documented to have increased each year from 2016 to 2019 (13.85 kg/hr to 19.67 kg/hr) (Figure 13.5). A combination of monitoring tools including an annual stock assessment and the production of an annual Habitats Regulations Assessment ensure that the growth in the fishery continues to occur sustainably and in line with achieving the conservation objectives of the Poole Harbour MPA.

The benefits of sustainable management are also being seen in the economics of the fishery. A study undertaken by the New Economics Foundation in 2018 looked at the local economic impact of commercial fishing in Poole Harbour for the top 11 species landed (Williams and Davies 2018). Information from landings data for 2016 and 2017, combined with information from local stakeholders, producers and industry experts, enabled an estimate to be made of the indirect and induced gross value added associated with each

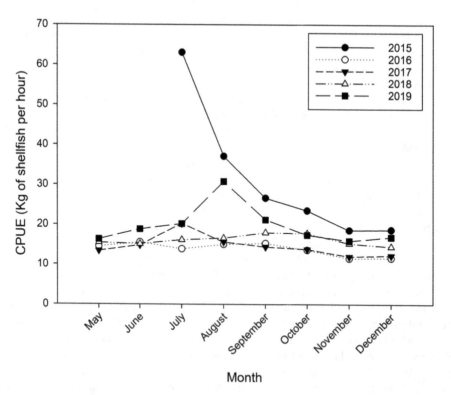

Figure 13.5 Average catch per unit effort (CPUE) (shown as Kg of shellfish harvested per hour) across all vessels for each month of the dredge season for 2015 to 2019. The Poole Harbour Dredge Permit Byelaw came into force on 1 July 2015. Therefore, there is no catch data for May and June 2015. The initial high level of CPUE seen in 2015 is a factor of the fishery being newly open at a time of year when Manila Clam had historically not been able to be fished in Poole Harbour. 2016 to 2019 are representative of the common pattern of fishing seen. High CPUE for August 2019 is due to an increase in landings for the Common Cockle *Cerastoderma edule*.

species and thus the overall economic impact (Williams and Davies 2018). For Manila Clam, the value for total economic activity was £1,436,216, making it the most economically valuable species landed into Poole (Williams and Davies 2018). The study also highlighted other future opportunities for the fishery, for example integration with aquaculture practices in Poole Harbour, to maintain markets throughout the year and increase spawning stock biomass (Williams and Davies 2018). These types of initiatives are now able to be explored as a result of the improvements in relationships between fishers and fisheries managers, engagement with the local community and an understanding of the relationship between the fishery and the needs of the protected site in which it operates. Although challenges will still remain (Williams and Davies 2018), there is a willingness across all parties to work together to ensure that the fishery continues to develop and lead the way as an example of progressive fisheries management.

References

Beninger, P.G. and Lucas, A. 1984. Seasonal variations in condition, reproductive activity, and gross biochemical composition of two species of adult clam reared in a common habitat: *Tapes decussatus* L. (Jeffreys) and *Tapes philippinarum* (Adams & Reeve). *Journal of Experimental Marine Biology and Ecology* 79: 19–37. https://doi.org/10.1016/0022-0981(84)90028-5

Bidegain, G., Bárcena, J.F., García, A., and Juanes, J.A. 2015. Predicting coexistence and predominance patterns between the introduced Manila clam (*Ruditapes philippinarum*) and the European native clam (*Ruditapes decussatus*). *Estuarine, Coastal and Shelf Science* 152: 162–72. https://doi.org/10.1016/j.ecss.2014.11.018

Bidegain, G. and Juanes, J.A. 2013. Does expansion of the introduced Manila clam *Ruditapes philippinarum* cause competitive displacement of the European native clam *Ruditapes decussatus*? *Journal of Experimental Marine Biology and Ecology* 445: 44–52. https://doi.org/10.1016/j.jembe.2013.04.005

Birchenough, S.E., Clark, R.W.E., Pengelly, S., and Humphreys, J. 2019. Managing a dredge fishery within a marine protected area: Resolving environmental and socio-economic objectives. In: Humphreys, J. and Clark, R.W.E. (eds) *Marine Protected Areas Science, Policy and Management.* Elsevier, Amsterdam, pp. 459–74. https://doi.org/10.1016/B978-0-08-102698-4.00023-X

Cabral, S., Carvalho, F., Gaspar, M., Ramajal, J., Sá, E., Santos, C., Silva, G., Sousa, A., Costa, J.L., and Chainho, P. 2020. Non-indigenous species in soft-sediments: Are some estuaries more invaded than others? *Ecological Indicators* 110: 105640. https://doi.org/10.1016/j.ecolind.2019.105640

Caldow, R.W.G., Stillman, R.A., leV. dit Durell, S.E.A., West, A.D., McGrorty, S., Goss-Custard, J.D., Wood, P.J., and Humphreys, J. 2007. Benefits to shorebirds from invasion of a non-native shellfish. *Proceedings of the Royal Society B* 274: 1449–55. https://doi.org/10.1098/rspb.2007.0072

Carregosa, V., Figueira, E., Gil, A.M., Pereira, S., Pinto, J., Soares, A.M.V.M., and Freitas, R. 2014a. Tolerance of *Venerupis philippinarum* to salinity: Osmotic and metabolic aspects. *Comparative Biochemistry and Physiology, Part A* 171: 36–43. https://doi.org/10.1016/j.cbpa.2014.02.009

Carregosa, V., Velez, C., Soares, A.M.V.M., Figueira, E., and Freitas, R. 2014b. Physiological and biochemical responses of three Veneridae clams exposed to salinity changes. *Comparative Biochemistry and Physiology, Part B* 177–178: 1–9. https://doi.org/10.1016/j.cbpb.2014.08.001

Chew, K.K. 1989. Manila clam biology and fishery development in Western North America. In: Manzi, J.J. and Castagna, M. (eds) *Clam Mariculture in North America, Developments in Aquaculture and Fisheries Science.* Vol. 19, Elsevier, New York, pp. 243–61.

Clark, R., Humphreys, J., Solandt, J., and Weller, C. 2017. Dialetics of nature: The emergence of policy on the management of commercial fisheries in English European marine sites. *Marine Policy* 78: 11–17. https://doi.org/10.1016/j.marpol.2016.12.021

Cordero, D., Delgado, M., Liu, B., Ruesink, J., and Saavedra, C. 2017. Population genetics of the Manila clam (*Ruditapes philippinarum*) introduced in North America and Europe. *Nature Scientific Reports* 7: 39745. https://doi.org/10.1038/srep39745

de Montaudouin, X. 1997. Potential of bivalves' secondary settlement differs with species: A comparison between cockle (*Cerastoderma edule*) and clam (*Ruditapes philippinarum*) juvenile resuspension. *Marine Biology* 128: 639–48. https://doi.org/10.1007/s002270050130

Drummond, L., Mulcahy, M., and Culloty, S. 2006. The reproductive biology of the Manila clam, *Ruditapes philippinarum*, from the North-West of Ireland. *Aquaculture* 254: 326–40. https://doi.org/10.1016/j.aquaculture.2005.10.052

Dyrynda, P. 2005. Sub-tidal ecology of Poole Harbour – An overview. In: Humphreys, J.

and May, V. (eds) *The Ecology of Poole Harbour*. Elsevier, Amsterdam, pp. 109–30. https://doi.org/10.1016/S1568-2692(05)80013-0

Elston, R.A., Cheney, D.P., MacDonald, B.F., and Suhrbier, A.D. 2003. Tolerance and response of Manila clams, *Venerupis philippinarum* (A. Adams and Reeve, 1850) to low salinity. *Journal of Shellfish Research* 22(3): 667–74.

Eno, C. and Sanderson, W.G. 1997. Non-native marine species in British waters: A review and directory. Joint Nature Conservation Committee, pp. 137.

Flassch, J.P. and Leborgne, Y. 1992. Introduction in Europe, from 1972 to 1980, of the Japanese Manila clam (*Tapes philippinarum*) and the effects on aquaculture production and natural settlement. *ICES Marine Science Symposia* 194: 92–6. https://archimer.ifremer.fr/doc/00037/14871/

Goulletquer, P. 1997. A bibliography of the Manila clam *Tapes philippinarum*, IFREMER, RIDRV-97.02/RA/La Tremblade, p. 122. Available from: https://archimer.ifremer.fr/doc/00000/3221/ (accessed 1 June 2020).

Herbert, R.J.H., Willis, J., Jones, E., Ross, K., Hübner, R., Humphreys, J., Jensen, A., and Baugh, J. 2012. Invasion in tidal zones on complex coastlines: Modelling larvae of the non-native Manila clam, *Ruditapes philippinarum*, in the UK. *Journal of Biogeography* 39: 585–99. https://doi.org/10.1111/j.1365-2699.2011.02626.x

Howson, C.M. and Picton, B.E. (eds). 1997. The species directory of the marine fauna and flora of the British Isles and surrounding seas, Ulster Museum publication no. 276, Ulster Museum, Ross-On Wye and the marine conservation society, p. 514.

Humphreys, J. 2005. Salinity and tides in Poole Harbour: Estuary or lagoon? In: Humphreys, J. and May, V. (eds) *The Ecology of Poole Harbour*. Elsevier, Amsterdam, pp. 35–48. https://doi.org/10.1016/S1568-2692(05)80008-7

Humphreys, J. 2010. The introduction of the Manila clam to British coastal waters. *The Biologist* 57(3): 134–9.

Humphreys, J. and May, V. 2005. Introduction: Poole Harbour in context. In: Humphreys, J. and May, V. (eds) *The Ecology of Poole Harbour*. Elsevier, Amsterdam, pp. 1–8. https://doi.org/10.1016/S1568-2692(05)80005-1

Humphreys, J., Caldow, R.W.G., McGrorty, S., West, A.D., and Jensen, A.C. 2007. Population dynamics of naturalised Manila clams *Ruditapes philippinarum* in British coastal waters. *Marine Biology* 151: 2255–70. https://doi.org/10.1007/s00227-007-0660-x

Humphreys, J., Harris, M.C., Herbert, R.J.H., Farrell, P., Jensen, A., and Cragg, S.M. 2015. Introduction, dispersal and naturalization of the Manila clam *Ruditapes philippinarum* in British estuaries, 1980–2010. *Journal of the Marine Biological Association of the United Kingdom* 96(6): 1163–72. https://doi.org/10.1017/S0025315415000132

Humphreys, J., Herbert, R.J.H., Roberts, C., and Fletcher, S. 2014. A reappraisal of the history and economics of the Pacific oyster in Britain. *Aquaculture* 428–429: 117–24. https://doi.org/10.1016/j.aquaculture.2014.02.034

Hurtado, N.S., Pérez-García, C., Morán, P., and Pasantes, J.J. 2011. Genetic and cytological evidence of hybridization between native *Ruditapes decussatus* and introduced *Ruditapes philippinarum* (Mollusca, Bivalvia, Veneridae) in NW Spain. *Aquaculture* 311: 123–8. https://doi.org/10.1016/j.aquaculture.2010.12.015

Jensen, A., Carrier, I., and Richardson, N. 2005b. Marine fisheries of Poole Harbour. In: Humphreys, J. and May, V. (eds) *The Ecology of Poole Harbour*. Elsevier, Amsterdam, pp. 195–204. https://doi.org/10.1016/S1568-2692(05)80021-X

Jensen, A., Humphreys, J., Caldow, R.W.G., and Cesar, C. 2005a. The Manila clam in Poole Harbour. In: Humphreys, J. and May, V. (eds) *The Ecology of Poole Harbour*. Elsevier, Amsterdam, pp. 163–74. https://doi.org/10.1016/S1568-2692(05)80018-X

Jensen, A.C., Humphreys, J., Caldow, R.W.G., Grisley, C., and Dyrynda, P.E.J. 2004. Naturalization of the Manila clam (*Tapes philippinarum*), an alien species, and establishment of a clam fishery within Poole Harbour, Dorset. *Journal of the Marine Biological Association of the United Kingdom* 84: 1069–73. https://doi.org/10.1017/S0025315404010446h

Koo, B.J. and Seo, J. 2020. Filtration rates of Manila clam, *Ruditapes philippinarum*, in tidal flats with different hydrographic regimes. *Plos One* 15(2): e0228873. https://doi.org/10.1371/journal.pone.0228873

Laing, I. and Child, A.R. 1996. Comparative tolerance of small juvenile palourdes (*Tapes decussatus* L.) and Manila clams (*Tapes philippinarum* Adams & Reeve) to low temperature. *Journal of Experimental Marine Biology and Ecology* 195: 267–85. https://doi.org/10.1016/0022-0981(95)00097-6

Laing, I.S.D. and Utting, S.D. 1994. The physiology and biochemistry of diploid and triploid clams (*Tapes philippinarum*) larvae and juveniles. *Journal of Experimental Marine Biology and Ecology* 184: 159–69. https://doi.org/10.1016/0022-0981(94)90002-7

Laruelle, F., Guillou, J., and Paulet, Y.M. 1994. Reproductive pattern of the clams, *Ruditapes decussatus* and *R. philippinarum* on intertidal flats in Brittany. *Journal of the Marine Biological Association of the United Kingdom* 74: 351–66. https://doi.org/10.1017/S0025315400039382

Magoon, C. and Vining, R. 1981. Introduction to Shellfish aquaculture in the Puget Sound region, State of Washington Department of Natural Resources Handbook, Olympia, p. 68.

Mitchell, R. 1974. Aspects of the ecology of the lamellibranch *Mercenaria mercenaria* (L.) in British waters. *Hydrobiological Bulletin* 8: 124–38. https://doi.org/10.1007/BF02254913

Moura, P., Garaulet, L.L., Vasconcelos, P., Chainho, P., Costa, J.L., and Gaspar, M.B. 2017. Age and growth of a highly successful invasive species: The Manila clam *Ruditapes philippinarum* (Adams & Reeve, 1850) in the Tagus Estuary (Portugal). *Aquatic Invasions* 12(2): 133–46. https://doi.org/10.3391/ai.2017.12.2.02

Nakamura, Y. 2004. Suspension feeding and growth of juvenile Manila clam *Ruditapes philippinarum* reared in the laboratory. *Fisheries Science* 70: 215–22. https://doi.org/10.1111/j.1444-2906.2003.00794.x

Natural England. 1990. Poole Harbour site of special scientific interest citation. Available from: https://designatedsites.naturalengland.org.uk/PDFsForWeb/Citation/1000110.pdf (accessed 1 June 2020).

Numaguchi, K. 1998. Preliminary experiments on the influence of water temperature, salinity and air exposure on the mortality of Manila clam larvae. *Aquaculture International* 6: 77–81. https://doi.org/10.1023/A:1009225921044

Ozbay, G., Blank, G., and Thunjai, T. 2014. Impacts of aquaculture on habitats and best management practices (BMPs). In: Sustainable Aquaculture Techniques, Hernandez-Vergara, M.P., and Perez-Rostro, C.I. (eds) *InTech, Croatia*. https://doi.org/10.5772/57471

Phillipson, J. and Symes, D. 2010. Recontextualising inshore fisheries: The changing face of British inshore fisheries management. *Marine Policy* 34(1): 1207–14. https://doi.org/10.1016/j.marpol.2010.04.005

Pranovi, F., Franceschini, G., Casale, M., Zucchetta, M., Torricelli, P., and Giovanardi, O. 2006. An ecological imbalance induced by a non-native species: The Manila clam in the Venice Lagoon. *Biological Invasions* 8: 595–609. https://doi.org/10.1007/s10530-005-1602-5

Rodríquez-Moscoso, E., Pazo, J.P., García, A., and Cortes, F.F. 1992. Reproductive cycle of Manila clam, *Ruditapes philippinarum* (Adams & Reeve 1850) in Ria of Vigo (NW Spain). *Scientia Marina* 56(1): 61–7.

Rodwell, L.D., Lowther, J., Hunter, C., and Mangi, S.C. 2014. Fisheries co-management in a new era of marine policy in the UK: A preliminary assessment of stakeholder perceptions. *Marine Policy* 45: 261–8. https://doi.org/10.1016/j.marpol.2013.09.008

Ruesink, J.L., Feist, B.E., Harvey, C.J., Hong, J.S., Trimble, A.C., and Wisehart, L.M. 2006. Changes in productivity associated with four introduced species: Ecosystem transformation of a 'pristine' estuary. *Marine Ecology Progress Series* 311: 203–15. https://doi.org/10.3354/meps311203

Scarlato, O.A. 1981. *Bivalves of Temperate Waters of the Northwestern Part of the Pacific Ocean*. Nauka Press, Leningrad, p. 408.

Sladonja, B., Bettoso, N., Zentilin, A., Tamberlich, F., and Acquavita, A. 2011. Manila clam (*Tapes philippinarum* Adams & Reeve, 1852) in the Lagoon of Marano and Grado (Northern Adriatic Sea, Italy): Socio-economic and environmental pathway of a shell farm. In: Sladonja, B. (ed) *Aquaculture and the Environment – A Shared Destiny, InTech, Croatia*, pp. 51–78. https://doi.org/10.5772/31737

Sorokin, Y.I. and Giovanardi, O. 1995. Trophic characteristics of the Manila clam (*Tapes philippinarum* Adams and Reeve). *ICES Journal of Marine Science* 52: 853–62. https://doi.org/10.1006/jmsc.1995.0082

Utting, S.D. 1995. Triploidy in the clam *Tapes philippinarum*. Workshop on the Environmental Impacts of Aquaculture Using Organisms Derived Through Modern Biotechnology. OECD, Paris, pp. 114–19.

Watanabe, S., Katayama, S., Kodama, M., Cho, N., Nakata, K., and Fukuda, M. 2009. Small-scale variation in feeding environments for the Manila clam *Ruditapes philippinarum* in a tidal flat in Tokyo Bay. *Fisheries Science* 75: 937–45. https://doi.org/10.1007/s12562-009-0113-1

Williams, C. and Davies, W. 2018. *A Tale of Three Fisheries: The Value of the Small-Scale Commercial Fishing Fleet, Aquaculture and the Recreational Charter Boat Fleet to the Local Economy of Poole*. New Economics Foundation Consulting Limited, London.

CHAPTER 14

Ecology and Exploitation of Poole Harbour Oysters

JOHN HUMPHREYS

Abstract

Two species of oyster are found in Poole Harbour: the native European Flat Oyster *Ostrea edulis*, which provided a thriving wild fishery in the early to mid-nineteenth century, and the Pacific Oyster *Magallana* (*Crassostrea*) *gigas*, which is the basis for the harbour's current pre-eminent position for oyster cultivation. In response to a Europe-wide decline in the Flat Oyster and many failed attempts at restoration, in 1890 a new species of oyster was introduced, initially from Portugal. Today's Poole Harbour production, much of which is exported to the Far East, can be traced back to a 1960s introduction of Pacific Oysters. Although a non-native species, the Pacific Oyster creates a dilemma for environmental agencies, especially in the context of globalisation and climate change.

Keywords: *Ostrea edulis*, *Magallana gigas*, *Crassostrea gigas*, Pacific Oyster, European Oyster, Poole Harbour, oyster fisheries

Correspondence: jhc@jhc.co

Introduction

The significance of Poole Harbour for oyster production is evidenced from both archaeological research and contemporary accounts. In the eighteenth century, Daniel Defoe reported Poole to be 'famous for the best and the biggest oysters in all this part of England . . . they are barrelled up here, and sent not only to London, but to the West Indies, and to Spain, and Italy, and other parts' (Defoe 1726). This pre-eminence continues to the present day, Poole Harbour being the largest oyster production site in Britain.

The basis of this operation was and is the shallow, warm and fertile waters of the harbour, combined with its double high tides. However, since the eighteenth century there have been fundamental changes in the industry triggered by the decline of native oyster stock, followed by a resurgence on the basis of innovative culture methods and a newly introduced species.

Today, the seabed of Poole Harbour provides habitat for two species of oyster: the European Flat Oyster *Ostrea edulis* referenced by Defoe and the Pacific Oyster *Crassostrea gigas* (now known as *Magallana gigas*). The first of these species is a native of British waters whereas the second is indigenous to eastern coasts of Asia. Figure 14.1 shows specimens of both species taken from Poole Harbour. In terms of external morphology, the two

John Humphreys, 'Ecology and Exploitation of Poole Harbour Oysters' in: *Harbour Ecology*. Pelagic Publishing (2022). © John Humphreys. DOI: 10.53061/XYMT2923

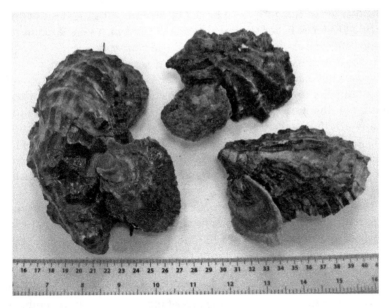

Figure 14.1 Poole Harbour oyster species: three Pacific Oysters (elongated and markedly corrugated), each with a (more rounded and smoother) native Flat Oyster attached.

species are easily distinguished: the Flat Oyster being broadly round or oval and growing to around 100 mm in diameter and the Pacific Oyster being elongated, more deeply sculptured and cupped and capable of growing much larger.

The biology of the Poole Harbour species

The habitats of the two Poole Harbour species usually differ with respect to position relative to intertidal shores: the native Flat Oyster is associated with subtidal areas whereas the Pacific Oyster is commonly considered an intertidal species, tolerating periods of exposure and immersion. However, the microtidal and somewhat lagoon-like characteristics of Poole Harbour (Humphreys 2005) blur these distinctions with the result that Flat Oysters are found growing attached to live Pacific Oysters, as Figure 14.1.

Oysters belong to the highly specialised Bivalvia group of molluscs, in which two distinct shells (valves), attached by a connecting ligament, are secreted around otherwise soft bodies. Whereas most bivalves, such as cockles and clams, follow the ancestral habit of burrowing in sediments, oysters are adapted for an epibenthic life attached to the surface of the seabed, which they achieve by secreting an organic adhesive, in effect committing them to a sedentary life lying on their left side. As a consequence, oysters have unequal valves, the left being more cup-shaped and the right valve above being flatter and forming a sort of lid.

In common with other bivalves, oysters have no head, the feeding functions of which have been lost over the course of an evolutionary history in which the structure of the gill has been enlarged and modified to add water filtering to the original respiratory function. Through this adaptation, a current of water is maintained over the gills from which food in the form of particulate organic matter such as microscopic plants (phytoplankton) is filtered and passed along by cilia to the mouth.

The anatomy of the gills and surrounding chambers along with differences in muscle physiology and shell shape determine the species' ability to cope with turbid waters without potential lethal clogging by sediment of the fine gill structures. Of the two species, the Pacific Oyster is better equipped to inhabit turbid waters: a point not without significance in estuarine environments such as Poole Harbour.

Although adult oysters are sedentary, dispersal is achieved through a lifecycle that includes the release of planktonic larvae which may be carried in tidal currents, but which have sufficient motility to control their vertical position in the water column so as to have some influence on the area they subsequently occupy and reasonable control over the substrate on which they settle. In this respect larvae are known to attach preferentially, but not exclusively, to calcium carbonate-rich substances especially when coincident with chemical signals released by adult oysters. This habit often results in the formation of oyster beds or reefs as generations of oysters settle on or near the shells or shell fragments of predecessors.

Despite such similarities, the reproductive biology of the two Poole species differs significantly: both are successive hermaphrodites (being a single sex but changing sex over time) whose sperm is released directly into the water. But whereas *C. gigas* releases its eggs for fertilisation in the water column, *O. edulis* retains its eggs inside its shell to be fertilised by inhalant water drawing sperm from outside and not released until they have developed into a later (veliger) larval stage. This behaviour can be seen as an adaptation for limiting dispersal range to increase the probability of staying near compatible habitat types. Despite this brooding habit the prodigious fecundity common among bivalves is maintained, with larger *O. edulis* females found to be brooding as many as 1.5 million larvae (Walne 1979).

For both species water temperature is a significant factor determining when gonad maturation and spawning can occur. Reflecting their natural geographical distribution, the European Flat Oyster is reported as capable of spawning at lower water temperatures (around 15°C) than the Pacific Oyster (roughly 20°C). However, such simple figures are misleading for various reasons: firstly, the various stages of gonad development and gametogenesis each require sustained periods of increasing spring temperatures and, secondly, as species of oysters are known through both natural selection and phenotypic physiological adaptation to form races whose tolerances vary.

Both Poole Harbour species are euryhaline, being tolerant of the varying salinities characteristic of estuarine waters. However, in this respect Pacific Oysters have the advantage, reportedly being able to tolerate salinities as low as 12 (parts salt per thousand parts water) in contrast to the Native Oyster, whose distribution is generally limited to salinities consistently above 20 (coastal seawater is typically around 34).

The rise and fall of the native oyster fishery

In the modern period the pattern of oyster production in Poole has broadly followed that of the country as a whole. Commencing in the 1830s, a boom in British native oyster production has been attributed to the provision of rapid transport via a growing railway system, which coincided with population growth, poverty and the consequent availability of labour. It has been estimated that in excess of 500 million oysters passed through London's Billingsgate Market in 1850 and a total British annual consumption of 1.5 billion oysters was reported by *The Times* newspaper as late as 1867 (*The Times* 15 October 1867).

At that time oysters were a protein staple – in contrast to the luxury food they are today – and Britain was thought to contain the richest natural oyster beds in Europe. Poole Harbour significantly contributed to a bonanza, which generated considerable socio-economic benefit, not least for Poole itself which was reported as having upwards of 40 vessels employed on oyster fishing each spring.

However, the boom was followed by a crash in stocks, a decline recognised in 1866 by a Royal Commission on Sea Fisheries. That year official oyster landings for England and Wales were first officially recorded; the data suggested a catastrophic fall, with a total production across England and Wales of only 40 million oysters (around 3,500 tonnes). Lovell (1867) reported that native oysters cost two-pence 'so scarce had the molluscs become'. Despite efforts by the Poole Corporation to preserve the fishery, by the late 1840s the oyster beds had been 'over dredged to extinction' (Cullingford 2003).

Various efforts at restoration of the fishery in Poole failed, including an 1867 oyster breeding venture based in shallow ponds in Sandbanks and an 1879 attempt to restrict oyster fishing from grounds in front of Brownsea Castle. In 1881 the 'Poole Oyster Fishery Company' succeeded in obtaining a Regulating Order from the Government Board of Trade covering 200 acres of the Wareham Channel. This enabled the management of stocks through the licensing of oyster dredgers with enforcement by a water bailiff (Philpots 1890). In 1890 Philpots was reporting the area to have been cleaned and stocked and 'confidently expected that our famous beds will be again plentifully stocked with luscious bivalves'.

But despite all such efforts *Ostrea edulis* continued to decline on a widespread scale, with a relentless downward trend across England and Wales – to the extent that the 1886 reported landings of 3,500 tonnes had dropped to below 500 tonnes by 1947.

Poole Harbour and Poole Bay continued intermittently to provide impoverished but commercially exploited native oyster beds for a few more decades. Hillier and Blythe (1994) describe various attempts in the 1950s to revive the Poole oyster industry including the laying of hundreds and thousands of oysters with the consequence that 'the first native Poole oysters to go to market for over a quarter of a century were lifted in 1959'. In support of such efforts, archive materials held by the Southern Inshore Fisheries and Conservation Authority (Southern IFCA) document a range of fishery management and regulation measures applied throughout the twentieth century, including varying closed season dates, increasing minimum landing sizes, maximum landing sizes designed to conserve the largest and most fecund specimens, maximum size of oyster dredges and total fishery closures. However, to this day native oyster stocks in Poole Harbour remain sparse and depleted, as they do elsewhere on the south coast and beyond.

Causes of decline and the prospect of restoration

The combination of both ecological and economic impacts of native oyster decline has led to recent restoration efforts involving collaborations between fishers, regulators, academics and environmental non-governmental organisations (eNGOs) among others. In Britain and Ireland, a Native Oyster Network was established in 2017 to promote such collaborations and facilitate knowledge transfer between the various restoration projects, which now exist from Dornoch Firth to the Solent. The Solent project combines the deployment of brood stock in bags and layings on the seabed,(supported by the Blue Marine Foundation eNGO), with the complete closure to dredging of the oyster beds and annual stock assessments by the Southern IFCA. The universities of Portsmouth, Southampton and Bournemouth are also involved in various ways.

The initiatives of the Poole Oyster Fishery Company and others show that efforts to restore native oyster stocks date back to the nineteenth century and continued in the twentieth century, both nationally and locally. In 1947 the UK fisheries ministry established a laboratory at Conwy to determine the causes of native oyster decline, with a view to elucidating the necessary measures for their restoration. For Poole Harbour, the period from 1919 to the 1990s is summarised in a document in the archives of what is now the

Table 14.1 File note from Southern Inshore Fisheries and Conservation Authority (formerly Southern Sea Fisheries District) archive

Poole Harbour regulated fishery – History of events	
1919/20	56 licences taken out. 892,500 oysters landed
1920/21	43 licences taken up. 1,000,000 oysters landed
December 1921	Oyster beds throughout district reported to be suffering serious mortality from a disease which was sweeping through Europe
December 1922	It is assessed that millions of oysters have been lost to disease between 1920 and 1922
1922 to 1929	Great efforts were made by cleaning the beds and relaying to revitalise the Poole oyster fishery but in December 1929 the CFO recommended no fishing that year
1929 to 1939	The very limited oyster fishery gradually declines until by 1939 everybody has given up
1948	Only one commercial oyster bed reported in the District and that is at Newtown
1958 to 1966	A series of individuals and organisations attempt to revive the oyster cultivation industry on layings in the harbour. Many setbacks and 1966 see the first real returns
1980	Poole Regulating Order Licences again issued
1986 February	*Bonamia* first detected in Pole Harbour
1988	Surveys showed virtually no oysters outside the layings
199? (illegible)	Further survey gave even more depressing results – hardly an oyster to be found

Attributable to Chief Fishery Officer M.A. Whitley. Undated.

Southern IFCA, which is transcribed here as Table 14.1. This gives an indication of both continued efforts at restoration and the significance of disease in the native oyster decline. It is a salutary lesson that all these efforts ultimately failed, even if some delivered benefits in the short term.

Therefore, it cannot be assumed that present-day restoration projects will succeed. Their success will depend on whether they sufficiently remedy the causes of a decline which has been attributed to multiple factors. Moreover, the extent to which such factors individually or in combination have contributed to the decline or are limiting factors in restoration is not well understood and will most likely vary geographically.

In Poole Harbour, while overfishing will have been a significant issue in the decline, other factors likely to be affecting stocks include competitors for space and food such as the non-native Slipper Limpet *Crepidula fornicata*, which is abundant, water quality issues such as turbidity and pollution, and most particularly the protozoan parasite *Bonamia ostreae*, first identified in Europe in 1979 and present along the south coasts of England. Direct or indirect effects of rising water temperatures as a consequence of climate change can also not be ruled out, not least as Poole Harbour's shallow bathymetry and southern position makes its water temperatures among the highest in Britain.

Restoration efforts must also address aspects of the biology of *O. edulis* such as the retention and brooding of eggs and larvae within their shell chambers and the settlement preference for conspecific culch: both of which add a density-dependent aspect to successful spawning and recruitment episodes. Essentially, juvenile Native Oyster recruitment is most successful on established oyster beds in which sufficient abundance is already established.

These complex multifactoral challenges suggest that the best chance of restoration will involve co-ordinated multi-agency efforts, in which the effective regulation of fishing and water quality is combined with significant long-term investment. Although there are international examples of apparently successful restoration projects, whether the UK projects will succeed long term is currently uncertain. Moreover, in the event that restoration projects do bear fruit, history suggests that for long-term benefit all stakeholders will need to recognise the need for strict and imaginative approaches to the management of the fishery. For example, if a mosaic of unexploited and fished oyster beds could be established, a 'Maximum Sustainable Yield' in the latter might be achievable if a minimum of half the biomass of the former were maintained on the commercial grounds (Laing *et al.* 2006).

The arrival of the Pacific Oyster

It is an indication of difficulties encountered restoring the native Flat Oyster stocks in the 1890s that the Poole Oyster Company decided to establish a contract for the supply of live non-native species of 'Tagus Oysters' *Ostrea angulata* from Portugal (Philpots 1890). In fact, *O. angulata* is now considered an out-of-date synonym for *C. angulata* which modern genetic analysis suggests may in fact be the same species as the Pacific Oyster *C. gigas*, both forms of which were introduced to Europe by human agency, although at different times and via different routes (Humphreys *et al.* 2014). If a conspecific relationship is accepted for these oysters, then *C. gigas* was first introduced into Poole Harbour in 1890.

In any event these non-native oysters struggled to become established in British water temperatures and efforts to sustain the industry through imported seed were ultimately unsuccessful. Continued industry decline was attributed to pollution, American pests introduced with imported stock, unusually severe winters (1939/40 and 1946/47) and occasional unexplained large-scale mortalities. Then in 1962 commercial importation of the Pacific Oyster was stopped in response to disease in source populations (Humphreys *et al.* 2014).

In response to more stringent regulation to prevent the introduction of pests and diseases the British government fisheries laboratory at Conwy produced pathogen-free Pacific Oyster brood stock from which UK hatcheries could produce seed oysters as a safe alternative to imported seed. This was the basis of a 1960s re-introduction from which the current Poole oyster industry derives, although not without continuing early difficulties.

In the mid-1960s the Poole Oyster Company cultivated imported seed oysters in mesh bags fixed to trestles. However, the viability of Pacific Oyster growing was again seriously compromised this time by tributyltin (TBT), a toxic agent in the water. From the early 1960s TBT was used as an additive to anti-fouling paints for vessel hulls. Slowly released into the water, its toxicity to marine organisms supresses attachment and growth, giving significant vessel performance and fuel-economy benefits (Dafforn *et al.* 2011). Even at low contamination levels, oysters are highly sensitive to TBT. By the early 1980s French oyster farms reported reduced oyster settlement, compromised larval development and shell malformations in 80–100% of oysters (Dafforn *et al.* 2011).

By virtue of thousands of recreational vessels, Poole Harbour's water quality was similarly compromised, jeopardising Pacific Oyster culture as a viable commercial enterprise. Recognition of the impact and extent of TBT contamination was followed in 1987 by the banning of the use of TBTs in anti-fouling paint for small (mainly recreational) UK vessels. This marked a turning point for Pacific Oyster culture and today Poole Harbour is again a pre-eminent oyster production site.

Othniel Oysters

The current significance of Poole Harbour for oyster production is the result of the establishment and growth of the locally owned and operated Othniel Oysters Ltd. The company was founded in 1987 by Poole fisherman Gary Wordsworth, whose move into aquaculture involved obtaining an initial lease for 5 acres of harbour seabed. Over the years the company has been innovative in terms of both aquaculture technology and species cultured. For instance, with the support of the British government fisheries ministry, Othniel was one of the first in the UK to develop the seabed culture of Manila Clams (Humphreys *et al.* 2015). Today, however, the company specialises in Pacific Oysters.

The centre of the operation is Othniel's 'Ferry No 3': the former Sandbanks-Studland chain ferry vessel now moored off the north coast of Brownsea Island and converted into

Figure 14.2 Ferry No 3, the original Sandbanks-Studland chain ferry vessel now converted to Othniel Oyster Ltd.'s facility in Poole Harbour. In the foreground a harvesting vessel is tied up alongside.

an oyster cultivation and processing facility (Figure 14.2). Each year, seed Pacific Oysters from hatcheries in France are grown alongside in racks, until they reach sufficient size to resist the predation of crabs, at which point they are transferred to the harbour bed where conditions provide for the fast growth of premium oysters.

Othniel's approach in Poole has been developed to suit the harbour conditions, which favour growth to saleable size directly on the seabed rather than the more common trestle culture used elsewhere. This along with the extended 'double' high-water stand in the harbour has led the company to innovate in terms of harvesting technology to bring oysters up from submerged beds. Figure 14.2 shows the company's harvesting vessel alongside Ferry No. 3.

By virtue of this local success story, Othniel, still run by Wordsworth, is the largest single oyster producer in the UK. From an initial 5 acres (2 ha), by 2020 the company was leasing over 40 ha (100 acres) of the Poole Harbour seabed from the Southern IFCA. Production has typically been around 400 tonnes per year, much of which is exported to the Far East and mainland Europe.

Pacific Oysters in the Poole Harbour ecosystem

At the time of the 1960s introduction of the Pacific Oyster, it was thought that the species when deposited as juvenile 'spat' supplied from elsewhere would fatten well to market size but would not successfully reproduce due to English water temperatures being too low for the temperature-sensitive gametic and larval stages of the lifecycle. The species was therefore assumed not to be able to naturalise (establish wild populations) in UK waters. Nevertheless, the species was known to occasionally spawn in Essex waters, and this led some to suggest that it be considered as part of the British marine fauna (Yonge 1960).

Such categorisations are not insignificant: In the 1980s, for example, the future of British Pacific Oyster cultivation could have been threatened by new conservation legislation in the form of the Wildlife and Countryside Act 1981, which included clauses on the introduction of non-indigenous species. In particular the act could have prevented release of juvenile Pacific Oysters for fattening. However, in 1982 a general licence to 'release or allow to escape into the wild' was issued (London Gazette 30 November 1982) under the terms of the act on the basis that the species was 'already resident' in British waters (Humphreys *et al.* 2014).

It might be supposed that this legal status for the species would resolve the question of the Pacific Oyster in Britain but in fact this is not the case. Repeated government-sponsored reviews designed to assess the ecological risk posed by the species (e.g. Herbert *et al.* 2012, 2016) have not yet resulted in clear policy or a settled position on management. Concerns among conservation organisations relate to the general idea that non-native species can change indigenous ecosystems and competitively exclude native species. Yet not all non-native species are 'invasive' in this sense, and there is little evidence that the Pacific Oyster, especially if properly managed as a fishery resource, is significantly detrimental. Conversely, ecological benefits are apparent.

Pacific Oysters when thriving form reefs within which biodiversity is greater than the surrounding sediment seabed. In the Colne estuary, Essex, areas of Pacific Oyster reef have become established on what would otherwise be intertidal mudflats. Comparisons of areas of oyster reef, cleared oyster reef and natural mudflat have shown that oyster reefs score highest in terms of both species richness and biomass (even excluding the oysters themselves), and cleared oyster reefs second. These results when modelled in relation to bird feeding behaviours suggested a conservation strategy in which a mosaic of the three area types may be optimal for bird populations. Since the clearance of oyster reefs was done by collaborating fishers with commercial oyster dredges, a virtuous circle of both economic and environmental benefit is implied (Herbert *et al.* 2018).

In any event, Pacific Oyster spat now introduced into Poole Harbour for fattening are first chromosomally reconfigured in such a way as to impede gametogenesis. These 'triploid' oysters direct energy normally used for reproduction into fattening: arguably, a double benefit as it also mitigates against wild settlement spawned from cultivated beds.

Pacific Oysters in Poole Harbour also contribute to the alleviation of water quality challenges. The most ecologically significant chemical pollutant in Poole Harbour is nitrogen, notably in the form of nitrates. Plants need nitrogen for growth, and it is therefore used in agricultural fertilisers in the harbour's river catchment, from where it is released into the harbour. Too much nitrogen in the water column creates 'eutrophic' conditions by enabling algal blooms, to an extent which degrades the natural environment. These blooms are most noticeable as green 'algal mats', which accumulate over much of the harbour's mudflats as the summer progresses. Less noticeable but equally important are the microscopic phytoplankton blooms which detrimentally impact the ecology of the water column.

Filter feeding shellfish such as the Pacific Oyster feed on phytoplanktonic algae, thereby removing them from the water and assimilating the nitrogen they contain into the oysters' own tissues. Commercially harvested shellfish such as the Pacific Oyster represent a significant 'ecological service' in terms of water purification, since they not only filter out phytoplankton reducing the turbidity of the water but also, when they are harvested, remove the nitrogen they contain entirely from the harbour system.

Gravestock *et al.* (2020) have examined the significance of this benefit in Poole Harbour by calculating the filtration rates of Pacific Oysters under conditions typical of the harbour environment. They conclude that the total volume of water filtered by Pacific Oysters on the harbour's aquaculture beds ranges up to 1.9×10^6 m³ per day, during the peak growth season. When combined with the other commercial aquaculture species (the Mussel *Mytilus edulis*) this represents over 61% of the harbour's total high-water volume on a neap tide when the water quality issues are most acute.

Phytoplankton abundance is often measured in terms of the amount of green pigment chlorophyll in the water column. On the basis of their filtration calculations, Gravestock *et al.* (2020) estimate a peak removal of 231 kg of chlorophyll per day from the harbour by the two aquaculture species. While these authors do not convert this figure to phytoplankton mass removed, studies in Chesapeake Bay have shown the related species *C. virginica* to be capable of removing up to 23–40% by mass of phytoplankton production. Approximate and localised as they are, such estimates nevertheless substantiate a significant water purification role for cultured Pacific Oysters in Poole Harbour.

Conclusion

The history of the Poole oyster industry reflects many of the marine environmental issues of the modern age. An industry which in ancient times is known to have thrived in the context of a natural environment relatively unmodified by human populations has, since the industrial revolution, struggled to exist against a series of human-made barriers. Many of the main categories of marine conservation concerns are represented in this history, from unregulated overfishing in the context of a growing metropolitan population in need of a protein staple, to TBT and nitrates in the water, to the anthropogenic arrival of new diseases and non-native predator species.

Yet ironically the solution has itself been found in the introduction of a non-native oyster, which, unlike its indigenous cousin, seems capable of persisting in the turbid, eutrophic and warming marine environment that human populations have produced. An anthropogenically introduced species is arguably a solution to an anthropogenically altered environment. The Pacific Oyster provides a viable industry while improving water quality, enhancing biodiversity and in all probability increasing the harbour's carrying capacity of important bird populations (Humphreys *et al.* 2021).

In this context it is of interest that the non-native Manila Clam, also introduced for aquaculture (Humphreys 2010), has, since it became naturalised, provided both a wild fishery and a food source for important Poole Harbour wader populations to an extent that has been calculated to reduce their overwinter mortality (Caldow *et al.* 2007).

All this creates questions for the orthodox assumptions of conservation. There are good reasons why the control of non-native species in open marine systems is regarded as one of the major challenges of environmental management. However, the Pacific Oyster demonstrates that the arrival of non-native species can be beneficial in terms of biodiversity and ecosystem (trophic) function. In demonstrating some of the benefits from Pacific Oysters in English waters, Herbert *et al.* (2018) have described their findings as creating a 'conundrum' for environmental agencies and managers.

Moreover, the Pacific Oyster elucidates the problems of a dogmatic reaction to non-native species, especially in the context of that most profound planetary scale impact, climate change. It has been shown for European breeding bird populations that their geographical range has shifted northwards on average 1 km per year for the last 30 years, with consequent changes to the breeding bird fauna of Britain as southern species move in (EBBA2 2020). The arrival of new species of breeding birds is seldom perceived as a problem.

It can be argued that the equanimity demonstrated in relation to the arrival of non-native feathered fauna could sometimes be usefully applied to the climate-related redistributions of benthic species: each new non-native arrival carefully examined, not with the loaded terms of 'aliens' and 'impacts', but rather with a more dispassionate and balanced assessment of ecological and where relevant socio-economic benefits and risks.

Rombouts *et al.* (2012) in modelling future changes in the distributions of four native bivalve species find a general northern displacement, predicting the disappearance of three of them from the English Channel. As warming British coasts become more conducive to successful Pacific Oyster spawning, the conventional concern that non-native species will damage British marine biodiversity needs to be carefully reconsidered on a case-by-case basis. The Pacific Oyster may be telling us that as indigenous species decline or retreat northwards, warmer water arrivals will be needed if the biodiversity and functional resilience of our benthic ecosystems, and their associated fisheries, is to be maintained.

References

Caldow, R.W.G., Stillman, R.A., dit Durell, S.E.A.L.V., West, A.D., McGrorty, S., Goss-Custard, J.D., Wood, P.J., and Humphreys, J. 2007. Benefits to shorebirds from invasion of a non-native shellfish. *Proceedings of the Royal Society B* 274: 1449–55. https://doi.org/10.1098/rspb.2007.0072

Cullingford, C.N. 2003. *A History of Poole*. Phillimore, Chichester.

Dafforn, K.A., Lewis, J.A., and Johnston, E.L. 2011. Antifouling strategies: History and regulation, ecological impacts and mitigation. *Marine Pollution Bulletin* 62: 453–65. https://doi.org/10.1016/j.marpolbul.2011.01.012

Defoe, D. 1726. *A Tour through the Whole Island of Great Britain*. London, Penguin Classics 1986 edition.

EBBA2. 2020. *European Breeding Bird Atlas 2 European Bird Census Council*. Bedfordshire, Sandy.

Gravestock, V.J., Nicoll, R., Clark, R.W.E., and Humphreys, J. 2020. Assessing the benefits of shellfish aquaculture in improving water quality in Poole Harbour, an estuarine Marine Protected Area. In: Humphreys, J. and Clark, R.G. (eds) *Marine Protected Areas: Science, Policy and Management*. Elsevier, Amsterdam. https://doi.org/10.1016/B978-0-08-102698-4.00037-X

Herbert, R.J.H., Davies, C.J., Bowgen, K.M., Hatton, J., and Stillman, R.A. 2018. The importance of non-native Pacific oyster reefs as supplementary feeding areas for coastal birds on estuary mudflats. *Aquatic Conservation: Marine and Freshwater Ecosystems* 28: 1294–307. https://doi.org/10.1002/aqc.2938

Herbert, R.J.H., Humphreys, J., Robert, C., Fletcher, S., and Crowe, T. 2016. Ecological impacts of non-native Pacific oysters (*Crassostrea gigas*) and management measures for protected areas in Europe. *Biodiversity and Conservation* 25: 2835–65. https://doi.org/10.1007/s10531-016-1209-4

Herbert, R.J.H., Roberts, C., Humphreys, J., and Fletcher, S. 2012. *The Pacific Oyster in the UK: Economic, Legal and Environmental Issues associated with its Cultivation, Wild Establishment and Exploitation*. Shellfish Association of Great Britain, London.

Hillier, J. and Blythe, M. 1994. *The Spirit of Poole*. Poole, Poole Historical Trust.

Humphreys, J. 2005. Salinity and tides in Poole Harbour. In: Humphreys, J. and May, V. (eds) *The Ecology of Poole Harbour: Proceedings in Marine Science 7*. Elsevier, Amsterdam. https://doi.org/10.1016/S1568-2692(05)80008-7

Humphreys, J. 2010. The introduction of the Manila clam to British coastal waters. *Biologist* 57(3): 134–8.

Humphreys, J., Harris, M.C.R., Herbert, R.J.H., Farrell, P., Jensen, A., and Cragg, S.M. 2015. Introduction naturalisation and dispersal of the Manila clam *Ruditapes philippinarum* in British estuaries 1980–2010. *Journal of the Marine Biological Association of the UK* 95(6): 1163–72. https://doi.org/10.1017/S0025315415000132

Humphreys, J., Herbert, R.J.H., Roberts, C., and Fletcher, S. 2014. A reappraisal of the history and economics of the Pacific Oyster in Britain. *Aquaculture* 428–9: 117–24. https://doi.org/10.1016/j.aquaculture.2014.02.034

Humphreys, J., Syvret, M., Horsfall, S., Williams, C., Woolmer, A., and Adamson, E. 2021. Why we should learn to love Pacific oysters. *The Marine Biologist* 20: 10–11.

Laing, I., Walker, P., and Areal, F. 2006. Return of the native – Is European oyster (*Ostrea edulis*) stock restoration in the UK feasible? *Aquatic Living Resources* 19: 283–7. https://doi.org/10.1051/alr:2006029

Lovell, M.S. 1867. *The Edible Mollusca of Great Britain and Ireland* (2nd edn). Reeve & Co, London. https://doi.org/10.5962/bhl.title.56462

Philpots, J.R. 1890. *Oysters and All About Them*. Richardson, London.

Rombouts, I., Beaugrand, G., and Dauvin, J.-C. 2012. Potential changes in benthic macrofaunal distributions from the English Channel simulated under climate change scenarios. *Estuarine, Coastal and Shelf Science* 99: 153–61. https://doi.org/10.1016/j.ecss.2011.12.026

Walne, P.R. 1979. *Culture of Bivalve Molluscs*. Fishing News Books Ltd., Farnham.

Yonge, C.M. 1960. *Collins New Naturalist Series. Oysters*, London.

PART IV

Water Quality

CHAPTER 15

Water Framework Directive Ecological Monitoring in Poole Harbour, 2007–2019

SUZY WITT

Abstract

The Water Framework Directive (WFD) is a piece of European Union legislation that requires member states to make plans to protect and improve the water environment. As a statutory body, the Environment Agency (EA) is required to monitor ecological and water quality elements within Poole Harbour under this legislation. This chapter provides a summary of monitoring results from this programme and briefly describes the tools used to classify the harbour. Given the ecological and economic importance of the harbour, anthropogenic pressures highlighted during this monitoring are considered.

Keywords: Water Framework Directive (WFD), monitoring, Environment Agency (EA), nutrients, EQR

Correspondence: suzy.witt@environment-agency.gov.uk

Introduction

The Water Framework Directive (WFD) was implemented into law in the UK in 2003. It requires the ecology of river basin catchments to be assessed to determine where action is needed to improve the water environment.

The Environment Agency (EA) assesses the status of rivers, lakes, coastal and estuarine water environments (waterbodies) by looking at ecological, chemical and physical elements. For each element, a variety of tools have been developed to ensure a consistent approach and to allow the classification of waterbodies according to the results obtained. Each tool has been developed and calibrated at a European level.

For ecological elements, data obtained are compared to those describing reference conditions (where a waterbody is minimally disturbed). This is reported as an Ecological Quality Ratio (EQR). An EQR with a value of one represents reference conditions and a value of zero represents a severe impact. The EQR is divided into five ecological status classes (High, Good, Moderate, Poor and Bad) that are defined by changes in the biological community in response to disturbance.

The ecological elements which have been monitored in Poole Harbour include angiosperms (saltmarsh and intertidal seagrass), opportunistic macroalgae, fish, phytoplankton and benthic invertebrates. Water quality and supporting elements have been assessed by measuring dissolved inorganic nitrogen (DIN), dissolved oxygen (DO) concentrations and

Suzy Witt, 'Water Framework Directive Ecological Monitoring in Poole Harbour, 2007–2019' in: Harbour Ecology. Pelagic Publishing (2022). © Suzy Witt. DOI: 10.53061/RUUU3584 the levels of specific pollutants and priority hazardous substances. Hydro-morphological supporting elements are also considered as part of classification. The presence of invasive non-native species is also taken into account and if any are present in a waterbody it cannot be classed greater than Good, even if individual elements are classed as High.

Poole Harbour is an ecologically and economically important harbour and is part of the surveillance programme of WFD monitoring. This means that all elements are monitored within the harbour, as opposed to other waterbodies where only certain elements are considered at a lower frequency. Monitoring is carried out on a risk basis, whereby higher risk waterbodies are prioritised. This chapter provides a summary of the monitoring that has been undertaken by the EA in Poole Harbour from 2007, when monitoring began, to 2019. Each ecological element is considered separately and a summary of WFD classification is provided.

The technical details of each individual monitoring methodology will not be discussed here, as further information and details on the development and application of each tool is available on the UKTAG website (www.wfduk.org).

Angiosperms (saltmarsh and intertidal seagrass)

Saltmarsh vegetation consists of a limited number of salt-tolerant species adapted to regular immersion by the tides. A saltmarsh system shows a clear zonation according to the frequency of inundation. Saltmarsh communities are affected by hydraulic factors such as wave climate, currents, sediment supply and transport, sea level and storm frequency. Anthropogenic factors include coastal squeeze, grazing, trampling, boat movements, pollution and eutrophication. The saltmarsh WFD index was developed to respond to these pressures.

Sampling is a combination of data obtained from aerial imagery and ground surveys carried out on foot. Five different aspects are considered, including the current and historical extent of the marsh, as well as species diversity and the number of zones present. Aerial imagery provides information on the extent of saltmarsh areas and the position of different zones on the marsh, and imagery is normally captured during the year of ground sampling. Images captured either the year before or after the survey can also be used for analysis and interpretation. The ground survey provides data for species diversity and abundance and assists in the desk-based interpretation of aerial photography. The survey window is between 1 June and 30 September and ideally coincides with when the plants are at their most floristic. Initial baseline surveys of the saltmarsh communities were carried out in 2007–8 and 2012–13. In 2016 a 'reduced transect' method was employed, which involved surveying the same parameters but at only four transects within the harbour. These transects were considered most representative of the species diversity of the saltmarsh. Further photo interpretation of aerial imagery is carried out following reduced transect surveys to ensure areas of marsh are correctly assigned to vegetation types.

Figures 15.1 and 15.2 illustrate the position of transects within the harbour. Transects were placed to cover the maximum area of saltmarsh while taking the potentially dangerous nature of this intertidal zone into consideration. Transects are normally carried out perpendicular to the shore, but not always, as monitoring is designed to encompass as many different community types as possible. Two 4 m² quadrats are surveyed in each saltmarsh zone while travelling along a transect line and also where there was a change in the community. Within each quadrat the species present and their percentage cover are determined, as well as recording bare ground or any negative indicators. Additional plant species encountered while walking between quadrats or transects are also logged as part of the survey.

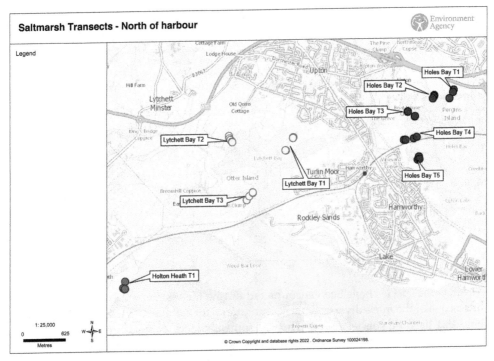

Figure 15.1 Position of saltmarsh transects in the north of Poole Harbour (©Bluesky International Ltd/Getmapping PLC ©OrdnanceSurvey © Environment Agency).

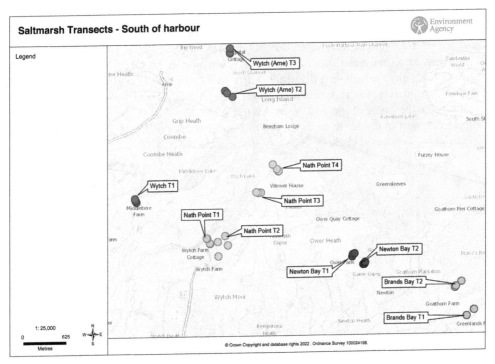

Figure 15.2 Position of saltmarsh transects in the south of Poole Harbour (©Bluesky International Ltd/Getmapping PLC ©OrdnanceSurvey © Environment Agency).

Table 15.1 Saltmarsh species encountered during Environment Agency surveys, 2007–16

Species	
Creeping Bent *Agrostis stolonifera*	Lax-flowered Sea-lavender *Limonium humile*
Bulbous Foxtail *Alopecurus bulbosus*	Common Sea-lavender *Limonium vulgare*
Thrift *Armeria maritima*	Hemlock Water-dropwort *Oenanthe crocata*
Sea Aster *Aster tripolium*	Parsley Water-dropwort *Oenanthe lachenalii*
Common Orache *Atriplex hastata*	Common Reed *Phragmites australis*
Sea Purslane *Atriplex portulacoides*	Sea Plantain *Plantago maritima*
Spear-leaved Orache *Atriplex prostrata*	Common Saltmarsh Grass *Puccinellia maritima*
Sea Clubrush *Bolboschoenus maritimus*	Common Glasswort *Salicornia europaea* agg.
Bostrychia scorpioides (red alga)	Perennial Glasswort *Sarcocornia perennis*
English Scurvygrass *Cochlearia anglica*	Grey Clubrush *Schoenoplectus tabernaemontani*
Common Scurvygrass *Cochlearia officinalis*	Common Cordgrass *Spartina anglica*
Sea Couch *Elytrigia atherica*	Greater Sea-spurry *Spergularia media*
Red Fescue *Festuca rubra*	Annual Seablite *Suaeda maritima*
Sea Milkwort *Lysimachia maritima*	Shrubby Seablite *Suaeda vera*
Saltmarsh Rush *Juncus gerardii*	Sea Arrowgrass *Triglochin maritima*
Sea Rush *Juncus maritimus*	

Thirty-one species have been encountered during these surveys, which places the harbour as one of the more diverse estuaries monitored for saltmarsh in England under WFD. Table 15.1 lists the species found. There is some evidence of pressure from erosion and coastal squeeze, however, in certain areas of the harbour. For example, looking at aerial imagery it is evident that in the centre of Holes Bay saltmarsh has eroded. Taking all metrics into account, however, Poole is currently classed as Good for saltmarsh under WFD.

During the opportunistic macroalgae survey in 2011, intertidal seagrass was discovered in the south of the harbour. Small beds of Beaked Tasselweed *Ruppia maritima* were mapped in surveys undertaken by hovercraft in 2012, 2015 and 2016 by using a GPS to delimit the edge of the beds. Quadrats were carried out within the beds to determine the percentage cover of the seagrass, but due to the sparse nature of the beds this element is not currently classified under WFD in this waterbody. The seagrass plants exist within the patches of macroalgae, and their extent and distribution have altered during monitoring, culminating in only single plants found during the 2019 opportunistic macroalgae survey. Further research is needed on the impact of macroalgae and other factors on intertidal seagrass beds to determine the cause of the decline of this element within the harbour. Subtidal seagrass beds are present in the north of the harbour, but these are not currently monitored by the EA under WFD but will likely be considered as a separate ecological element in the future.

Fish

Fish species are surveyed annually at six sites in the harbour, using both a small beam trawl and a seine net. This was originally carried out biannually in both spring and autumn; however, from 2016 this has only been carried out in the autumn. From 2019 these surveys were paused to allocate resources to other monitoring programmes. Our coastal survey vessel undertakes an otter trawl in the Middle Ship Channel in the autumn and this has been carried out in all years since 2007.

All fish caught during the surveys are identified to species, measured and returned at each site. The sites monitored have altered since surveying began in the harbour in 2007. This has been due to the availability of boats and equipment and also the need for local information. From 2009 the same six sites have been monitored with trawling and seine netting carried out from the shore to obtain a comparable dataset. At Redcliffe, a site on the River Frome, only seine netting is carried out due to the impracticalities of trawling in the river. At the Middle Ship Channel site trawling is the only methodology used. The location of all sites is shown in Figure 15.3. Figures 15.4, 15.5 and 15.6 show the photographs taken on various surveys.

Since 2007 over 60 species of fish have been recorded in the harbour. A list of encountered species is shown in Table 15.2.

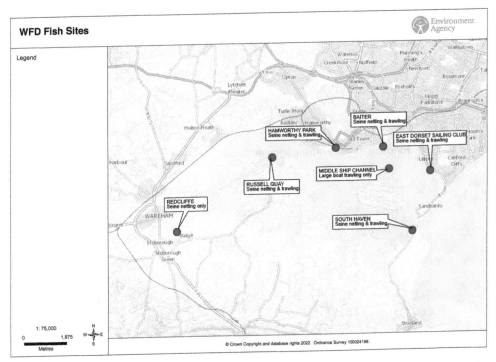

Figure 15.3 Position of Water Framework Directive (WFD) fish monitoring sites from 2009 onwards (©Bluesky International Ltd/Getmapping PLC ©OrdnanceSurvey © Environment Agency).

Figure 15.4 Otter trawling aboard the Environment Agency coastal survey vessel.

Figure 15.5 Environment Agency officer seine netting.

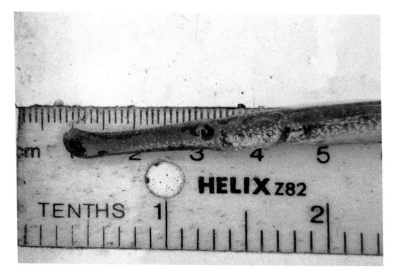

Figure 15.6 Measuring a Deep-snouted (Broad-nosed) Pipefish.

The harbour is classified using the Transitional Fish Classification Index. This index looks at a variety of measures including species diversity and composition, species abundance, nursery function and trophic integrity. Each aspect is assessed by comparing the observed values with those expected under reference conditions.

Poole Harbour is currently at Good WFD status for the fish element. It is an important nursery area for fish, and this is demonstrated by the high numbers of juvenile fish caught during our surveys.

Opportunistic macroalgae

An opportunistic algal species is able to take advantage of conditions in which other species often struggle to survive or compete. Blooms form principally of species of Sea Lettuce *Ulva* (this includes taxa formerly known as *Enteromorpha*), *Chaetomorpha* or *Cladophora*, although other green, red and brown algae may reach nuisance proportions.

Table 15.2 Fish species encountered during Water Framework Directive (WFD) surveys, 2007–18

Fifteen-spined (Sea) Stickleback	Deep-snouted (Broad-nosed) Pipefish	Lesser Sand Eel	Sea Bass
Two-spotted Clingfish	Dover Sole	Lesser Weever	Shanny
Two-spotted Goby	Dragonet	Long-spined Sea Scorpion	Short-snouted Seahorse
Three-spined Stickleback	European Eel/Elvers	Minnow*	Snake Pipefish
Five-bearded Rockling	Flounder	Motherless Minnow/Sunbleak**	Solenette
Baillon's Wrasse	Garfish	Painted Goby	Spiny Seahorse
Ballan Wrasse	Gilt-head Bream	Pollack	Sprat
Black Goby	Golden Grey Mullet	Red Mullet	Straight-nosed Pipefish
Black Sea Bream	Goldsinny	Reticulated Dragonet	Thick-lipped Grey Mullet
Brill	Grayling*	Roach*	Thin-lipped Grey Mullet
Brown/Sea Trout*	Greater Pipefish	Rock Cook	Tompot Blenny
Coley/Saithe/Coalfish	Greater Sand Eel	Rock Goby	Transparent Goby
Common Goby	Herring	Rudd*	Tub Gurnard
Corkwing Wrasse	Hooknose/Pogge	Sand Goby	Turbot
Cuckoo Wrasse	Lamprey*	Sand Smelt	Worm Pipefish
Dab	Lemon Sole	Sand Sole	
Dace*	Lesser (Nilsson's) Pipefish	Scaldfish	

*Species recorded only at River Frome site Redcliffe **Invasive species (recorded only at Redcliffe)

The formation of opportunistic macroalgal blooms is considered indicative of anthropogenically elevated nutrient levels. Deleterious effects include blanketing of the surface causing a hostile physico-chemical environment in the underlying sediment, effects on birds including changes in the feeding behaviour of waders and the smothering of seagrass and saltmarsh beds. Furthermore, high levels of nutrients have been found to reduce the biomass of the root systems of saltmarsh plants (Alldred *et al.* 2017), making them more susceptible to erosion. Extensive proliferation of algae can also affect the economic viability of this waterbody, which is currently used for amenity purposes, boating and water sports as well as being a designated EC shellfish water and harvesting area. The impact on commercial enterprises and public interactions within the harbour must not be overlooked.

To monitor opportunistic algae, aerial photography and digital imaging are combined with ground surveys. Aerial imagery is either captured using EA planes or data from other organisations, such as the Channel Coastal Observatory. The extent, percentage cover and biomass of algal species are derived from this methodology and values are then compared with reference conditions.

In Poole Harbour ground surveys have been carried out by hovercraft in 2008, 2011, 2015 and 2019. Figure 15.7 shows the distribution of macroalgae in 2015 and Figure 15.8 shows the distribution in 2019. Algal patches occur in most intertidal areas of the harbour and some patches have high biomass levels, particularly Holes Bay, where biomass regularly exceeds 3 kg per m^2. This amount of algae, although providing habitat for invertebrate grazers, can produce deleterious conditions within the sediment as well as affecting other species and habitats.

Poole Harbour is currently classed as Moderate under WFD for the macroalgae element.

Further information and discussion on the distribution and density of macroalgae within the harbour are included as an additional chapter in this book (see 'Nuisance Macroalgae in Poole Harbour').

Phytoplankton

The EA collects samples for phytoplankton analysis at five sites in the harbour throughout the year. These sites are shown in Figure 15.9. Phytoplankton are an ideal indicator of

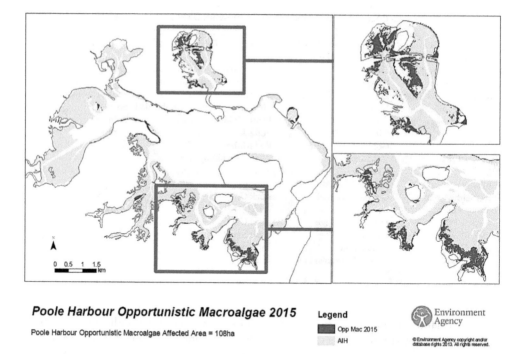

Poole Harbour Opportunistic Macroalgae 2015

Poole Harbour Opportunistic Macroalgae Affected Area = 108ha

Legend

■ Opp Mac 2015

AIH

Environment Agency

© Environment Agency copyright and/or database rights 2013. All rights reserved.

Figure 15.7 Distribution of opportunistic macroalgae in 2015 as shown by Environment Agency aerial imagery.

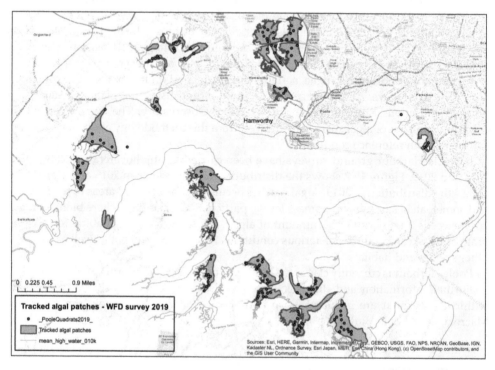

Figure 15.8 Distribution of opportunistic macroalgae in 2019 as shown by Environment Agency ground survey using a hovercraft and hand-held GPS unit.

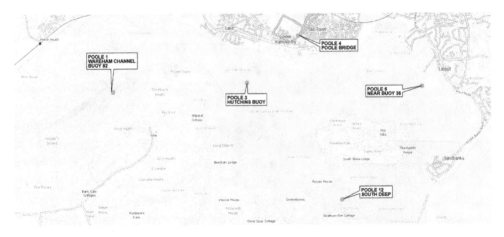

Figure 15.9 Position of sites monitored for phytoplankton and water quality elements in Poole Harbour (©Bluesky International Ltd/Getmapping PLC ©OrdnanceSurvey © Environment Agency).

changing nutrient conditions as they are short-lived and derive their nutrients from the water. Phytoplankton samples are collected from close to the water surface, avoiding any surface film and without disturbing bottom sediments.

The concentration of chlorophyll a in the water column is measured in order to determine phytoplankton biomass. In addition, the number of sampling occasions when phytoplankton counts (both individual taxa cells and total phytoplankton cells) exceed a threshold is also considered. These two measures are compared with those expected under reference conditions in order to classify the waterbody.

Poole is classed as Good under WFD in this ecological element. Phytoplankton growth is not currently considered a problem in Poole Harbour. Phytoplankton blooms above 5×10^5 cells per litre have occurred in every year from 2007; however, classification under the WFD has never fallen below good status since it was first classified in 2010.

Benthic invertebrates

Benthic invertebrates are an important component of marine ecological systems and are effective indicators of certain types of disturbance or pressure.

Data are obtained for Poole Harbour through intertidal core sampling to determine invertebrate abundance data, particle size and the salinity of the habitat sampled. Core samples are taken down to a depth of 15 cm using a 0.01 m² core and three replicates are taken at each site and combined to produce one sample per site. Sampling has been carried out in the spring of 2008, 2011 and 2015. There are 30 sites (shown in Figure 15.10) distributed throughout the harbour.

The WFD IQI index is used to classify the harbour and is specific to muddy or sandy habitats, therefore other habitats are excluded from monitoring to ensure consistency.

The fauna encountered is dominated by polychaete and oligochaete worms, hydrobid gastropods and larger bivalves. Over 80 species have been found over the three surveys. A species list is shown in Table 15.3. The fauna is typical of this harsh environment, where fine sediments are prone to anoxia and tidal cycles result in variations in salinity.

The current WFD classification is Good for this ecological element.

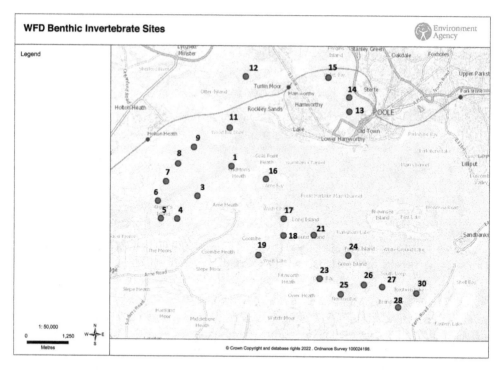

Figure 15.10 Position of WFD benthic invertebrate sites in Poole Harbour (© OrdnanceSurvey and © Environment Agency)

Table 15.3 Benthic species encountered during Water Framework Directive (WFD) surveys in 2008, 2011 and 2015

Spio armata	Caulleriella killariensis	Praunus flexuosus	Macoma balthica
Spio martinensis	Eteone longa	Eusarsiella zostericola	Cerastoderma edule
Tubificoides amplivasatus	Malacoceros fuliginosus	Crassicorophium bonnellii	Ruditapes philippinarum
Tubificoides benedii	Heterochaeta costata	Apocorophium acutum	Mysia undata
Melinna palmata	Arenicola marina	Carcinus maenas	Mytilus edulis
Paranais litoralis	Exogone naidina	Corophium volutator	Mya arenaria
Pseudopolydora paucibranchiata	Cossura pygodactylata	Balanus improvisus	Scrobicularia plana
Streblospio shrubsolii	Nephtys hombergii	Cyathura carinata	Lucinoma borealis
Phyllodoce mucosa	Ampharete grubei	Idotea chelipes	Lepidochitona cinerea
Chaetozone gibber	Euclymene oerstedii	Microdeutopus anomalus	Hydrobia ulvae
Sphaerosyllis tetralix	Anaitides mucosa	Gammarus salinus	Retusa obtusa
Polydora cornuta	Brania limbata	Gammarus locusta	Alderia modesta
Aphelochaeta marioni	Polydora quadrilobata	Microprotopus maculatus	Akera bullata
Brania pusilla	Gattyana cirrosa	Aora gracilis	Limapontia depressa
Sphaerosyllis taylori	Pygospio elegans	Monocorophium insidiosum	Peringia ulvae
Ampharete lindstroemi	Desdemona ornata	Ampelisca brevicornis	
Notomastus latericeus	Manayunkia aestuarina	Microdeutopus gryllotalpa	Amathia lendigera
Glycera alba	Neoamphitrite figulus		Anguinella palmata
Tubificoides pseudogaster	Neanthes virens		Electra crustulenta
Cossura longocirrata	Alitta virens	Timoclea ovata	Amphipholis squamata
Galathowenia oculata		Abra tenuis	

Invertebrates recorded at genus level have been excluded.

Table 15.4 A summary of determinants measured at water quality sites within the harbour

SPECIFIC POLLUTANTS	Arsenic
	Copper
	Iron
	Unionised ammonia
	Zinc
PHYSICO-CHEMICAL QUALITY ELEMENTS	Dissolved inorganic nitrogen (DIN)
	Dissolved oxygen (DO)
PRIORITY SUBSTANCES	Lead
	Nickel
	Trichloromethane
PRIORITY HAZARDOUS SUBSTANCES	Cadmium
	Mercury
	Nonylphenol
	Tributyltiun (TBT)*

*Tributyltin (TBT) data is not used for classification because concentrations are below the limit of quantification.

Water quality

Water quality is monitored regularly at five sites in the harbour. These sites are the same ones that are monitored for phytoplankton and are shown in Figure 15.9. This monitoring is undertaken to satisfy the requirements of several directives as well as the WFD, including the Environmental Quality Standards Directive, the Nitrates Directive and the Urban Wastewater Treatment Directive (UWWTD). A summary of determinants measured is provided in Table 15.4.

For classification purposes under the WFD, the environmental quality standards for specific pollutants and physico-chemical quality elements are used to assess ecological status, whereas the standards for priority substances and priority hazardous substances are used to assess chemical status.

The EA has now changed the approach to chemical status. This has resulted in an apparent significant decline in chemical classification (as shown by Figure 15.12). This, however, has not been driven by a sudden deterioration in water quality but reflects a change in approach to classifying the chemical status of English waterbodies. This means that classifications for 2019 are not directly comparable to those for previous years. The revised approach includes improvements to biota monitoring capabilities (monitoring for the presence of substances in fish and shellfish as well as in water) as well as to the interpretation of data.

The current physico-chemical status in Poole is Moderate. This is due to elevated levels of dissolved inorganic nitrogen. Figure 15.11 illustrates DIN levels over time at all sites within the harbour. Nitrogen and phosphate inputs include both point discharges, such as sewage treatment works (STW), and diffuse land run-off sources from within the catchment. The EA and Natural England have drawn up a Nitrogen Reduction Strategy and management plan (Bryan and Kite 2013) to try to address nutrient loading within the harbour. This identifies the sources of nitrogen and sets a target to bring the harbour to Good status. It also suggests options that might achieve this.

Poole Harbour was designated a Sensitive Area (Eutrophic) and Polluted Water (Eutrophic) under the EC Urban Waste Water Treatment and Nitrates Directives in 2002. Nutrient stripping was installed at Poole STW in 2008. This has resulted in a subsequent reduction in DIN levels in Holes Bay, illustrated by site 4 (Poole Bridge) in Figure 15.11. Nutrient levels in the Wareham Channel (which captures diffuse sources of nitrogen

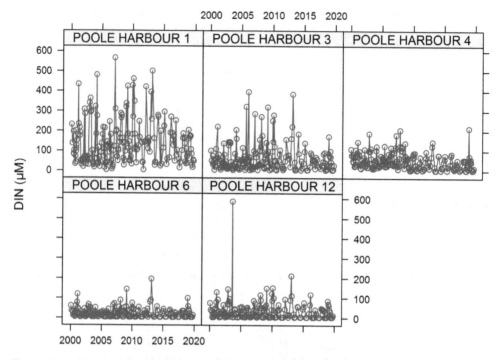

Figure 15.11 Dissolved inorganic nitrogen (DIN) levels over time at five sites within the harbour. (Note: All Environment Agency monitoring data has been used, not just that collected for WFD purposes.)

entering the harbour from the rivers Frome and Piddle) have increased steadily since 1995 (Environment Agency 2016). Since 2000 levels have stabilised but do not appear to have reduced significantly. The slow travel time through the Chalk aquifer in the catchment means that nitrate leaching remains high, despite regulation to reduce nitrogen losses from agriculture to water (Environment Agency 2016).

DIN levels over the winter period exceed WFD standards and indicate that Poole Harbour is hyper-nutrified throughout. The current WFD classification for physico-chemical quality elements is Moderate.

Summary

All data for this programme are fed into classifications of Poole Harbour, and these are updated to correspond to River Basin Management Plan cycles. River Basin Management Plans set out how partners will work together to protect and improve the water environment. The classifications are used within these plans to drive and develop programmes of measures to improve or maintain the ecological status of waterbodies.

The WFD classification cycle results for Poole Harbour are shown in Figure 15.12. The overall classification for the waterbody is Moderate and is derived from the worst class for all elements and cannot be higher than this – even if an individual element is classed as high. Elements that are currently failing WFD targets are the biological quality element macroalgae and the water quality element dissolved inorganic nitrogen. Both of these are interrelated, as elevated nutrient levels lead to the proliferation of nuisance algae.

Although Poole Harbour is not currently meeting objectives under the WFD, collaborative working through catchment initiatives and projects has advanced the study of the

	2009 Cycle 1	2015 Cycle 2	2019 Cycle 3
Overall Waterbody Classification	*Poor*	*Moderate*	*Moderate*
ECOLOGICAL	Poor	Moderate	Moderate
Biological quality elements	Poor	Moderate	Moderate
Angiosperms	-	Good	Good
Fish	Good	Good	Good
Invertebrates	Good	Good	Good
Macroalgae	Poor	Moderate	Moderate
Phytoplankton	-	Good	Good
HYDROMORPHOLOGICAL SUPPORTING ELEMENTS	Supports good	Supports good	Supports good
PHYSICO-CHEMICAL QUALITY ELEMENTS	Moderate	Moderate	Moderate
Dissolved Inorganic Nitrogen (DIN)	Moderate	Moderate	Moderate
Dissolved oxygen	High	High	High
Specific pollutants	High	High	High
Supporting elements (Surface Water)	Moderate	Moderate	Moderate
CHEMICAL	Good	Good	Fail*
Other Pollutants	Good	Does not require assessment	Does not require assessment
Priority hazardous substances	Good	Good	Fail*
Priority substances	Good	Good	Good

Figure 15.12 A summary of WFD classification for Poole Harbour. *The apparent deterioration in WFD status in 2019 has not been driven by a deterioration in water quality but is due to a change in the Environment Agency's approach to classifying the chemical status of waterbodies. For this element, 2019 results are therefore not directly comparable with previous classifications.

harbour and the pressures on it. The continuation of partnership working and involving local stakeholders and the community throughout the area will ensure that improvements can be made to this important environment.

Acknowledgements

Richard Acornley (Senior Environmental Monitoring Officer, EA).

References

Alldred, M., Liberti, A., and Baines, S.B. 2017. Impact of salinity and nutrients on salt marsh stability. *Ecosphere* 8(11). https://doi.org/10.1002/ecs2.2010

Bryan, G. and Kite, D. 2013. Strategy for managing nitrogen in the Poole Harbour catchment to 2035. With contributions from Money, R., Jonas, P. and Barden, R. Published by the Environment Agency.

Environment Agency. 2016. DATASHEET: Nitrate vulnerable zone (NVZ) designation 2017 – Eutrophic waters (Estuaries and coastal waters). NVZ name: Poole Harbour NVZ ID: ET1.

CHAPTER 16

Wessex Water's Environmental Improvement Work in the Poole Harbour Catchment, 2000–20

RUTH BARDEN

Abstract

Wessex Water Services Ltd. is the water and sewerage supplier within the Poole Harbour catchment. Increase in nitrogen loading over decades is impacting the conservation status of the Poole Harbour Special Area of Protection for Birds and Special Site of Scientific Interest, as well as the groundwater abstracted for public water supply. This has led to a range of interventions by Wessex Water to address these issues and in compliance with national and European legislation. This chapter highlights the legislative drivers for investment and describes the actions undertaken by Wessex Water at sewage treatment assets, drinking water sources and more widely in the catchment.

Keywords: nitrogen, sewage treatment, water treatment, auction, Water Framework Directive

Correspondence: Ruth.Barden@wessexwater.co.uk

Introduction

Wessex Water is the regional water and sewerage company with an operating area of 10,000 km^2, providing 2.8 million people with sewerage and 1.4 million people with water services. The company provides sewerage services across the whole Poole Harbour catchment and drinking water for the vast majority of the area; however, Poole itself is served by South West Water.

Water companies are regulated from an environmental and economic perspective to ensure that we fulfil our duties in conserving and enhancing the environment and that customers receive a good service which is value for money. This chapter focuses on the environmental improvements which the company has delivered across the catchment during the first 20 years of the twenty-first century, highlighting a key wastewater and supply challenge.

Ruth Barden, 'Wessex Water's Environmental Improvement Work in the Poole Harbour Catchment 2000–20' in: *Harbour Ecology*. Pelagic Publishing (2022). © Ruth Barden. DOI: 10.53061/UZJT9025

Poole Harbour catchment: A Wessex Water perspective

We describe the catchment as the watershed of the rivers Frome, Piddle, Corfe and Sherford, all of which flow into Poole Harbour. Within this catchment we have a mix of sewerage, sewage treatment and water supply assets.

We operate 21 water recycling centres (WRC, formerly called sewage treatment works), 172 sewage pumping stations and 18 combined sewer overflows. Sewage from domestic properties and business is connected to our sewerage network, where it flows by both gravity and pumping to our water recycling centres for treatment and discharge back into the environment. Our waste assets are illustrated in Figure 16.1.

Wessex Water also operates 13 water treatment works within the catchment, 3 river flow supports (where we discharge groundwater to support flows in rivers) and 65 distribution sites. The drinking water supplied in this catchment is abstracted from the Chalk aquifer, which underlies much of Dorset. The abstracted groundwater is treated and put into the supply network to serve our customers.

Environmental legislation

The majority of the environmental improvement work undertaken results from compliance with UK and European legislation. In some cases, this legislation is very prescriptive in the way that these improvements are delivered and technologies used, focusing on outputs. However, more recently, environmental legislation is aiming to deliver wider outcomes, often at a catchment scale. This approach enables Wessex Water to be more innovative, searching for interventions which will deliver the widest environmental benefits at scale. These approaches are also often cheaper for our customers.

The key legislation which has influenced our investment across the catchment is shown in Table 16.1.

Further details of the individual sites, interventions, costs and outcomes are detailed in Appendix 16.1.

Key issue: Nitrogen

This section describes a key issue for the Poole harbour catchment: nitrogen. This is a parallel issue facing both the provision of drinking water and our sewage treatment processes. Nitrogen contributions to the environment derive from animals and people, and therefore the biggest contributors are agriculture and sewage treatment.

Nitrogen and sewage

Nitrogen is of particular concern as it affects algal growth and the protected habitats and species within Poole Harbour. Nitrogen is found in domestic sewage in a number of forms but primarily as ammonia, which is a breakdown product from urine derived from our diet. Most sewage treatment works which discharge to watercourses will have a consented limit for the ammonia concentration as it is toxic to fish. In order to achieve this limit, biological treatment is used to convert the ammonia to nitrate which is then discharged to the receiving watercourse.

Standard sewage treatment works will reduce the overall nitrogen load, in addition to converting the incoming ammonia to nitrate. The level of removal depends on the process type. A small amount of nitrogen will be removed through the settlement process and transferred into the sludge. Process data from the sewage treatment works within the Poole Harbour catchment suggest nitrogen removal rates as indicated in Table 16.2.

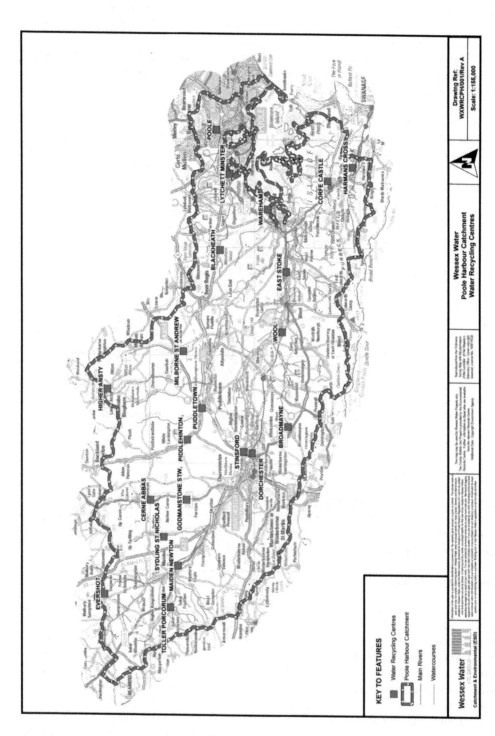

Figure 16.1 Wessex Water water recycling centres (WRCs) in the catchment.

Table 16.1 A summary of relevant environmental legislation

Legislation	Date	Intervention
EU Bathing Water Directive	1976	This directive introduced designated bathing waters and bacteriological standards to protect public health and improve environmental quality. There were seven bathing waters designated in the Poole Harbour catchment, primarily around the harbour itself and Studland. The asset improvements included screening on inlet works and overflows, increased levels of stormwater storage and improvements to overflows.
EU Shellfish Waters Directive	1979	Directive to designate and safeguard shellfish growing areas from bacterial pollution. This included the designation of areas of Poole Harbour and Wareham Channel. Improvements were made to the sewerage system to reduce the number of spills into these waters and those with an influence. Typically, this has involved constructing storm storage at WRCs and increasing the hydraulic capacity of sewers.
EU Urban Waste Water Treatment Directive	1992	The initial focus was the 'proper collection, treatment and discharge of wastewater, and correct disposal or re-use of the resulting sludge'. This included a large programme of improving and permitting sewage overflows, providing sewerage to smaller villages and secondary (biological) treatment at water recycling centre. In the Poole Harbour catchment from 2000, this has included: • Phosphorus removal at Dorchester and Wool WRCs (to 2 mg/L total phosphorus), with flow transfers from Lulworth, Warmwell and Bradford Peverell • Upgraded secondary (BAFF) treatment at Poole WRC. Poole Harbour was designated a 'Sensitive Area' for nitrogen in 2002, requiring nitrogen removal to be installed at our largest Water Recycling Centre (WRC), Poole, in 2009.
EU Habitats and Birds Directives	1994	Poole Harbour was designated as a Special Protection Area for Birds, although already a Site of Special Scientific Interest. In 2010, the Environment Agency confirmed that nitrogen removal at Poole WRC is sufficient to satisfy the water quality requirements associated with this directive. However, since 2015, 40 tonnes/yr of nitrogen has been offset via catchment management, with respect to forecast growth at Dorchester WRC.
EU Water Framework Directive	2000	This outcome-focused directive captures both waste and supply activities, identifying improvements to ensure that all waterbodies (rivers, groundwaters, estuaries and seas) achieve good status by 2027. Further phosphorus removal has been constructed at Dorchester and Wool WRCs (to 1 mg/L) and a number of investigations undertaken to inform our previous (2015–20) and current (2020–5) Business Plans. The WFD introduced Drinking Water Protection and Safeguard Areas to limit, and ideally reverse, rising trends impacting groundwater quality abstracted for water supply. This has focused on both nitrate and pesticide levels in the groundwater sources. This has driven an extensive programme of catchment management, working with farmers to reduce leaching of agri-chemicals impacting raw water supplies.
EU Revised Bathing Water Directive	2006	An update to the 1976 directive, which tightened the previous bacteriological standards, introduced a new classification system with a minimum 'Satisfactory' requirement and enhanced the monitoring programme. This resulted in further sewerage improvements to reduce the frequency of spills from overflows into bathing waters and rivers flowing to them. Holton Heath WRCs was converted to a pumping station with flows treated at Wareham WRC, where an upgraded secondary treatment and UV plant was installed.

Table 16.2 Water Recycling Centre (WRC) nitrogen removal rates

Process type	Average nitrogen eemoval	Catchment data source
Biological filter works	23.5%	Data from 14 WRC
Activated sludge plant	56.5%	Data from 6 WRC
Poole WRC	85%	Data from Poole WRC

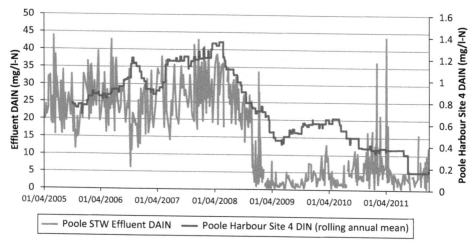

Figure 16.2 Changes in Poole STW and Poole Harbour Site 4 DAIN (Dissolved Available Inorganic Nitrogen) concentrations. Note: Site 4 is the sampling point at the mouth of Holes Bay.

Poole WRC is the only sewage works within the catchment to (currently) provide proactive nitrogen removal.[1] This nitrogen removal plant costing £12 million was commissioned in 2008 to ensure that the treated discharge from the works does not exceed 10 mg/L of total nitrogen (N) on an annual average basis. The treatment process involved the use of methanol, which has a relatively large carbon footprint and accounted for around 1.5% of Wessex Water's annual sewage and sludge treatment CO_2 emissions when constructed. Annual operating costs were around £0.9 million.

Due to the high running and carbon costs of this treatment process, alternative options were sought. In 2013, an additional treatment process was installed, which is the DEMON (or de-ammonification) process used more commonly in Europe to provide nitrogen reduction. This process uses anammox bacteria which are anaerobic and oxidise ammonium (hence 'anammox'), reducing the chemical dosing and level of aeration required. The DEMON process represents an estimated £140 k/yr saving in chemical dosing costs and a 50% reduction in energy use.

The N removal scheme at Poole took the N input to Holes Bay down to 55 tonnes/yr when the works came on stream in 2008. During the first 18 months, the site outperformed the 10 mg/L consent, typically achieving a 5 mg/L concentration (see Figure 16.2). This has now stabilised at a discharge of 114 tonnes/yr in 2015 (9 mg/L average discharge concentration) following commissioning and development in the catchment, resulting in increased flows and an increase in the level of sludge treatment provided by the site. The site currently removes 927 tonnes N/yr.

Nitrogen removal is expensive and chemically and energy intensive. Therefore, in addition we are delivering a catchment nitrogen offsetting scheme as a more sustainable alternative, working with farmers and landowners to deliver a further 40 tonnes/yr nitrogen reduction to offset some of the nitrogen load discharged from Dorchester WRC. This is delivered through our EnTrade nutrient trading platform. Following a pilot auction in 2016, further auctions have been held in 2017, 2018 and 2019. EnTrade has saved an estimated 275 tonnes of nitrogen from entering Poole Harbour by funding 65 farmers, who placed 702 bids on EnTrade and have received £500,000 funding to change land management practices.

---

[1] Wareham WRC will have nitrogen removal from 2022 to achieve a 15 mg/L total nitrogen permit.

We recognise that EnTrade is a way to engage with farmers and to deliver wider environmental, or natural capital, benefits than just nitrogen reduction. From 2020, we have been working with other partners to deliver nitrogen offsetting, biodiversity habitat creation, carbon offsetting through tree planting and natural flood management interventions within the Poole Harbour catchment. We see this as a blossoming environmental market to deliver natural capital outcomes for the catchment, beyond nitrogen, to satisfy current legislation and the future 25 Year Environment Plan, outlined by the government in 2018.

Nitrogen and water supply

Rising nitrate levels in groundwater is also of concern for public water supplies. Wessex Water relies on groundwater for public water supply in Dorset. Drinking water regulations place a limit on the level of nitrate, and some supply sources have shown a steady increase over time to a point where action has had to be taken to avoid breaches of the regulations. Nitrate removal plants are expensive to build, operate and maintain and are unsustainable. Accordingly, Wessex has looked for more sustainable means of tackling nitrates in groundwater by means of catchment management.

Under the European Water Framework Directive, many water supply catchments have been designated a drinking water safeguard zone by the Environment Agency. Within these zones certain substances, such as nitrate fertilizers, must be managed carefully to prevent the pollution of raw water in sources that are used to provide drinking water.

We work with the farmers inside the safeguard zone to:

- raise awareness of groundwater quality issues;
- share the results of water, soil, crop and manure testing that we have carried out for them;
- provide advice and information about ways to improve the efficiency of how their crops use key inputs such as fertilisers and pesticides;
- compensate them (where appropriate) for adopting alternative practices. An example of this is encouraging farmers to grow cover crops (also known as green manures) to lock up nutrients such as nitrogen and thus prevent them from leaching.

There are 13 public water supply sources in the Poole Harbour catchment, some of which exceed current Drinking Water Standards (DWS), and others have rising trends. Table 16.3 indicates the sources within the catchment where ongoing monitoring is undertaken to establish he status of the source.

Table 16.3 Nitrate trends in drinking water sources

Source	Current Status
Winterbourne Abbas	DWS exceedance
Belhuish	Rising
Briantspuddle	Rising
Alton Pancras	Rising – not current threat
Milbourne St Andrew	High – falling
Hooke	Stable – below DWS
Forston	Stable/falling
Dewlish	Falling
Eagle Lodge	Falling
Maiden Newton	Low
Cattistock	Low
Litton Cheney	Falling – Source Protection Zone inside catchment but actual source is outside
Langdon	Not used

Figure 16.3 Two key elements of the nitrogen removal plant at Poole WRC under construction: (A) the sequencing batch reactor which reduces ammonia levels, (B) the DEMON anaerobic sand filter.

Table 16.4 Work plan for the Poole Harbour catchment, 2020–5

(i)	**New phosphorus removal at WRCs**
	• Cerne Abbas WRC (removing 0.3 tonnes per annum by 2021
	• Corfe Castle WRC (0.48 tonnes per annum by 2021)
	• Piddlehinton WRC (0.13 tonnes per annum removed by 2025)
	• Dorchester WRC (tighten existing permit to remove 18.54 tonnes by 2025)
(ii)	*Nitrogen removal at WRCs and catchment offsetting, target for removal of 100 tonnes of nitrogen/yr through:*
	• new nitrogen removal process for Wareham WRC (to achieve 15 mg/L total nitrogen), removing approx. 9 tonnes of nitrogen/yr by 2021.
	• continuing nitrogen offsetting from Dorchester WRC – 40 tonnes per annum reduction to 2025.
	• additional nitrogen offsetting in line with our Performance Commitment – 51 tonnes/yr to 2025.
(iii)	**Installation of Ultraviolet (UV) treatment to reduce bacteria to improve shellfish waters**
	• Corfe Castle WRC (by 2021)
(iv)	**Environmental Investigations**
	• Dewlish boreholes – implement abstraction licence change to reduce abstraction in the Devils Brook.
	• Dorchester WRC Seasonal Permitting – investigate an innovative seasonal phosphorus permitting approach at Dorchester WRC.
	• Dorset Frome SSSI Water Quality – assess the contribution of our WRCs, CSOs, water resources and catchment management have on water quality of the river Frome SSSI.
	• Poole Harbour catchment WRCs – understand the N and P contributions from WRCs.
	• Poole Harbour Shellfish Waters – assess our discharges and their impact on shellfish waters and climate change impacts.
	• Poole WRC Options Appraisal – assess improvements for discharge quality or outfall relocation.
	• Holes Bay – Investigation of N & P loads from Wessex Water discharges to Holes Bay, Poole.
(v)	**Biodiversity**
	• Briantspuddle and Litton Cheney DrWPA – biodiversity opportunity investigation and catchment nitrate reduction measures
	• Poole Harbour catchment Biodiversity Project – 72 ha of habitat to be improved to enhance biodiversity and reduce nutrients to rivers.

Nitrates are so high at the Langdon public water supply source that it is no longer used. The Milborne St Andrew source is not used in winter when the nitrate level in the water is at its highest.

We work with the Drinking Water Inspectorate (DWI) to ensure that all water supply sources comply with the drinking water standard. Many of these have been identified as Safeguard Zones, under the Water Framework Directive, where measures to reverse rising nitrate trends need to be implemented. At these water supply sites within the catchment the DWI has supported Wessex Water implementing the catchment management approach instead of installing treatment. Figure 16.3 illustrates the level of success which catchment management has delivered in one catchment.

Future investment

In December 2019 Wessex Water received Ofwat's Final Determination of Our Business Plan for 2020–25. This means that we have a clear idea of our investment plan in Poole Harbour over this period. This includes investment to meet regulatory outputs, satisfying requirements identified under the Urban Waste Water Treatment, Water Framework and Bathing Water Directives primarily.

We have also articulated different ways of working to address these challenges, which include using environmental markets to work collaboratively to deliver resilient and sustainable outcomes. We will be continuing to deliver nitrogen reductions through EnTrade but are an eager partner to the evolving Poole Harbour Nitrogen Management Scheme. In addition, we are working to improve biodiversity through our catchment management work but also linking up with other partners to deliver Nature Recovery Networks. Table 16.4 highlights Wessex Water's main areas of planned work within the Poole Harbour catchment.

Appendix 16.1: Wessex Water Investment in Poole Harbour Catchment, 2005–20

Scheme name	Sites/Description	Outcome	WW cost or contribution
AMP 4 (2005–10)			
Sewage Treatment Works Improvement – Phosphorus Removal	Bradford Peverell STW (abandon works and transferred to Dorchester STW) Wool STW – 1 mg/L Dorchester STW (stricter consent) – 1 mg/L	Centralisation of continuous discharges to Wool and Dorchester with 1 mg/L total phosphorus discharges at both sites	£11 m
Sewage Treatment Works Improvement – Nitrogen Removal	Nitrogen removal to achieve a total discharge permit of 10 mg/L at Poole STW	Reduction in total nitrogen discharged of 250 tonnes N/yr (total removal approx. 900 tonnes N/yr)	£12 m
STW Improvement	Evershot STW (Flows) Wool STW (transfer of flows from West Lulworth STW) (Bathing Water) Warmwell STW (abandon works and transfer to Dorchester STW) (Growth)	Improved treatment and hydraulic capacity Removal of continuous discharge from West Lulworth Tighter consent and P removal provided at Wool STW Removal of continuous discharge at Warmwell Tighter consent and P removal provided at Dorchester STW	£1.6 m (inc above) (inc above)
Water Treatment Works Nitrate Removal	Eagle Lodge WTW Empool WTW Winterbourne Abbas WTW	Installation of nitrate removal plant	£2.5 m
Pesticide Removal	Ulwell Water Treatment Works	Removal of a disused sheep dip impacting water quality	£4 k
Flow/Abstraction Investigations	Tadnoll Brook Warmwell Heath	Understanding impacts on river flows in sensitive locations	
AMP5 (2010-15)			
SSSI Investigation – Effect of Sewage Treatment Works Discharges of Phosphorus Levels	Bere Stream SSSI (phosphorus and flow) Poole Harbour SSSI River Frome SSSI	Understanding of Wessex Water nutrient contributions within the catchment and recommendations for AMP 6 action	

(Continued)

Scheme name	Sites/Description	Outcome	WW cost or contribution
Intermittent Event Duration Monitoring	Nine sites across the catchment	Installation of monitoring on overflows discharging to high amenity watercourses, e.g. bathing waters	
Sewage Treatment Works Improvements – Storm Storage	Holton Heath STW	Sewage treatment works converted to a pumping station with increased storm capacity. Flows pumped to Wareham STW for UV disinfection prior to discharge	(see below)
Sewage Treatment Works Improvements	Holton Heath STW UV (abandon and transfer flows to Wareham STW for UV) Wareham STW Puddletown STW	Improvements at Wareham STW to provide additional treatment to cater for flows from Holton Heath and Cold Harbour. Improved UV disinfection	£4.96 m £0.8 m
Inflow Reduction	Cerne Abbas STW	To reduce the ingress of groundwater into foul sewers to minimise the risks of sewage flooding during wet weather or high-water table events	
Shellfish Water Investigation	Poole Harbour Shellfish Waters	Investigation to understand the impact of Corfe and Studland STWs on Poole Harbour (south) shellfishery	£50 k
Catchment Management – Source Protection for Nitrates	Winterbourne Abbas WTW Empool WTW Eagle Lodge WTW Hooke WTW	Provision of agronomic advice, sampling and monitoring to farmers within these Safeguard Zones to reduce nitrogen leaching	
AMP6 (2015–20)			
Sewage Treatment Works Improvement – Phosphorus Removal	Phosphorus removal at Maiden Newton STW to 1 mg/L	Reduction in phosphorus loading	approx. £0.7 m
Catchment Management – Source Protection for Nitrates	Trialling a nitrogen offsetting approach with farmers	Reduction in nitrogen leaching/run-off to Poole Harbour by 40 tonnes/yr	approx. £1.9 m
Chemical Investigation Programme 2	Chemical investigations at Dorchester STW to include upstream, downstream and final effluent sampling	Understanding the levels of specific chemicals in the discharge, watercourse and removal rates	£10 m (regional programme)
Bathing Water Investigation	Poole Harbour (Rockley Sands) impacts of sewerage assets on bathing water quality	To investigate the sewerage system and frequency of spills, which may impact bathing water quality	£200 k
Catchment Management – Source Protection for Nitrates	Continuation of catchment management work at five existing catchments and at new sites including Alton Pancras, Belhuish and Forston	Reduction in nitrogen leaching from land management practices	
Supporting work			
Poole Harbour Catchment Initiative	A collaborative catchment initiative, working in partnership, to deliver water quality and environmental improvements	Production of the Catchment Plan and Action Plan, co-ordination of supporting groups covering agriculture, monitoring and delivery	£80/yr

(Continued)

Scheme name	Sites/Description	Outcome	WW cost or contribution
Stour Catchment Initiative	A collaborative catchment initiative, working in partnership, to deliver water quality and environmental improvement. Co-hosted by Dorset Wildlife Trust and Wessex Water	Commenced autumn 2013, the Steering Group has prepared a Catchment Plan which will be web-based and launched summer 2015	£50k/yr
Strategy for Reducing Nitrogen in Poole Harbour by 2035	WW are supporting the development of both Implementation Plans – Diffuse Pollution and Nitrogen Neutral Development	Supporting Environment Agency and Natural England	
Poole Harbour Aquatic Management Plan	WW sit on the Steering Group for this Management Plan – Poole STW, etc.		
Dorset Wild Rivers	WW part-funding and supporting project through Partners Programme (£20,000 funding from 2010 to 2015) and sit on Steering Group. Funding of £20,000 confirmed for 2015–20	Since 2010 successfully delivered over 14 km of chalk stream restoration, 29.5 ha of wet woodland planting, 13 new scrapes or ponds, 4,000 volunteer hours pulling Himalayan balsam and 58 farm visits (resulting in 32 ha of improved farm habitat through HLS)	£100k (over five years)
University of Bath Water Research Programme	Sustainable wastewater treatment technologies, including algal and reed-bed research. Two related projects will aim to provide a robust design using reedbeds for phosphorus removal, and to trial the use of algae for nitrogen and phosphorus removal from sewage effluent	A pilot high rate algal pond for phosphorus and nitrogen removal will be constructed summer 2015	£1 m
University of Southampton PhD	Sediment coring throughout the harbour and tributaries to date and analyse the palaeo-ecology to understand nutrient fluctuations over time		£40 k

CHAPTER 17

Managing Poole Harbour Water Quality through a Catchment-based Approach

FIONA BOWLES

Abstract

In 2009, Natural England determined that the Poole Harbour Special Protection Area was in unfavourable condition due to eutrophication. Various consequences have been identified including loss of seagrass, change in the functioning of saltmarsh, phytoplankton blooms and dense growth of opportunistic macroalgae. This chapter reviews the history of eutrophication in Poole Harbour and how it has been tackled through the management of its terrestrial and freshwater catchment in a partnered approach.

Keywords: European marine site, eutrophication, macroalgae, nitrogen neutral, Poole Harbour, nitrogen, Natural Capital, restoration

Correspondence: fiona.bowles@btopenworld.com

Introduction

In 2009, Natural England (the government's adviser for the natural environment which protects England's protected sites) determined that the Poole Harbour Special Protection Area (SPA) was in unfavourable condition due to eutrophication. Various effects have been identified including loss of seagrass, change in the functioning of saltmarsh, phytoplankton blooms and dense growth of opportunistic macroalgae. The increasing incidence of algal mats leads to increased incidence and severity of low-oxygen conditions, reduced invertebrate biodiversity and effects on the availability of invertebrate food sources for estuarine birds (Gooday et al. 2018). For the review of consents under the Habitats Directive (European Council,1992), the source of this eutrophication was investigated by the Environment Agency (EA) in 2008 (Acornley et al. 2008), and by Wessex Water in 2010–13 (Cascade 2012). It was found to be due to increased nutrient loads from:

- point source discharges, primarily treated sewage into both the catchment and the harbour directly;
- diffuse sources throughout the catchment;
- increased marine loads on incoming tides from the Solent.

The harbour's 830 km² catchment (Gooday et al. 2018) is drained primarily by two chalk rivers, the Dorset Frome and the Piddle, with a series of smaller local waterbodies draining to the harbour across the tertiary geology (clays, sands and gravel). About 75% of the harbour catchment is in agricultural holdings, mostly grassland for sheep, dairy and other cattle, and much of the rest is used for arable cropping. The eastern shores are more urban, with the growing population in Poole.

This chapter reviews the historic changes in nutrient levels and the role of partnership working in developing measures to restore the water quality of the Poole Harbour European marine site (EMS) through its catchment. Due to the varied forms of both nitrogen and phosphorus that are monitored or permitted throughout the catchment, those generic terms are used except where a specific standard is noted.

Water quality issues in Poole Harbour

In addition to the condition assessment of SSSI and European marine sites, the state of all waterbodies in the UK was established under the Water Framework Directive (European Council 2000) through analysis of water quality, channel morphology and ecological indicators of flow, eutrophication and chemistry. This directive includes transitional and coastal waters. The results published for both 2009 and 2015 demonstrated that the Poole Harbour transitional water was Moderate, while the freshwater rivers and tributaries were Moderate or Good, with the main stems of the rivers Frome, Piddle and Sherford being in Poor condition. The overall condition status was reduced further from 2009 to 2019 due to changes in the chemical criteria. Looking at the Ecological Class, 5 out of 20 waterbodies deemed Good in 2019. Figure 17.1 illustrates the most recent Ecological Class of the waterbodies of the Poole Harbour catchment.

The Dorset Frome has a long-term monitoring record, collected by the Freshwater Biological Association, Centre for Ecology and Hydrology and Wessex Water at East Stoke, which shows a clear increase in nitrogen and phosphorus from the 1960s (CEH/FBA 2011). The groundwater records go back to 1976 and demonstrate a rising trend in nitrogen levels in the Chalk aquifer from the 1980s to 2010. The water quality within the harbour is harder to monitor because of the effect of changing salinity on nitrogen detection, but changes in algal mats were observed from the early 1980s when the Wessex Water

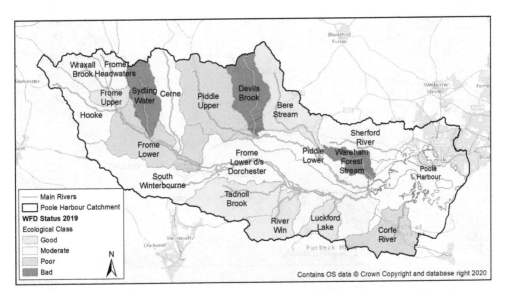

Figure 17.1 Poole Harbour catchment 2016 Water Framework Directive status and failing elements.

Authority contracted aerial surveys. More recently, the effect of nutrients on changes in saltmarsh has been reported (Deegan *et al.* 2012).

A paleoenvironmental study was therefore undertaken between 2013 and 2018 to assess the effects of long-term nutrient changes in the estuarine sediment record. This study (Crossley *et al.* 2020) demonstrated that temporal changes in water quality and ecology had occurred but varied across the four sites assessed (Figure 17.2).

Holes Bay, which is the prime area of impact from the Poole sewage treatment works, showed increased sedimentation rates between 1940 and the 1970s, with an increase in nutrients and algal activity. Post-1970s, the sedimentation rate slowed, but nutrients continued to rise with eutrophication after about the year 2000. This reflects the development and loss of heaths until the 1970s and the continued population growth and discharges from then on. The silt record here did not pick up the relatively recent reduction in nitrogen from the sewage treatment works from 2008.

The estuaries of the Frome and Piddle demonstrate similar changes in sedimentation accumulation rates after the Agriculture Act 1947, although the timings vary slightly and, as with Holes Bay, the rates do not return to pre-1900 levels. Earliest anthropological change is evident in the 1840s to the 1880s, and nutrient isotopes reflect a greater influence of farming rather than sewage impacts in these rivers. A water quality tipping point was reached in the 1970s here.

At Arne, a site less impacted by direct river or sewage inputs, the core showed a smaller increase in sediment accumulation rates and a return to 1900 rates. Nutrient change has also been less, as the productivity is more from marine algae here, with more terrestrial input from the 1920s.

This sediment history supports the conclusions from modelling of nitrogen sources carried out by Wessex Water and the EA from 2008 to 2013, which indicates that the freshwater contributions are now primarily from land run-off, with 14% from sewerage and 5% from aerial deposition (see Figure 17.3). Non-sewered housing is not a significant source of nitrogen (1%) (RSPB 2013). However, marine input is also significant, especially around the southern channel, but contributes less than 25% when compared to freshwater sources. This is affected by nitrogen increase from the Solent and its catchment rivers.

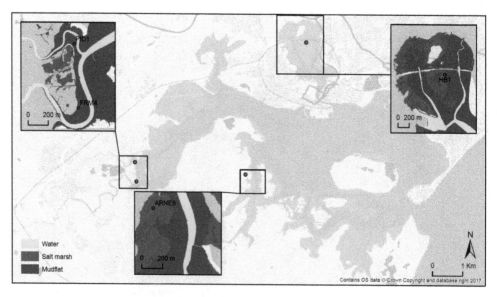

Figure 17.2 Paleoenvironmental study site map (Crossley 2019, In: Humphreys and Clark (eds) *Marine Protected Areas: Science, Policy and Management*, Amsterdam: Elsevier).

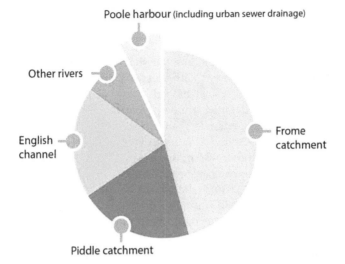

Figure 17.3 Estimated percentage contribution of sub-catchment areas to total annual inorganic nitrogen load on Poole Harbour 2006–2010, post nitrogen reduction at Poole sewage treatment works (redrawn from Kite *et al.* 2012).

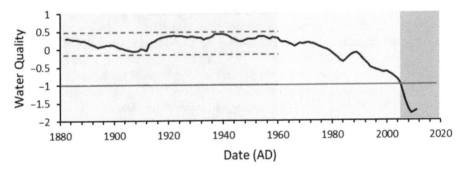

Figure 17.4 Diagram showing the envelope of good water quality for Poole Harbour (dashed red lines) and transition to critical system state (red line). The colours represent the safe (green), cautious (yellow) and dangerous (red states). Note that the Water Quality Index shown is an average of the four cores (L. Crossley, In: Humphreys and Clark (eds) *Marine Protected Areas: Science, Policy and Management*, Amsterdam: Elsevier.

While nitrogen is historically deemed to be the limiting nutrient for algal growth in marine habitats, the recent judicial review study determined that phosphorus was also a significant contributor to the excess algal mass (Kite and Nicolson 2018). A Water Quality Index calculated from the combined analysis of the paleoecology study indicates that the decline is continuing despite measures to reduce the nutrient inputs (Crossley 2018) and suggests that if the estuary is a step change system, then it is outside of its safe operating system as shown in Figure 17.4.

Catchment-wide issues

In addition to the need to reduce nutrients in the harbour, nitrogen levels in groundwater exceeded the safe limit for drinking water some of the time, requiring either mixing of sources or denitrification of raw water by Wessex Water. The non-favourable condition of Poole Harbour effectively stopped further development in the catchment without an action plan to resolve the issue, so the Local Authorities, with Wessex Water, the EA and Natural England, developed their 'Strategy for managing nitrogen in the Poole Harbour catchment to 2035', referred to as the 'Nitrogen Management Plan' (Bryan and Kite 2013), which both allocated target reductions on a fair shares principle and instigated the concept of 'nitrogen neutral' development with Wessex Water and developers offsetting the nitrogen arising from the increased sewage load.

The Lower Frome SSSI is deemed unfavourable for raised phosphorus levels, and excess phosphorus and algal parameters are a key reason for failure throughout the Poole Harbour waterbodies, despite recent improvements in sewage treatment. This is in part related to the higher than natural sediment losses both recorded in water quality sampling and demonstrated by the paleoenvironmental study. The latter and the REFORM study (Grabowski and Gurnell 2016) indicate that the Dorset Frome was able to adjust initially to increased sediment (soil) run-off in the 1940s, when agriculture increased for the war effort, through natural geomorphological processes, before reaching capacity in the 1950s.

The rivers of the Poole catchment are no longer natural. Our need of water for drinking, to irrigate our crops and livestock and to provide our fisheries for food and recreation has meant that rivers have been managed since the Bronze Age. A recent review of the history of the Dorset Stour has revealed how rivers were once a central feature in villages and flood waters were harnessed for fertilising land in water meadows. However, rivers were then gradually marginalised and channelised to make room for farming on their rich floodplains and for housing development. This in turn led to more fear of flooding and storm channels were developed by Dorset and Avon River Authority to deliver floods to sea quickly during the 1950s and 1960s, carrying a heavy load of silt and nutrients washed off an increasingly arable landscape. This general modification of river systems has led to physical modification being the highest reason for failure of English rivers (EA 2019). Loss of floodplain and natural wetlands is a key cause of this increased nutrient and sediment transfer to the estuary. Restoration of rivers and their flood plains in the north-west United States (Cluer and Thorne 2014) has shown water

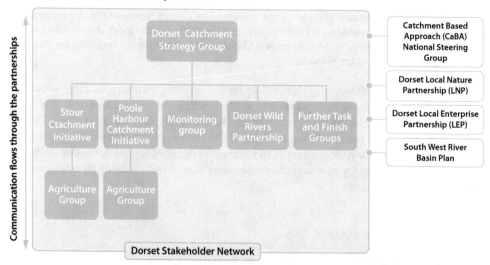

Figure 17.5 Dorset Management Catchment Governance Structure, 2018 (Bowles 2020, In: Humphreys and Clark (eds) *Marine Protected Areas: Science, Policy and Management*, Amsterdam: Elsevier).

quality benefits downstream, as well as improved groundwater levels and summer flows and improved biodiversity and fish populations. The judicial review identified that 250–300 ha of wetlands would contribute to achieving the nitrogen and phosphorus targets for Poole Harbour EMS.

It was in acknowledgement of the challenges of delivering measures around diffuse pollution in the River Basin Management Plan 2009 that led DEFRA to develop the catchment-based approach, known as CaBA. The Poole Harbour Catchment Initiative (PHCI) developed from the pilot Frome & Piddle CI (2011–13) through which partner organisations work together to restore the groundwater, rivers and Poole Harbour for the benefit of all of those who live in, work in and enjoy the catchment (Wessex Water 2014). The partners represent statutory, industry and environmental organisations with an interest in, or an impact on, the catchment. Experience from 2011 to 2018 showed that engagement is effective at different geographical scales within the partnership and the PHCI evolved to form a Poole Harbour delivery group, supported by a Dorset Strategy Group and specialist sub-groups (see Figure 17.5).

The partnership identified five key issues for action (see Table 17.1).

Legislation on water and waterbodies in the UK currently splits responsibility between the Environment Agency (water quality, flood, modification and abstraction), the local councils (flood and channel planning permission) and water companies (trade waste and some surface water drainage). This, with short-term economic decision-making, has historically led to the issues being resolved in a way that is at worst conflicting and commonly at increased carbon cost. For example, increasing storm flow capacity of rivers and flood plain drainage has reduced summer flows. Sewage treatment improvements for tighter discharge consents for phosphorus and/or nitrogen are generally at a high energy cost. However, since 2005 wider consideration has been given to treating problems at source and a stakeholder workshop for the Frome and Piddle Catchment Plan 2013 identified that measures for nutrient reduction on the River Piddle that delivered multiple benefits were preferred. Since then the catchment action plan and the prioritisation of projects for funding have looked to achieving multi-benefits. Thus, the harbour issues are typically resolved within the river catchment to benefit freshwater habitats too.

Table 17.2 illustrates the measures considered and implemented in reducing eutrophication and river modification impacts. Initially, outcomes of measures were identified in terms of tonnes of nutrient release avoided, or removed, to estimate simple cost benefits.

The catchment partnership both influences the project choice for delivery via EA and Natural England programmes and has directly delivered project work through its partners, individually and jointly, to implement the catchment action plan. Delivery has been

Table 17.1 Table of issues identified for the Frome and Piddle Catchment Action Plan 2013 and Poole Harbour Catchment Plan 2014

Agreed key issues	Impacts
Nitrogen	Excess causes growth of algae (estuary) and pollutes raw groundwater supply
Phosphorus	Excess causes growth of algae, affecting plants and fish, primarily in rivers (later found to affect estuary as well)
Sediment	Excess when washed off land, smothering invertebrates and fish and can carry phosphorus too
River channel and habitat alterations	Historic weirs and barriers can limit fish migration and impound reaches. Changing the natural river profile by widening, deepening and straightening affects flooding and natural processes
Water quantity	Low flows can affect the aesthetics and ecology of the river and high flows increase the risk of flooding

Table 17.2 Examples of restoration measures used in the Poole Harbour catchment

Issue	Measure	Partner/s	Comment on outcomes
Nitrogen in raw groundwater supply	Denitrification in water treatment	WW	Achieves drinking water standards, high cost for carbon and £££
Nitrogen in raw groundwater supply	Catchment management	WW, CSF, FWAG	Reduces N losses, plus soil and P loss to surface waters £
Increased algal mats in Poole Harbour	Algal harvesting	EA	Reviewed but deemed so far to be non-significant in terms of N removal, temporary benefit for birds, possible negative impacts on invertebrates
Increased algal mats in Poole Harbour	Denitrification at Poole STW	WW	Reduction of N to Holes Bay. Increased carbon usage and £££
Phosphorus in Frome SSSI	Tertiary P removal	Wessex Water	Reduces P load, increased ££ and carbon cost
Phosphorus in Frome SSSI	Catchment management	CSF, FWAG-SW, WW	Reduces surface water P run-off directly and sediment
Flood risk to village	Storm channels and embankments	Environment Agency	Increases flow to sea in storms, reduces summer flows in other channels, requires weirs which impound, reduces wet floodplain
Hatches and weirs-barriers to fish	Fish Pass (salmon and eel)	FP & WDFA & EA	Allows salmonid/eel passage. Impoundment issues remain for WQ and other species movement
River restoration	River narrowing /flow deflectors	EA, WW, FP & WDFA	Improves habitat diversity and thus biodiversity and fish, extent of wetland benefit depends on degree of floodplain reconnected
Natural flood management	Log jams and flood plain reconnection	DC, FWAG-SW, EA	Slows the flow, reduces flood peak, improves summer flow in streams and heath habitat, biodiversity

Legend

WW: Wessex Water, FP & WDFA: Frome, Piddle & West Dorset Fisheries Assoc., EA: Environment Agency, FWAG-SW: Farming and Wildlife Advisory Group, South West, CSF: Catchment Sensitive Farming, DC: Dorset Council

through changes to permitted discharges and through working with landowners to reduce diffuse pollution. The significant changes in practice needed to reduce nutrient and sediment lost through farming have significant economic implications. Through the period 2011–19 the key progress was on trialling measures for nutrient retention and in engaging farmers in their uptake, both to meet their industry targets and to deliver nitrogen neutral development for the planning authorities.

Table 17.3 illustrates the timeline for the evolution of engagement and economic processes for water quality in Poole Harbour catchment. Recent funding for piloting natural flood management allows measures to be trialed in the River Piddle, and in the Frome headwaters at Hooke, through Dorset Council, FWAG SW, Dorset Wildlife Trust and the Environment Agency working with landowners. These options, which are supported in principle by the DEFRA 25-year plan (DEFRA 2018), bring a range of alternative funding sources to water quality mitigation.

Until recently, a significant proportion of the cost of water quality improvement has been borne by the water industry to achieve tighter treatment levels for sewage or by DEFRA-funded projects. In Poole Harbour, nitrogen reduction has been driven by legislation and funded by government or by Wessex Water. The progress listed in Table 17.3 illustrates how the more recent opportunities have developed to fund restoration of the European marine site water quality.

Table 17.3 The timeline for the evolution of engagement and economic processes for water quality in Poole Harbour catchment

Year	Initiative
2005	Catchment Sensitive Farming (CSF) priority area established
2005	Wessex Water catchment work commenced on nitrogen in drinking water sources. WAGriCo project engaged farmers in PH on nitrogen issues and experience in Germany
2008	Environment Agency undertook review of consents for PH European marine site
2009	Poole Harbour European marine site determined to be in unfavourable condition, due to Nitrogen
2010–13	Wessex Water investigation of Poole Harbour and the influence of sewage inputs
2010	PH sewage treatment works discharge consent tightened to 10 mg/L total nitrogen, treatment change 2009
2011	Frome & Piddle Catchment Initiative pilot start
2012	PH Agriculture group formed by CSF
2013	Frome & Piddle Catchment Initiative extended to become Poole Harbour Catchment Initiative (PHCI)
2014	CSF Agriculture group remit extended to act for PHCI
2015	WWF and Angling trust challenge of MPA protection and start of PH judicial review of consent order
2015	WW start of the Dorchester Water Recycling Centre N offsetting project
2016	WW Inception of EnTrade, an online platform to distribute open and transparently funding to farmers for undertaking nutrient reduction measures
2017	Local Authority published supplementary planning guidance to achieve nitrogen neutrality
2017	DEFRA introduces farming rules for water
2018	NFU develop PH nutrient management scheme with PH agriculture group
2020	EnTrade trial natural capital auction for partnered measure delivery by landowners

Funding restoration

Looking just at reducing leaching rates to groundwater nitrogen, there have been four types of projects:

1. Legislative – implementation of a nitrate vulnerable zone (NVZ) that requires farmers to maintain records of nitrogen application and places limits on farm practices. Compliance with the new farming rules for water is required for farms in the NVZ or receiving government grant payments.
2. Advisory – free agri-advice is run by the Catchment Sensitive Farming team, which provides training and advice, through one-to-ones, group events and clinics, on best practice farming, approximately two-thirds of the catchment having been engaged with their activities up to 2017. Advice has been given to over 260 farms, which has encouraged the implementation of over 8,300 different mitigation methods or changes to farm management. Wessex Water and other non-governmental organisations including FWAG-SW operate similar schemes co-operatively and Wessex Water also offers practical support testing leaching rates at a field scale. Modelling of the catchment topography, hydrology and soil types has provided maps of high-risk areas to support joint prioritisation of interventions.
3. Wessex Water funds nitrogen reduction in three ways:
   - Direct funding of nitrogen removal from sewage treatment works and reduction in spills from consented sewage overflows. The removal of an additional 243 tonnes N p.a. at Poole sewage treatment works through denitrification had a capital cost of £12 million and an initial revenue cost of £0.9 million p.a. Following further works the total removal by 2015 was 927 tonnes p.a. at a reduced revenue cost (see chapter 26).
   - Funding for a team of agri-advisors to advise farmers in water supply safeguard zones to reduce nitrogen losses to groundwater. Trials in the WAGriCo project have demonstrated that this is more cost-effective than treating raw drinking water directly.

- Funding farmers to offset the output of sewage works (Wessex Water 2017) At Dorchester sewage treatment works the 40 tonnes p.a. has a budget of £8 per Kg of N reduced, which is equivalent to the revenue cost of nitrate removal treatment works, delivered through a reverse auction system. By 2017, the offsetting schemes had improved nutrient management on over 10,000 ha of land within the catchment across almost 40 farms, with a further 4,000 ha within the safeguard zones.

4. For nitrogen neutral development, the Poole Borough Council (now BCP) Supplementary Planning Document (Borough of Poole 2017) requires that a community infrastructure levy payment is paid for new development. It assumes that each person contributes 35 kg of nitrogen per year, of which Wessex Water is required by the Urban Waste Water Treatment Directive (European Council 1992) to remove 75%. The residual must be offset by changing the management of high input land to low input uses, for example 100 houses would require 6.2 ha of arable land to be reverted, in perpetuity (80–120 years). The community infrastructure levy is collected from developers by BCP and Dorset Council. Land purchase for wetland creation or potentially conservation covenants offer permanent offsetting of nitrogen discharge for new properties built within the catchment.

These projects illustrate three key drivers of change: legislative, voluntary and cost benefit to beneficiaries. There are three funding models: polluter pays (or their customers), beneficiary pays or taxpayer pays, with some additional support from charities.

The work to date has contributed to the understanding of what the cost and benefit to the farmer is in changing practice. The reverse auction platform helps to provide both buyers and sellers with a market price/value for the 'goods' that they wish to sell or buy. As the amount of N leaching is currently higher than the demand for offsetting, the price bid to undertake measures is low, whereas the amount of P offsetting available is low, compared to the demand, so the price bid to reduce phosphorus run-off is high. However, Wessex Water also found that as experience of the scheme benefits grew, the cost/kg of nitrogen reduced to less than the actual cost to the farmer (Wessex Water pers. comm.).

In 2018, engagement through the Poole Harbour agricultural group has resulted in the development of a farmer-led nutrient management scheme. This Poole Harbour Nitrogen Management Scheme, hosted by the National Farmers Union proposes to create one entity with the tools and support for reducing nitrogen, owned by its farmer members. This would deliver catchment governance and management and a nitrogen trading tool through a community interest company. This mechanism may be tested for the new Environmental Land Management Scheme that DEFRA is piloting to replace the European-funded grant schemes.

The fact that the 'Nitrogen Management Plan' (and management plans for other European Marine Protected Areas) relied on voluntary measures was challenged by various NGOs and resulted in a Judicial Review Order to review the management plan. The review report for that order, currently in draft, may recommend statutory measures to achieve the required outcomes. The technical report (Kite and Nicholson 2018) has confirmed the target nutrient reductions to meet favourable condition. Overall reductions in discharge are required for nitrogen (1,200 tonnes p.a.), phosphate and sediment, and the measures modelled range from the existing farming rules, through improvements to arable practice such as cover crops and fertilizer spreader calibration, to arable reversion and reduction of stocking levels (35%). These measures would reduce phosphorus and sediment losses, as well as nitrogen leaching, and reduce faecal coliforms, nitrous oxide and methane emissions from farming (Gooday et al. 2018).

The financial implications of this of scale of land use change are significant, with 250–300 ha of wetlands indicated in addition to meet the total reduction in nitrogen needed. There is a clear benefit in developing a scheme which funds a range of goods and services, through multiple buyers and sellers, to widen the funding opportunities.

Future evolution of a natural capital approach for resolving water quality problems in Poole Harbour catchment and European marine site

The experience to date has shown that the water quality problems can only be resolved through reducing the inputs (pollutants) and improving the resilience of waterbodies to allow natural processes to contribute to clean up. The project focus has been on the catchment since this has the greater impact, but the increasing marine input is influenced by raised nitrogen in the Solent, and this too needs to be resolved in the Test and Itchen catchments. While markets have been established for nitrogen, sellers should ideally be paid for the other goods provided, such as biodiversity, phosphorus and sediment reduction and flow modification. Wessex Water designed a natural capital auction for 2020 to test willingness to sell biodiversity (hectares of habitat) and natural flood management benefits as well as nitrogen. The data gained may influence the future Environmental Land Management Scheme and provide data for developing a catchment or county-wide environmental funding scheme, building on previous studies on natural capital in Dorset (Aecom 2015; Ash 2015). The SW River Basin Management Plan (2016) indicated that the value of Poole Harbour catchment at good status was £114.3 million and that the costs for implementing measures for cycle 2 (2015–21) would be £43.3 million, showing a 2.6× cost–benefit ratio (EA 2014). Catchment partnerships themselves have been shown to offer a benefit of 3× the DEFRA funds in delivering nationally (Caba 2020), largely through the partnering illustrated in this catchment.

Projects currently under consideration for the catchment action plan which can contribute to the EMS restoration include the statutory Biodiversity net gains and Nature recovery network schemes plus Stage Zero river restoration (flood plain reconnection), constructed wetlands, beaver re-introduction (and management) and other natural flood management measures. Given the apparent tipping point that Poole Harbour water quality has reached (Crossley 2018), significant cross-funding of such schemes will be needed to restore its ecology.

Conclusions

1. Partnership started within the catchment of Poole Harbour in advance of the trialing and implementation of the government's catchment-based approach. It has been a major contributory factor in understanding the problems that have affected the European marine site and its freshwater catchment. The data indicates that despite reductions in polluting activity, the water quality is in a 'critical state'.
2. The catchment-based approach brought discussion on the key problems affecting the entire catchment (eutrophication), which has since been widened further to consideration of channel morphology and flow issues. It also influenced the move from water quality monitoring and modelling to considering a wider weight of evidence, including palae-oenviromental data.
3. After the reduction in direct discharge of nitrogen into the harbour from Poole sewage treatment works, in 2008, the key reason for the site's unfavourable condition

arises from its freshwater catchment, so measures are more effectively targeted there. By prioritising measures in the headwaters, water quality benefits potentially improve a greater length of waterbody.

4. The European marine site condition has been a key driver for action for the whole catchment and has enabled, along with raw ground water quality, a significant development of schemes and projects to reduce nitrogen pollution. Unlike sewage treatment solutions, the land management measures for nitrogen generally have beneficial outcomes for sediment and phosphorus loss too, which are also key polluters of the freshwater habitats.

5. Wetland creation has been recommended as part of the 'Nitrogen Management Plan', and this would contribute to future actions for natural flood management and biodiversity gain within the catchment.

6. Working in partnership has enabled better data sharing and modelling, opportunity mapping and more effective use of time in communicating problems to the water and farming industries. It has also supported a range of offsetting and payment for ecosystem service schemes to be developed and trialed.

7. The scale of change needed to reverse eutrophication is significant both to the farming industry and to its customers, as well as in terms of landscape change. A series of funding mechanisms have been used to influence or fund restoration measures between 2005 and 2019. These could inform the development of wider environmental management funding scheme to meet the aspirations of the DEFRA 25-year plan (DEFRA 2018), the Dorset Local Nature Partnership and the catchment partners.

References

Acornley, R., Jonas, P., and Witt, S. 2008. Poole Harbour SPA: Habitats directive assessment. Environment Agency Report, p. 69.

AECOM. 2015. Developing ecosystem accounts for protected areas in England and Scotland; Dorset AONB report. Report for DEFRA. Available from: http://sciencesearch.defra.gov.uk/Default.aspx?Module=More&Location=None&ProjectID=19271 (accessed 5 April 2022)

Ash, S. 2015. Dorset's green economy: A report for Dorset County Council. Available from: https://jurassiccoast.org/wp-content/uploads/2016/02/Dorsets-Environmental-Economy-Final-Report-Dec-2015.pdf (accessed 5 April 2022)

Borough of Poole. 2017. Nitrogen reduction in Poole Harbour Supplementary Planning Document 2017. Available from: https://www.google.com/url?sa=t&rct=j&q=&esrc=s&source=web&cd=&ved=2ahUKEwiDnYv-9v32AhUHHcAKHUwVDO4QFnoECAQQAQ&url=https%3A%2F%2Fwww.poole.gov.uk%2FEasySiteWeb%2FGatewayLink.aspx%3FalId%3D42779&usg=AOvVaw1JVjBTUdPBkQMO4Mg-Sgl3 (accessed 5 April 2022)

Bowles, F. 2020. Nitrogen pollution in coastal marine protected areas: A river catchment partnership to plan and deliver targets in a UK estuarine Special Protection Area. In: Humphreys, J. and Clark, R.G. (eds) Marine Protected Areas: Science, Policy and Management. Elsevier, Amsterdam. ISBN 978-0-08-102698-4.

Bryan, G. and Kite, D. 2013. Strategy for managing nitrogen in the Poole Harbour catchment to 2035. Environment agency report. Available from: http://webarchive.nationalarchives.gov.uk/20140328084622/http://www.environment-agency.gov.uk/research/library/publications/148450.aspx (accessed 5 April 2022)

CaBA Benefits Working Group. 2020. Catchment-based approach (CaBA) monitoring and evaluation report. Available from: https://catchmentbasedapproach.org/wp-content/uploads/2020/02/CaBA-Benefits-Report-2018-19-FINAL.pdf (accessed 5 April 2022)

Cascade. 2012. AMP5 Wastewater nutrient investigations: Poole Harbour SSSI. Report for Wessex Water. DM#1388653v3 (accessed 5 April 2022)

CEH/FBA. 2011. CEH_phosphorus_data_River_Frome. © Database Right/Copyright NERC – Centre for Ecology & Hydrology, & FBA (Freshwater Biological Association). All rights reserved.

Cluer, B. and Thorne, C. 2014. A stream evolution model integrating habitat and ecosystem benefits. River Research and Applications 30. https://doi.org/10.1002/rra.2631

Council Directive 91/271/EEC on Urban Waste-Water Treatment. 1991. Available from: https://

eur-lex.europa.eu/legal-content/EN/TXT/?uri
=CELEX:31991L0271 (accessed 5 April 2022)

Council Directive 92/43/EEC Conservation of Natural Habitats and of Wild Flora and Fauna. 1992. Available from: https://eur-lex.europa.eu/legal-content/EN/TXT/?qid=1582572861456&uri=CELEX:31992L0043 (accessed 3 April 2022)

Council Directive 2000/60/EC Establishing a Framework for the Community Action in the Field of Water Policy. 2000. Directive 2000/60/EC of the European Parliament and of the council establishing a framework for the community action in the field of water policy.

Crossley, L.H. 2018. Palaeoenvironmental reconstruction of Poole Harbour water quality and the implications for estuary management. University of Southampton, Faculty of Social, Human and Mathematical Sciences, PhD thesis, 1–293. Available from: https://eprints.soton.ac.uk/429023/(accessed September 2018)

Crossley, L.H., Langdon, P., Sear, D., Dearing, J., and Croudace, I. 2020. Palaeoenvironmental determination of biogeochemistry and ecological response in an estuarine marine protected area. In: Humphreys, J. and Clark, R.G. (eds) *Marine Protected Areas: Science, Policy and Management*. Elsevier, Amsterdam. ISBN 978-0-08-102698-4. https://doi.org/10.1016/B978-0-08-102698-4.00034-4

Deegan, L.A., Johnson, D.S., Warren, R.S., Peterson, B.J., Fleeger, J.W., Fagherazzi, S., and Wollheim, W.M. 2012. Coastal eutrophication as a driver of salt marsh loss. *Nature* 490: 388–92. https://doi.org/10.1038/nature11533

DEFRA. 2017. *Farming Rules for Water-getting Full Vale from Fertilisers and Soil*. DEFRA, London. https://www.gov.uk/government/publications/farming-rules-for-water-in-england (accessed 5 April 2022)

DEFRA. 2018. *A Green Future: Our 25 Year Plan to Improve the Environment*. DEFRA, London. https://assets.publishing.service.gov.uk/government/uploads/system/uploads/attachment_data/file/693158/25-year-environment-plan.pdf

Environment Agency. 2018. Poole Harbour consent order technical document, recommendations to deliver favorable status across Poole Harbour catchment (draft for consultation). Report for DEFRA.

Environment Agency. 2021. River basin management plan physical modifications challenge. 2019. Available from: https://prdldnrbm-data-sharing.s3.eu-west-2.amazonaws.com/Challenge+narratives/Physical+modification+Challenge+RBMP+2021.pdf (accessed 6 April 2022).

Environment Agency Consultation Plan: A summary of information about the water environment for the Dorset management catchment. 2014. Available from: https://circabc.europa.eu/webdav/CircaBC/env/wfd/Library/framework_directive/implementation_documents_1/2012-2014%20WFD%20public%20information%20and%20consultation%20documents/UK/UK08%20South%20West/Dorset.pdf (accessed 5 April 2022).

Gooday, R., Newell-Price, P., and Cao, Y. 2018. Poole Harbour scenario modelling, 2018. RSK ADAS report to Natural England.

Grabowski, R.C. and Gurnell, A.M. 2016. Diagnosing problems of fine sediment delivery and transfer in a lowland catchment. *Aquatic Sciences* 78: 95–106. https://doi.org/10.1007/s00027-015-0426-3

Kite, D.J., Bryan, G., and Jonas, P. 2012. Nitrogen in the Poole Harbour catchment technical report. In Bryan and Kite 2012. https://webarchive.nationalarchives.gov.uk/ukgwa/20140328111551mp_/http:/www.environment-agency.gov.uk/static/documents/Leisure/Strategy_for_Managing_Nitrogen_in_the_Poole_Harbour_Catchment_Final_06_06_13.pdf (accessed 7 April 2022).

Kite, D.J. and Nicolson, A. 2018. Background information for understanding the catchment situation on nitrogen nutrient enrichment in Poole Harbour Natura 2000 site. Report for WWF-UK et al and secretary of state (DEFRA and EA Consent order). MMO (2016) Evidence Supporting the Use of Environmental Remediation to Improve Water Quality in the south marine plan areas. A report produced for the Marine Management Organisation, p. 158. MMO Project No: 1105. ISBN: 978-1-909452-44-2.

RSPB. 2013. The feasibility of a nitrogen PES scheme in Poole Harbour catchment, RSPB 2013. Report to DEFRA. Available from: https://www.cbd.int/financial/pes/unitedkingdom-poole.pdf (accessed 3 April 2022).

Wessex Water. 2014. Poole Harbour catchment initiative; final catchment plan update May 2014. Available from: https://www.wessexwater.co.uk/pooleharbour/ (accessed 3 April 2022).

Wessex Water. 2017. Poole Harbour nitrogen offsetting project [Online]. Available from: https://www.wessexwater.co.uk/Nitrogen-offsetting-project (accessed 6 October 2017).

Wessex Water. 2019. Poole Harbour catchment partnership 'what palaeoecology tells us about the history of Poole Harbour'. Information note. Available from: https://www.wessexwater.co.uk/environment/catchment-partnerships/poole-harbour-catchment-partnership/meetings-and-documents (accessed 5 April 2022).

CHAPTER 18

Using Drone Surveys to Assess Opportunistic Green Algae in Poole Harbour

ANDREW HARRISON

Abstract

Estuarine environments can be difficult to survey using standard 'on the ground' sampling techniques because of potentially treacherous soft sediment substrate and dynamic tidal regimes. Until relatively recently, aerial surveys were restricted to manned aircraft due to the limiting size of camera technology. The rapid growth of drone technology and the ever-decreasing scale of camera sensors, however, mean that multiple different imaging sensor options can now be mounted on relatively small commercial drone platforms, making these technologies much more accessible for smaller, low-budget projects.

Drone aerial surveys were undertaken in four bays within Poole Harbour to assess the distribution, extent and seasonality of opportunistic green algae on intertidal mudflats. This chapter highlights the benefits of using drones to undertake aerial surveys within intertidal habitats and discusses current developments in imaging technology and ongoing research aimed at developing tools to enable quantification of algal biomass from high-resolution aerial imagery.

Keywords: drone, aerial survey, opportunistic green algae, intertidal, mudflats

Correspondence: andyharrison@bournemouth.ac.uk

Introduction

As a result of both historical and currently elevated levels of nitrate, Poole Harbour suffers from excessive growth of opportunistic green algae during the summer months (Figure 18.1). Aside from impacts on human amenity and recreation value, this has led to increased concerns in recent years with regard to the potential impact on the feeding behaviour of overwintering wildfowl, for which Poole Harbour is designated as a Site of Special Scientific Interest and Special Protection Area.

Monitoring the distribution and extent of opportunistic green algae on mudflats using traditional ground-based survey methods poses a number of challenges. Aside from the dangers of accessing and walking on soft sediments, many of these areas are highly sensitive from a bird disturbance and conservation perspective. Furthermore, the manpower

Figure 18.1 Holes Bay, Poole Harbour, with extensive opportunistic green algae coverage, August 2017.

and resource requirements to adequately cover the wide spatial extent necessary for a comprehensive survey are often prohibitive.

As technology advances, small unmanned aerial vehicles, or drones, are fast becoming a quick and cost-effective way to collect large amounts of high-resolution aerial mapping data that, until recently, was prohibitively expensive. With a suitably qualified and trained operator, drone surveys can provide a cost-effective and efficient option to collect valuable data beyond that which can be obtained from traditional ground-based surveys.

Funded by Natural England, Bournemouth University Global Environmental Solutions (BUG) undertook monthly drone aerial surveys of four bays within Poole Harbour: Arne Bay, Brands Bay, Holes Bay and Newton Bay. Surveys were conducted from August 2017 to February 2018 inclusive to monitor the growth and die-back of opportunistic green algae throughout the season. Standard RGB imagery was captured and processed for viewing and interrogation in Geographic Information System (GIS) software.

Undertaking drone aerial surveys

An aerial photographic survey was undertaken within each survey area using either a DJI Phantom 3 Professional quadcopter with 12-megapixel camera (August to December surveys) or a DJI Inspire 2 quadcopter with 20-megapixel camera (January and February surveys) (Figure 18.2).

In advance of undertaking a drone aerial mapping survey, the flight path for the area in question is pre-programmed into flight control software. The drone's Intelligent Flight Positioning system then uses GPS/GLONASS, combined with a built-in compass, to enable accurate automatic tracking of the pre-defined flight path during the survey.

Image capture parameters are also pre-defined using the flight control software to ensure sufficient coverage of the survey area. For the surveys in question, images were captured from an altitude of 50 mAGL (Above Ground Level), with a front and side overlap of

Figure 18.2 Dr Andy Harrison from Bournemouth University Global Environmental Solutions undertaking a drone aerial survey of opportunistic green algae in Holes Bay, Poole Harbour, using a DJI Phantom 3 Professional quadcopter drone.

70%. This provided high-resolution imagery with a ground resolution of 2.1 cm/pixel (DJI Phantom 3 Professional) or 1.3 cm/pixel (DJI Inspire 2). Using these pre-programmed settings, a total of between 450 and 700 images were captured during each survey, depending on the location and survey area, which ranged from approximately 11 ha to 23 ha.

One major advantage of drone surveys in these sensitive areas is that the 'remote-sensing' nature of the work means that there is minimal disturbance to sensitive wildlife features (e.g. compared to accessing the area on foot); in this case, there was a need to avoid disturbance of overwintering wildfowl feeding on intertidal invertebrates.

Previous trials undertaken to investigate the potential of drone flights to disturb overwintering birds found insignificant to no disturbance of birds when flying at a minimum altitude of 50 mAGL (unpublished trials on Brownsea Island Lagoon in conjunction with Dorset Wildlife Trust and Natural England). To ensure birds are not disturbed during the surveys, the drone should take off as far as possible from any bird activity and ascend vertically to a minimum altitude of 50 mAGL before flying over the survey area in question.

Image processing

Once a drone survey flight is complete, collected images must be downloaded and processed – in this case, to produce a single large aerial image of the survey area (an 'orthomosaic'). On completion of the Poole Harbour algal mat aerial surveys, post-processing of the data was conducted in Bournemouth University's dedicated GIS suite. Images were initially checked manually and any unsuitable images were removed (e.g. out of focus). For each survey, images were then combined using photogrammetry software (Agisoft Photoscan Professional) to produce a single large geo-referenced aerial orthomosaic.

Orthomosaics were exported as GeoTiff files for import into GIS software. This allows direct visual comparisons of spatial distribution and coverage between each location and over time within each location. Using this approach, large areas of intertidal mudflat (or any other feature of interest) can be surveyed within a relatively short time period, considerably more efficiently than undertaking a mapping exercise on the ground. The geo-referenced aerial orthomosaics produced allow for easy calculation of the distribution and extent of any features of interest (in this case, opportunistic green algae).

Example outputs

An example of an orthomosaic shown at full extent for Holes Bay in August 2017 is provided in Figure 18.3. This orthomosaic is a combination of approximately 700 images stitched together and covers an area of 23 ha. The total flight time to complete the aerial survey was approximately 40 minutes.

A partially zoomed image of Holes Bay from the same survey can be seen in Figure 18.4. The distribution and extent of opportunistic green algae coverage can be clearly seen in both images at two different scales. From these images, it is a simple process to accurately calculate algal coverage using measuring tools within the GIS software and track growth/die-back throughout the season.

Benefits of drone surveys

These surveys and associated GIS outputs highlight the efficacy of drone surveys for cost-effective collection of large amounts of data that could not otherwise have been collected with conventional 'on-the-ground' techniques. Of course, there will be occasions where ground surveys will still be required – such as 'ground-truthing' species identification in

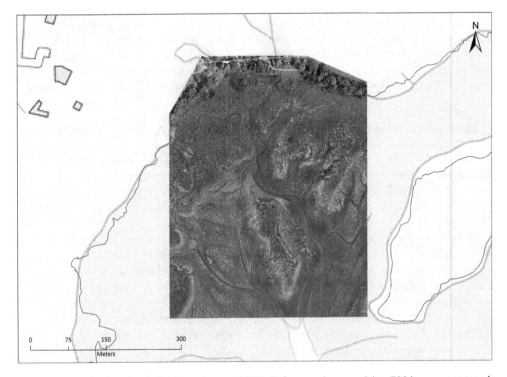

Figure 18.3 Holes Bay, Poole Harbour, August 2017. Orthomosaic comprising 700 images captured from a drone, shown at full extent.

Figure 18.4 Example of a partially zoomed orthomosaic image from Holes Bay, Poole Harbour, in August 2017.

aerial imagery. However, overall survey effort and time in the field can be considerably reduced by combining drone surveys with minimal targeted ground-truth field surveys, while collecting vastly more data than would otherwise have been possible using traditional ground-based field survey methods alone. In addition, the relatively low cost associated with drone surveys, particularly when set against the amount of useful data collected, means that multiple repeat surveys are possible, even for relatively low-value projects.

Future potential for drone surveys of opportunistic green algae

To date, drone surveys of opportunistic green algae distribution within Poole Harbour have been undertaken using standard RGB or 'True colour' imagery, which is comprised of red, green and blue bands, and attempts to replicate what is seen by the human eye. This is the format used for image capture by most everyday cameras, including those used with most drones.

RGB imagery provides an easily accessible visual overview and allows for manual interrogation of the data to calculate spatial distribution and density of the subject of interest within GIS software. However, RGB imagery is heavily influenced by environmental parameters, particularly light conditions. For example, low light conditions, shallow sun angles (e.g. during winter) and changing cloud cover (e.g. sunny intervals) will all influence the data captured by an RGB camera, resulting in differences in image quality and colour. This can confound attempts to assess temporal changes in the subject of interest (e.g. opportunistic green algal mats).

One option to overcome this is to capture imagery using near infrared (NIR) wavelengths in combination with RGB imagery. Reflection of NIR is very sensitive to the chlorophyll content in vegetation. By capturing NIR imagery, therefore, the presence/absence of vegetation and its relative health (less chlorophyll = stressed/dying vegetation) is readily distinguishable. However, similar to RGB imagery, environmental conditions (e.g. sun angle, cloud cover, atmospheric haze) during image capture can also alter NIR reflection.

To overcome the influence of environmental conditions and allow for tracking temporal changes in vegetation we can calculate numeric indices for the collected imagery, based on the ratio of NIR/RGB reflectance. The most commonly applied index (used extensively in the farming industry to assess crop health) is the Normalised Difference Vegetation Index (NDVI) (Weier and Herring 2000). Healthy vegetation (high chlorophyll content) will reflect a relatively higher ratio of NIR light to red light when compared to unhealthy vegetation (low chlorophyll concentration). This results in a higher NDVI value in healthier (greener) vegetation than unhealthy (less green) vegetation (Figure 18.5).

Importantly, because the NDVI is calculated as a ratio of reflected NIR/RGB wavelengths, it is unaffected by environmental conditions, allowing for reliable tracking of temporal trends. In addition, NDVI values are calculated at a single pixel resolution; therefore, the overall relative 'greenness' of an image can be quantified in numerical terms, thus providing empirical values associated with, for example, the relative amount of opportunistic green algae in an image. This removes the subjective element of manually interrogating an RGB image.

Figure 18.6 shows an RGB image from Holes Bay, with opportunistic green algae visible on the mudflats, while Figure 18.7 shows the same image but with NDVI values overlaid. Here we can see that the high chlorophyll areas, such as the shoreline trees, fields and opportunistic green algae, are represented by higher NDVI values than the low chlorophyll areas, such as the exposed mudflats, which have low (or negative) NDVI values. By selecting any area of the image, we can calculate an average (or total) NDVI value for all pixels within that area and, therefore, provide a numerical value for the 'greenness' of the area.

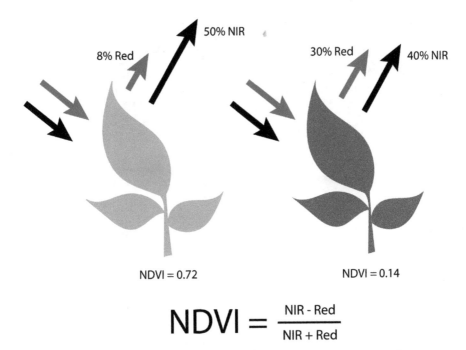

$$NDVI = \frac{NIR - Red}{NIR + Red}$$

Figure 18.5 Normalised Difference Vegetation Index (NDVI) calculation. **Note**: Healthy vegetation (high chlorophyll content) will reflect a relatively higher ratio of NIR light to red light when compared to unhealthy vegetation (low chlorophyll concentration). This results in a higher NDVI value in healthier (greener) vegetation than unhealthy (less green) vegetation.

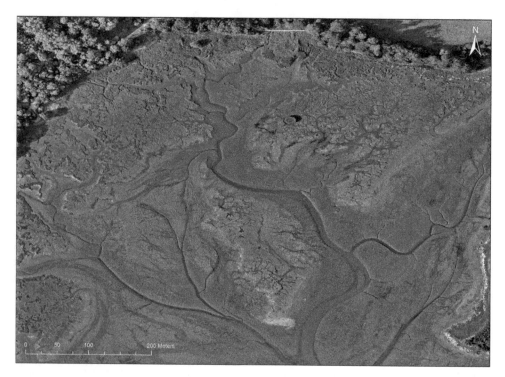

Figure 18.6 Partially zoomed RGB orthomosaic image of Holes Bay, Poole Harbour.

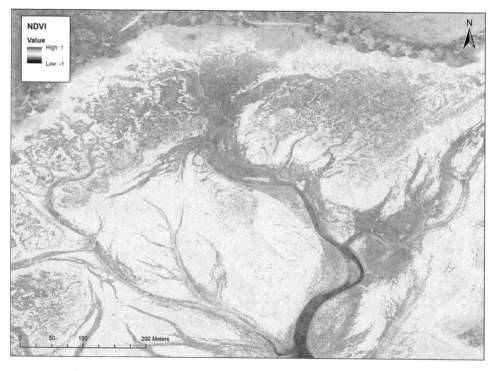

Figure 18.7 The same image as Figure 18.6, overlaid with Normalised Difference Vegetation Index (NDVI) values.

Until relatively recently, the size of multi-spectral sensors (including NIR) meant that aerial surveys using these instruments could only be undertaken from manned aircraft. This often meant that, particularly for small-scale projects, the costs of undertaking surveys were prohibitively expensive. However, the rapid growth of drone technology and the ever-decreasing scale of camera and imaging sensors mean that multiple different imaging sensor options can now be mounted on relatively small commercial drone platforms, making these technologies much more accessible for smaller, low-budget projects.

Using a combination of RGB and NIR imagery collected from drone aerial surveys, the distribution and extent of algal mats can be monitored both spatially and over time. Furthermore, in combination with new PhD research being undertaken on the impacts of opportunistic green algae in Christchurch Harbour, we hope to be able to ground-truth NDVI values obtained from drone aerial surveys with actual algal biomass by conducting concurrent ground surveys in defined areas. It is hoped that this may help to develop a technique whereby not only the extent and distribution but also the actual algal biomass could be assessed using drone aerial surveys alone.

Reference

Weier, J. and Herring, D. 2000. Measuring vegetation (NDVI & EVI). NASA Earth Observatory. Available from: http://earthobservatory.nasa .gov/Features/MeasuringVegetation/ (accessed 5 April 2022).

CHAPTER 19

Nuisance Macroalgae in Poole Harbour

SUZY WITT

Abstract

As a polluted eutrophic waterbody, Poole Harbour has suffered the proliferation of nuisance algae. The distribution and density of algae have been studied in the harbour since the 1980s. An overview of the studies undertaken is provided here, with details on the extent and biomass recorded. Trends in nutrient loading are considered, as well as initiatives to improve the ecological status of this estuary.

Keywords: macroalgae, Environment Agency (EA), nutrients, CASI, aerial imagery, DIN

Correspondence: suzy.witt@environment-agency.gov.uk

Introduction

The increase in the distribution of nuisance macroalgae is the result of the eutrophication of waterbodies, whereby excessive levels of nutrients (nitrogen and phosphorous) accumulate in the water column and sediment. Nitrogen and phosphate inputs include both point discharges, such as sewage treatment works (STW), and diffuse land run-off sources from within the catchment.

The impact of high nutrient concentrations to a sheltered estuary such as Poole Harbour, including the proliferation of macroalgal species such as Sea Lettuce *Ulva* spp., has been recognised for a number of years. These species do occur naturally, however, their profusion can produce a number of deleterious ecological effects. These include making the underlying sediments anoxic with the subsequent impact on benthic invertebrate species richness, abundance and biomass. Further deleterious effects include reducing the available feeding areas for birds and affecting feeding behaviour, deoxygenation of the water column and the smothering of saltmarsh and seagrass beds. The build-up of algal mats can also interfere with water-use activities, through the presence of rafts of floating detatched weed. Enjoyment of the coast can also be impacted by the aesthetic effects of the algae, such as decaying algae producing an odour nuisance or depositing on sites such as bathing waters.

Poole Harbour was designated a Sensitive Area (Eutrophic) and Polluted Water (Eutrophic) under the EC Urban Wastewater Treatment (UWWTD) and Nitrates Directives in 2002. Nutrient stripping was installed at Poole STW, which discharges to Holes Bay, in 2008. Urban Wastewater Treatment guidance suggests that excessive algal growth is considered to be areas of greater than 25% of the available intertidal habitat (AIH) in which the average macroalgal cover exceeds 25% (Environment Agency 2016). When considering

Suzy Witt, 'Nuisance Macroalgae in Poole Harbour' in: *Harbour Ecology*. Pelagic Publishing (2022). © Suzy Witt.
DOI: 10.53061/QVOS9745

the integrity of conservation sites under the Habitats Directive, Natural England guidance indicates that a problem area is likely to be one where cover by algae exceeds 15–25% of available habitat.

Poole Harbour consists of a large proportion of intertidal mudflats and marshes, which support high numbers of nationally and internationally important overwintering wild-fowl and waders. The site is designated a Marine Protected Area under the EC Marine Strategy Framework Directive, a Special Area of Conservation (SPA) under the EC Birds Directive, a Ramsar (wetland of international importance designated under the Ramsar Convention) and a Site of Special Scientific Interest designated under the Wildlife and Countryside Act 1981.

Algal monitoring in Poole Harbour

The Environment Agency (EA) undertakes assessment of opportunistic macroalgal blooms, including evaluating the extent and biomass of algae. This information is used to assess the impact of nutrient enrichment and chemical pollution and to classify waterbodies for the Water Framework Directive (WFD) according to their ecological status. Poole Harbour is currently at Moderate status under the WFD for macroalgae. The latest WFD classification for all waterbodies can be found on the data.gov.uk website (UK Government 2021).

The methodology for studying the distribution of macroalgae includes aerial photography and Compact Airborne Spectographic Imager (CASI), involving overflights of the area and the subsequent interpretation of the images and data created. Ground surveys are also important, to provide data to quality assure images and provide information on algal densities, biomass and any evidence of entrainment. The EA and its predecessors have carried out these methods in the harbour to produce an overview of trends in macroalgal cover over a number of years. In 1980, the Wessex Water Authority derived cover values of algal species for different areas of the harbour using aerial photography. The Channel Coastal Observatory has also carried out aerial imagery surveys in Poole. Table 19.1 illustrates all investigations carried out to try to capture the full extent of macroalgae in the harbour. The monitoring season is ideally between 1 June and 30 September and should aim to coincide with the peak of the algal bloom.

An example of the methods employed and the classification process utilised can be seen in a separate report for the 2005 CASI survey (Environment Agency Science Group 2006).

Affected area (AA) is a measure of the total area affected by the algal bloom and is measured in hectares (Ha). AA has been mapped by the EA using four different methods:

- Compact Airborne Spectographic Imager (CASI), as used in 1995, 1996, 1998 and 2005
- Digitized aerial photography, as used in 2008, 2014 and 2015
- Ground survey track, as used in 2008, 2009, 2011, 2015 and 2019
- Satellite imagery from Sentinel 2 in 2019

Macroalgal cover at the waterbody level is related to the available intertidal habitat (AIH). AIH occurs only between mean high water springs and mean low water springs and consists of mud, muddy sand, sandy mud, sand, stony mud and mussel beds. Unsuitable habitat for macroalgal growth (which is excluded from all surveys) includes permanent rock, cobbles, saltmarsh (excluding pioneer saltmarsh) and highly mobile sediments. The AIH for Poole Harbour is estimated to be 1,272 ha.

Ground surveys involve mapping the boundaries of algal patches using a hand-held GPS. As the intertidal zone in Poole Harbour consists of soft mud, a hovercraft is used to

Table 19.1 Investigations carried out since 1980 to try to capture the full extent of macroalgae in the harbour

Date	Comments	Source	Affected area (ha)	% AIH affected*	Density (% cover) measured?
1980	Aerial photography	Wessex Water	274	–	Yes: 41 ha 'thick algae'
1995	CASI	EA Geomatics	87	–	Yes: 87 ha > 20% cover
1996	CASI	EA Geomatics	282	–	Yes: 260 ha > 20% cover
1998	CASI	EA Geomatics	260	–	Yes: 195 ha > 20% cover
August 2005	CASI	EA Geomatics	242	–	Yes: 242 ha > 5% cover
August 2008	Aerial imagery +	CCO EA Ground	202	16	No
July 2008	Ground survey	survey	384	30	Yes
July 2009	Ground survey	EA Ground survey	420	33	Yes
August 2011	Ground survey	EA Ground survey	157	12	Yes
2013	Aerial imagery	CCO	Not available	–	No
July 2014	Aerial imagery	EA Geomatics	141	11	No
2015	Aerial imagery +	EA Geomatics		–	
	Ground survey**	EA Ground survey	109	9	Yes
2016	Aerial imagery	CCO	Not available	–	No
2019	Satellite imagery +	Sentinel + EA	Not available	–	
	Ground survey	Geomatics	345	27	Yes
		EA Ground survey			

AIH – Affected intertidal habitat, CASI – Compact Airborne Spectrographic Imager, EA – Environment Agency, ha – Hectares, CCO – Channel Coastal Observatory

*% AIH-affected figures are not derived prior to 2008 as this measure was not employed until the onset of WFD monitoring.

**Partial survey only in 2015.

help record the extent of the patches. Within each patch a number of 0.25 m² quadrats are placed. Quadrat locations follow a random or randomly stratified approach, depending on the variation in cover. In a patch of homogenous (uniform) cover, quadrats are placed randomly. Where percentage cover is more variable, quadrats are placed to ensure they are representative of the whole affected area – with areas of high and low density captured. Within each quadrat the percentage cover of algal species is recorded, as well as redox depth and the presence of any invertebrates or bare ground noted. Algae within the quadrat is then removed and taken back to the lab, where it is washed and the wet weight determined. This data is used to derive algal biomass per square metre. While removing algae, the presence of entrained algae can be investigated. Entrained algae is that which is growing within the sediment – usually at depths greater than 3 cm. An example of entrained algae is shown in Figure 19.1.

Entrainment occurs where algal density is high; it is not common when there is only low algal cover or density. Such algae can continue growing, even after it is covered by sediment. This can prevent the resurfacing of underlying invertebrates and restrict bird feeding activity. Entrained algae can also cause nutrient enrichment itself within the sediment through its decomposition (Wells *et al.* 2007). Algal spores can survive the winter period within the sediment, and this can mean that algal proliferation is triggered earlier in the growing season, increasing the amount of time deleterious effects can persist. Entrained algae has been found during EA ground surveys.

A further study carried out by the EA investigated macroalgal cover and biomass throughout the year, not just during the peak of growth in the summer. Sampling was carried out bimonthly and took place from April 2010 to October 2015. Sites were located in three areas – Holes Bay, Brands Bay and Arne Bay. These sites were chosen to cover the harbour as a whole, as well as for ease of access and the confirmed presence of significant summer algal mats. At each site five random quadrats were placed and the percentage

Figure 19.1 Left – quadrat showing entrainment, Right – survey quadrat illustrating algal density.

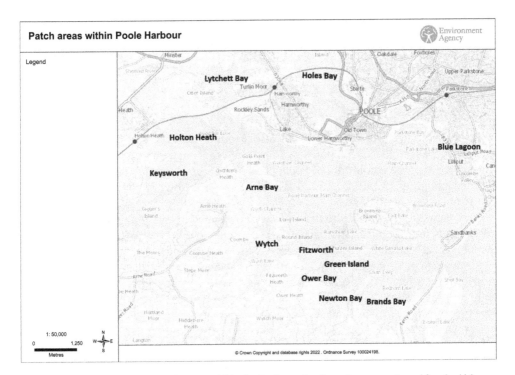

Figure 19.2 Map illustrating patch areas within the harbour that have been monitored for algal blooms (© Bluesky International Ltd/Getmapping PLC © Ordnance Survey and © Environment Agency).

cover of algae recorded, along with observations of redox depth and invertebrates present. Algae from within the quadrat was removed and taken back to the lab, where it was washed and the wet weight determined to derive algal biomass per square metre.

Results of algal mapping

Data collected during EA ground surveys is summarised in Figures 19.3–19.6 and Table 19.2. It must be noted that in 2015, only a partial survey was completed on a number of large algal patches and consequently only 33% (435 ha) of the available intertidal

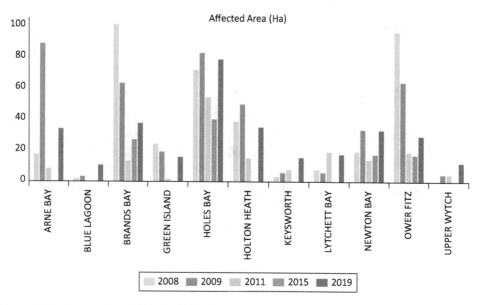

Figure 19.3 Affected Area (AA) for individual patch areas as determined by Environment Agency Water Framework Directive (WFD) ground surveys (hectares). *Note*: Surveys in 2011 and 2015 were partial surveys only.

habitat was surveyed. In 2011 there was also a reduced coverage of the AIH and fewer quadrats were sampled within patches. These two years must therefore be considered not representative of the whole AIH of Poole Harbour. The data for these two years has been included in Figures 19.3, 19.4, 19.5 and 19.6 and Table 19.2, however, to provide a complete picture of surveys undertaken.

Patch names are derived from the area they are located in and remain the same throughout the study period. The location of patch areas within the harbour is shown in Figure 19.2. The AA for each patch is recorded in hectares (ha) and these are detailed in Figure 19.3.

The average percentage cover within quadrats in each patch is illustrated in Figure 19.4, and Figure 19.5 details the average biomass (g per m²) of algae within each patch over the study period. Table 19.2 provides a summary of biomass and AA data for the five WFD algal ground surveys carried out by the EA. This is shown in graphical form in Figure 19.6.

The AA derived from each survey method from 1980 to the present day is shown in Figure 19.7 (excluding the figure using satellite imagery as this is awaiting confirmation). This graph gives a summary of algal cover changes throughout the last few decades. Each technique of study is different and as such caution must be used in direct comparisons between years. Even when comparing aerial imagery and CASI data there are issues in interpretation, as different resolutions were used and the amount of intertidal area exposed also fluctuated between years. Methods in aerial mapping have evolved over time; however, the best available methodology was used to collect data in each instance. The variations in intertidal area captured between years are largely due to weather conditions and the timings of surveys.

As well as taking account of the AA, it is important to consider the percentage of the AIH that is covered with algal mats. This figure has been calculated post-2008 and is part of the WFD classification process. Values for the percentage of AIH affected are included in Table 19.1.

The original CASI quantitative classification derived in 1998 is included as Figures 19.8 and 19.9. The images produced in the 2005 CASI survey are included as Figure 19.10

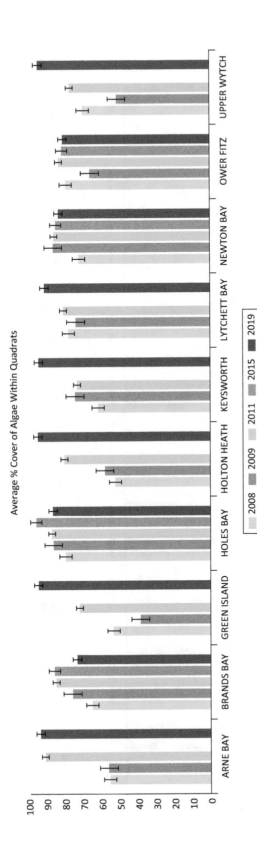

Figure 19.4 Average percentage cover of algae within quadrats (including Standard Error bars) over study period for individual patch areas as shown by Environment Agency Water Framework Directive (WFD) ground surveys. *Note:* The following averages were based on fewer than ten quadrats: 2008 – Green Island, Keysworth and Upper Wytch. 2009 – Keysworth, Newton Bay and Upper Wytch. 2011 – Arne Bay and Green Island. 2019 – Holton Heath and Keysworth. Surveys in 2011 and 2015 were partial surveys only.

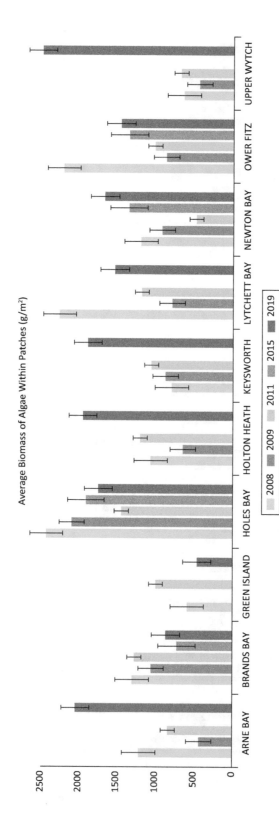

Figure 19.5 Average biomass (g/m2) (including Standard Error bars) over study period for individual patch areas as shown by Environment Agency Water Framework Directive (WFD) ground surveys. *Note:* The following averages were based on fewer than ten quadrats: 2008 – Green Island, Keysworth and Upper Wytch. 2009 – Keysworth, Newton Bay and Upper Wytch. 2011 – Arne Bay and Green Island. 2019 – Holton Heath and Keysworth. Surveys in 2011 and 2015 were partial surveys only.

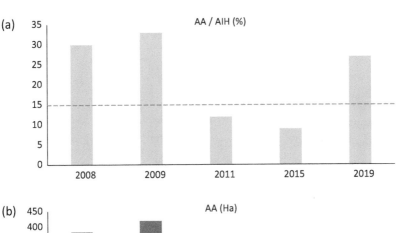

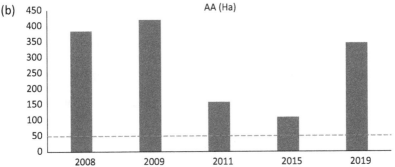

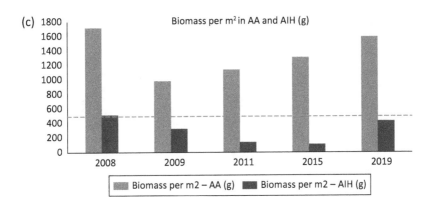

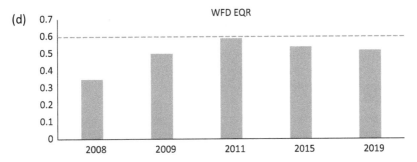

Figure 19.6 Summary data for Environment Agency ground surveys. *Note:* The Water Framework Directive (WFD) Good/Moderate boundary is included for each graph as a dashed line.

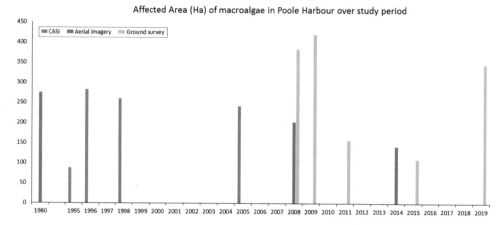

Figure 19.7 Affected Area (AA) in hectares of macroalgae in Poole Harbour as shown by different survey methods.

Table 19.2 Summary data for Environment Agency Water Framework Directive (WFD) ground surveys

Year	Average % cover in AIH	Biomass per m² – AIH (g)	Biomass per m² – AA (g)	AA (ha)	AA/AIH (%)	WFD EQR
2008	21.0	519	1,721	384	30	0.35
2009	22.8	325	985	420	33	0.50
2011	10.4	141	1,141	157	12	0.59
2015	7.3	112	1,315	109	9	0.54
2019	23.6	433	1,595	345	27	0.52

AIH – Available intertidal habitat, AA – Affected Area, EQR – Ecological Quality Ratio (which is derived by comparing observed results with that which is expected under reference conditions)

(Thematic Classification) and Figure 19.11 (Quantitative Classification). Maps derived from aerial imagery are also included for the 2008 and 2015 surveys as Figures 19.12 and 19.13. A map of algal patches seen in 2019, as collected using a ground GPS survey only, is included as Figure 19.14.

During the EA bimonthly algal mapping carried out from April 2010 to October 2015, it was evident that algae persisted throughout the year at Arne Bay, Holes Bay and Brands Bay. Figure 19.15 illustrates average biomass (g/m²) in each area over the study period. Holes Bay experienced the highest density of algae throughout the year and the highest percentage cover during winter months. The percentage cover of algae in Holes Bay has been consistently high over the study period, indicating the prevalence of the algae patch throughout the year – at every sampling occasion algal patches were found. There was also a persistence of algae beyond October in each year studied. The maximum average biomass found in Holes Bay was 3,671 g/m², which was observed in May 2015.

The persistence of macroalgae throughout the year was also evident in Brands Bay, although winter biomass figures were lower than for Holes Bay. Entrained algae was also found in Brands Bay on a number of occasions, ranging from 1 g to 201 g in a single quadrat. High values for percentage cover and biomass were observed here from June to October throughout the study. The highest average biomass observed in this bay was in August 2014 (3,704 g/m²).

The macroalgae cover throughout the year was less evident in Arne Bay than at the other two sites. The highest average biomass value in this bay was observed in November 2014,

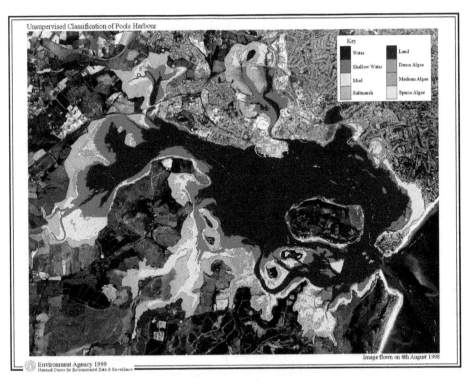

Figure 19.8 Classification output from 1998 Environment Agency Compact Airborne Spectrographic Imager (CASI) survey.

Figure 19.9 Classification output from 1998 Environment Agency Compact Airborne Spectrographic Imager (CASI) survey – Holes Bay and Wareham Channel only.

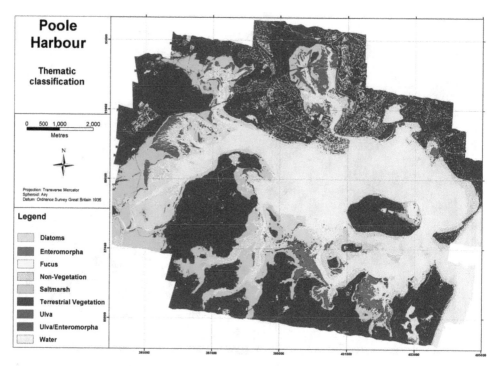

Figure 19.10 Thematic Classification output from 2005 Environment Agency Compact Airborne Spectrographic Imager (CASI) survey. Note: *Enteromorpha* is now known as *Ulva*.

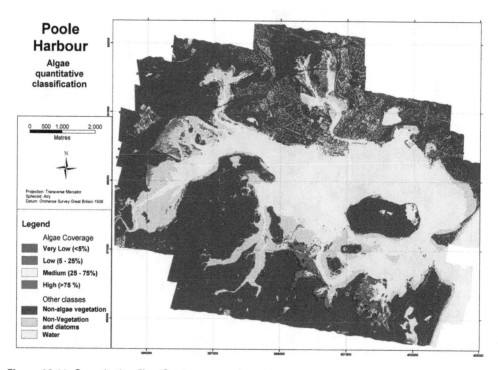

Figure 19.11 Quantitative Classification output from 2005 Environment Agency Compact Airborne Spectrographic Imager (CASI) survey.

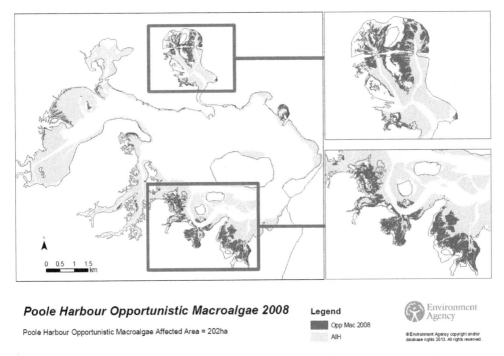

Poole Harbour Opportunistic Macroalgae 2008

Poole Harbour Opportunistic Macroalgae Affected Area = 202ha

Legend
- Opp Mac 2008
- AIH

Environment Agency

© Environment Agency copyright and/or database rights 2013. All rights reserved.

Figure 19.12 Distribution of opportunistic macroalgae in 2008 as shown by aerial imagery. Green areas are patches of algae within the intertidal zone. Available intertidal habitat (AIH) is shown in buff.

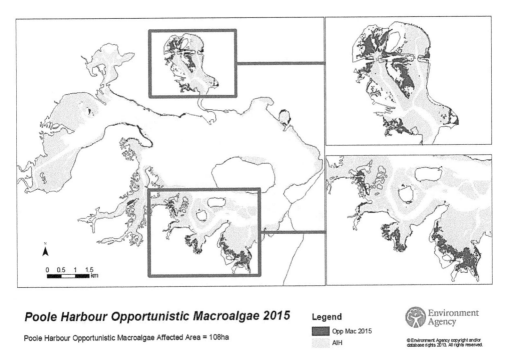

Poole Harbour Opportunistic Macroalgae 2015

Poole Harbour Opportunistic Macroalgae Affected Area = 108ha

Legend
- Opp Mac 2015
- AIH

Environment Agency

© Environment Agency copyright and/or database rights 2013. All rights reserved.

Figure 19.13 Distribution of opportunistic macroalgae in 2015 as shown by Environment Agency aerial imagery.

Figure 19.14 Distribution of opportunistic macroalgae in 2019 as shown by Environment Agency ground survey using a hovercraft and hand-held GPS unit.

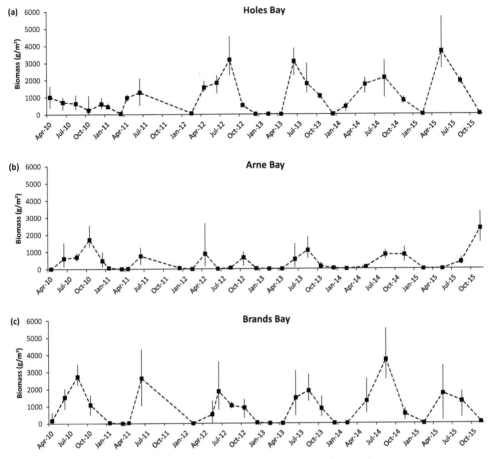

Figure 19.15 Average biomass (g/m²) of macroalgae observed during Environment Agency bimonthly survey at Holes Bay, Arne Bay and Brands Bay, 2010–15.

with 2,348 g/m² recorded, although the summer biomass value for this year was the second lowest for this site.

Water quality in the harbour

Water quality is measured by the EA at five sites within the harbour. These sites are shown in Figure 19.16. Figure 19.17 shows dissolved inorganic nitrogen (DIN) concentrations over time at these sites. The highest concentrations of DIN occur in the Wareham Channel to the east of the harbour, reflecting inputs from the River Frome and River Piddle, and the lowest concentrations are at the harbour entrance. There is a seasonal pattern at all sites, with the lowest concentrations occurring in summer and the highest in the winter (Acornley et al. 2006). Nitrogen and phosphate inputs include both point discharges, such as sewage treatment works (STW), and diffuse land run-off sources from within the catchment. Diffuse sources of nitrogen are considered the most significant in river catchments (Environment Agency 2016).

DIN levels in the Wareham Channel have increased steadily since 1995. However, since 2000 the rate of increase has slowed (Environment Agency 2016). Overall, there does not appear to have been a significant reduction in levels over time, which reflects surface

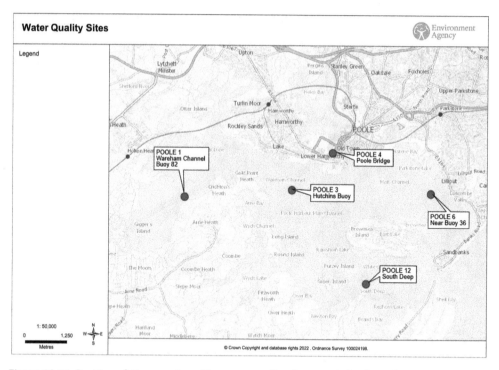

Figure 19.16 Position of sites monitored for water quality elements in Poole Harbour (including dissolved inorganic nitrogen, DIN) (© Bluesky International Ltd/Getmapping PLC © Ordnance Survey © Environment Agency).

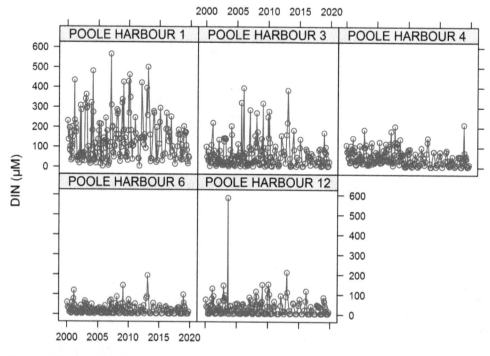

Figure 19.17 Dissolved inorganic nitrogen levels over time at five sites within the harbour. *Note*: All available Environment Agency monitoring data has been used.

water trends (Environment Agency 2016). The slow travel time through the Chalk aquifer in the catchment means that nitrate leaching remains high, despite regulation to reduce nitrogen losses from agriculture to water (Environment Agency 2016).

Nutrient levels in the vicnity of Poole STW have decreased since nutrient stripping was installed in 2008 (this is illustrated by the site Poole Harbour 4 in Figure 19.17, which is located at Poole Bridge).

Poole Harbour is currently failing WFD objectives for nutrients and is at Moderate status for this element. Nutrient levels measured over the winter period indicate that Poole Harbour is hyper-nutrified throughout.

Discussion

Increased macroalgal abundance, which is a result of eutrophication within the harbour, causes adverse ecological change to this important site. The adverse changes are listed in the introduction and include affecting bird feeding behaviour and the availability of food and the smothering of other habitats and causing aesthic nuisance.

From investigations throughout the study period it is apparent that the main areas of the harbour where algal mats occur tend to be the more sheltered zones, specifically Brands Bay, Newton Bay and Ower Bay as well as the western regions of the Wareham Channel and Holes Bay. Filamentous forms of Sea Lettuce *Ulva* (previously recored as Gut weed *Enteromorpha*) and other species such as *Cladophora* and *Chaetomorpha* are generally concentrated in the south of the harbour. Flat forms of Sea Lettuce tend to occur in the Wareham Channel and to the north of the harbour in Lytchett and Holes Bay.

When the Affected Area (AA) is considered, there has been variation throughout the study period. Figure 19.7 illustrates the AA in hectares and covers surveys from 1980 to 2019. Values are derived from three different methods of capture and as such caution must be used when comparing between years. In 2011 and 2015 a large decrease in AA was observed compared to other ground surveys. In 2015 only a partial survey was completed due to time constraints and only 33% (435 ha) of the AIH was surveyed. In 2011 there was also a reduced coverage of the AIH and fewer quadrats were sampled. The reduced coverage of these surveys explains the lower AA figures compared to other years of study. If AA values for the first ground survey in 2008 and the most recent in 2019 are compared, there is a reduction of 39 ha – from 384 ha in 2008 to 345 ha in 2019. If CASI and aerial imagery are taken in isolation there appears to be a greater reduction of AA across the years of study, from 241 ha in 1980 to 141 ha in 2014. Values captured by CASI and aerial imagery should ideally be combined with ground surveys to ensure the interpretation of imagery reflects that which is observed in the environment. Due to the small number of data points across the years of study and the differing methods of data capture, the statistical significance of these variations over the study period is difficult to determine.

When considering individual patches of algae within the harbour, as measured by ground surveys, there have also been variations in extent throughout the study period. Figure 19.3 illustrates AA for individual patch areas as determined by all EA ground surveys. In 2008 the highest values for AA were in Brands Bay and Ower Bay and Fitzworth. Holes Bay, Holton Heath, Arne Bay and Newton Bay saw maximum AA values in 2009. Holes Bay has shown consistently high algal coverage over time.

Figure 19.4 shows percentage cover values for each algal patch. With the exception of Green Island in 2009, average percent cover in quadrats has been above 50% in all areas and years of study. Holes Bay average percentage cover figures have always exceeded 75%. All patches except Brands Bay appear to have seen increases in percentage cover between 2008 and 2019. If the percentage cover within the AIH is considered, average cover has exceeded 20% in all years, except the reduced surveys of 2011 and 2015.

Biomass figures are more variable across the study period and within the different patch areas. Figure 19.5 illustrates average biomass per m² within the AA. Holes Bay has seen consistently high average biomass values (the majority greater than 1500 g/m²). Average biomass values throughout all algal patches remain high and in the majority of cases exceed 500 g per m².

Table 19.2 and Figure 19.6 summarise the data for EA ground surveys. In terms of AA, all values exceed the WFD Good/Moderate boundary of 50 ha – this is still the case in the partial surveys of 2011 and 2015. When AA as a percentage of AIH is considered, all full surveys exceed the WFD requirement for Good status of 15%. For biomass, the Good/Moderate boundary is 500 g per m². In terms of biomass per m² in the AA this value is substantially exceeded in every EA ground survey. This is even the case for the reduced surveys of 2011 and 2015. If biomass values are averaged across the AIH then only the 2008 survey results exceed 500 g per m². Essentially, where there are algal patches there is a high density of algae within them.

In terms of WFD EQR values, Poole Harbour was classed as Poor in 2008 and has been Moderate for all other ground surveys. The boundary between the Moderate and Good WFD classes is 0.6, and the EQR in 2011 was very close to this figure. The EQR values have increased over the study period, from a very low value of 0.35 in 2008, to 0.52 in the most recent survey.

Ground survey results for Poole Harbour indicate that the percentage cover of the AIH exceeds UWWTD guidelines (i.e. excessive growth is considered areas greater than 25% of AIH in which cover exceeds 25%) in 2008, 2009 and 2019. These figures are illustrated in Table 19.1 in the column '% AIH Affected'. Deleterious effects, such as those detailed in the introduction, are therefore likely.

There is evidence to show that algae also persists throughout the winter months within sheltered bays in the harbour, as seen in the bimonthly surveys carried out between 2010 and 2015. Both Holes Bay and Brands Bay exhibited algal biomass in winter months and Brands Bay also recorded entrained algae.

It is apparent that algal growth within the harbour is still extensive and remains above WFD and UWWTD guidelines. This is observable in the field as well as through the results given here, and is not particularly surprising given that there has not been a significant reduction in nutrient levels over the last few decades.

The variances in algal biomass between sites, even in the same month or year of study, indicates the complicated nature of algal growth patterns throughout the harbour and the problems inherent in data interpretation. Although temperature, turbidity, light and nutrient levels and bed stability influence the distribution of algae, the total surface area of the intertidal zone available for growth is an important limiting factor (Wells *et al.* 2007). This is why the calculation of AA and the AA as a percentage of the AIH are crucial in establishing the potential for deleterious effects within a waterbody.

A reduction in the extent of saltmarsh in the central part of Holes Bay has been observed between 1998 and 2005. The area previously taken up by saltmarsh is now largely occupied by algae. This change in habitat distribution is evident in Holes Bay if aerial images are examined. However, a similar reduction on a smaller scale cannot be ruled out in other areas of the harbour, highlighting a further deleterious effect of excessive macroalgal cover.

The proliferation of algal mats within the harbour has been observed and measured over the last few decades. Actions to try to reduce algal patches have included in 2001 Poole Harbour being designated both a Sensitive Area (Eutrophic) and Polluted Water (Eutrophic). This led, in 2002, to the designation of the area as a nitrate vulnerable zone (NVZ). This designation requires farmers to reduce nitrate pollution and follow mandatory rules. These are based on good agricultural practice and require the use of fertiliser and manure to be managed carefully (Environnment Agency 2016). The Environment Agency and Natural England devised a strategy to reduce nitrogen loading in the Poole Harbour catchment from both point and diffuse sources (Bryan and Kite 2013). This set out a range of options to address pollution from nitrogen and considered future urban developments, as well as the management of farming practices. To ensure that conservation objectives can be met, it was put forward in this strategy that the non-marine load of nitrogen coming into the harbour should be reduced to around 1,730 tonnes of nitrogen (N) per year (Bryan and Kite 2013).

The slow travel time of nitrate through the Chalk aquifer in this catchment, which can be 30–35 years on average (Bryan and Kite 2013), hinders the efforts to reduce the impact of algal blooms within the harbour. A mitigation measure that has been suggested is the removal of algal mats from the mudflats (Capuzzo and Forster 2014). This may provide a shorter term solution to the physical presence of algal blooms, and investigations into the impact and feasibility of removal and the subsequent use or disposal of algae are continuing. This method would not, however, remove a significant amount of nitrogen from within the harbour (Capuzzo and Forster 2014).

In terms of point source contributions, nutrient stripping was installed at Poole STW in 2008, which considerably reduced nutrient loading into Holes Bay. This had a smaller effect, however, on loadings in the harbour as a whole (Environment Agency 2016).

The monitoring of water quality and nutrient inputs within the harbour has been carried out by the EA for over 20 years. Algal mats and phytoplankton have also been studied following WFD legislation. This data has been used for modelling and to inform mitigation measures. Natural England and local universities have also collected data and carried out extensive studies within the harbour. The Catchment Sensitive Farming initiative is active within the Poole Harbour catchment and works with farmers and other partners to try to improve water and air quality in high priority areas such as this. The scheme is a partnership between Defra (Department for Environment, Food, and Rural Affairs), the EA and Natural England and offers farmers free training, advice and support for grant applications (UK Government 2020). Collaborative working through catchment initiatives and projects has advanced the study of the harbour and the pressures on it. A collaboration between Defra, the EA, the Esmée Fairbairn Foundation with the support of the Triodos Bank UK has been initiated to encourage sustainable private sector investment in the environment (Esmée Fairbairn Foundation 2020). The Poole Harbour Nutrient Management Scheme, which has been put together by the NFU and farmers and seeks to create a solution to the nutrient issue within the harbour, has been chosen to receive grant funding in a pilot set up by the collaboration. The scheme is developing a nutrient accounting tool, and it is hoped this will result in considerable reductions in nutrients entering the harbour.

It is clear that due to the valuable and sensitive environment of Poole Harbour and also its economic and social importance to the local community, further studies and investigations are needed. This is the case for algal blooms, but also for other species and habitats, to understand the ecosystem as a whole. The continuation of partnership working and involving local stakeholders and the community throughout the catchment will ensure that improvements can be made to this important environment.

Acknowledgement

Richard Acornley (Senior Environmental Monitoring Officer, EA)

References

Acornley, R., Jonas, P., and Witt, S. 2006. Poole Harbour SPA: Habitats directive assessment. Published by the Environment Agency.

Bryan, G. and Kite, D. 2013. Strategy for managing nitrogen in the Poole Harbour catchment to 2035. With contributions from Money, R., Jonas, P. and Barden, R. Published by the Environment Agency.

Capuzzo, E. and Forster, R.M. 2014. Removal and subsequent uses of opportunistic green macroalgae (*Ulva lactuca*) in Poole Harbour. Cefas contract report C6285. Issued 12 April 2014.

Environment Agency. 2016. DATASHEET: Nitrate vulnerable zone (NVZ) designation 2017 – Eutrophic waters (Estuaries and coastal waters). NVZ name: Poole Harbour NVZ ID: ET1.

Environment Agency Science Group. 2006. Poole Harbour CASI survey. Project PM_0442. Project Report.

Esmée Fairbairn Foundation. 2020. Available from: https://www.esmeefairbairn.org.uk (accessed 4 June 2020).

UK Government. 2020. Available from: www.gov.uk/guidance/catchment-sensitive-farming-reduce-agricultural-water-pollution (accessed 4 June 2020).

UK Government. 2021. Available from: https://data.gov.uk/dataset/41cb73a1-91b7-4a36-80f4-b4c6e102651a/wfd-classification-status-cycle-2 (accessed 12 February 2021).

Wells, E., Best, M., Scanlan, C., and Foden, J. 2007. Water Framework Directive development of classification tools for ecological assessment, opportunistic macroalgae blooming. Water Framework Directive – United Kingdom Technical Advisory Group (WFD-UKTAG).

CHAPTER 20

The Potential Economic Exploitation of Macroalgal Mats in Poole Harbour and Other Channel Estuarine Systems

SINEAD E. MORRIS, GORDON J. WATSON, SOPHIE RICHIER,
IAIN D. GREEN, ANNESIA L. LAMB and DANIEL J. FRANKLIN

Abstract

The nuisance growth of green macroalgal mats, mostly composed of the green seaweed *Ulva*, is a long-standing issue in Poole Harbour. Mat accumulations can have a variety of impacts, and there is increasing interest in the potential exploitation of mats as part of the environmental management of the harbour. We outline the knowledge gaps in the potential harvesting methods for green algal mats and explore ongoing research that is attempting to reduce these knowledge gaps. Once harvested, *Ulva* has many potential economic uses, proven and otherwise. We consider some of the applications of harvested *Ulva*, such as in the bioenergy and agriculture sectors. Sustainable harvesting and use of *Ulva* from Poole Harbour have the potential to promote economic growth while conserving the health of the local ecosystems. *Ulva* also has the potential to capture atmospheric carbon. We explore how these factors provide an opportunity for the advancement of Poole Harbour, its industries and its citizens within the framework of the 'blue economy'.

Keywords: macroalgae, economic use, management, carbon capture, blue economy, algal harvesting

Correspondence: smorris1@bournemouth.ac.uk

Introduction: Defining the problem and the nature of the ecological impacts in Poole Harbour and other similar ecosystems in the channel region

Macroalgal (seaweed) mats, typically composed of several types of green seaweed (e.g. *Ulva*, 'sea-lettuce', *Chaetomorpha* and others), are thought to be increasing as a consequence of increased nutrient concentrations in estuarine and coastal ecosystems (Smetacek and Zingone 2013). Large accumulations of macroalgae in coastal systems ('green tides') can have negative ecological, economic and amenity consequences (i.e. negative effects on ecosystem services; Watson *et al.* 2018). Among many ecological effects, macroalgal mats

Sinead E. Morris, Gordon J. Watson, Sophie Richier, Iain D. Green, Annesia L. Lamb and Daniel J. Franklin, 'The Potential Economic Exploitation of Macroalgal Mats in Poole Harbour and Other Channel Estuarine Systems' in: *Harbour Ecology*. Pelagic Publishing (2022). © Sinead E. Morris, Gordon J. Watson, Sophie Richier, Iain D. Green, Annesia L. Lamb and Daniel J. Franklin. DOI: 10.53061/HDDQ6575 can alter sediment oxygen dynamics, thereby changing the composition of the benthic fauna, inhibit the growth and persistence of saltmarsh or seagrass communities, and produce unsightly and malodorous conditions involving potentially harmful gases such as hydrogen sulphide (H_2S). Macroalgal mats, long identified as a problem in Poole Harbour (e.g. Langston et al. 2003) have increased (Thornton 2016), and their extent and abundance are currently monitored by the Environment Agency (EA). The EA has classified Poole Harbour as Poor/Moderate under the Water Framework Directive 'opportunistic macroalgae' methodology (e.g. Witt, this volume). For conservation and management purposes two potentially harmful thresholds for macroalgal mat biomass are recognised in Poole Harbour (and nationally within the UK), between 0.5 and 2 kgm² (wet weight), and these thresholds are regularly exceeded in Poole Harbour (Thornton et al. 2020). In nearby Brittany, where nuisance mat accumulations have become especially problematic over the last two decades (Ménesguen et al. 2006) the mats are regarded as harmful enough to warrant harvesting and disposal. Without removal, algal mats decompose and can emit toxic gases. In particular, emissions of hydrogen sulphide (H_2S) can cause health risks to those exposed. In June 2011, the French Agency for Food, Environmental and Occupational Health & Safety (ANSES) described the health risks associated with these emissions and issued recommendations for the protection of workers and the public. Thousands of tonnes of algal mat are collected each year along the French Channel coast. Up to 74,000 m³ of mat was collected in Brittany in 2010 (Perrot et al. 2014). Depending on shore sediment type, and local tidal conditions, mat collection is not always technically feasible. Tractor-based removal appears to be the best option on sandy beaches, whereas algal removal on mudflats with soft sediments is more difficult and expensive, requiring special techniques.

In Poole Harbour there are several economic consequences of the prolific growth of green mat-forming seaweed. The shellfish that are cultivated in the harbour may need pressure washing and boat hulls may require more frequent anti-fouling treatment in mat-rich systems. It is also possible that the potential yield of bait worms, which globally are among the most valuable of marine bioresources (Watson et al. 2017), from the harbour's intertidal sediments has changed over time though there is no evidence for this. Interactions between mats and worms are currently under investigation. Elsewhere in the Channel region, especially in Brittany, tensions have arisen against some farming industries, as green tides have been severely impacting tourism for the past 20 years. Action plans have been implemented by communities since the nineties and two major programmes (e.g. Plan de Lutte contre les Marées vertes; PLAV) specific to 'green tides' have been launched by the French government since 2010 to tackle at source Breton mat issues. Consistent efforts and progress have been made by all Breton players over the past 20 years and decreases in nitrate have been recorded, especially in surface waters.

Potential removal and/or harvesting of green macroalgal mats in Poole Harbour

Removal, or harvesting, of mats is being explored as an immediate short-term solution in Poole Harbour. In Jersey, washed-up seaweed is gathered from beaches using excavators, tractors and trailers (Jersey Royal Company 2015) with potential negative effects on tourism being the principal driver. A CEFAS (the Centre for Environment, Fisheries and Aquaculture Science) report noted that boat-based mechanical harvesting is the most viable option for seaweed removal in Poole Harbour (Capuzzo and Forster 2014). This conclusion was based on harvesting costs of ~£75/tonne with an initial target for removal of ~1,000 tonnes, over up to 40 days of harvesting. Although it seemed unlikely that harvested mat would have a marketable value, proposed beneficial uses include agricultural fertiliser and soil conditioner, fuel for anaerobic digestion and compost. For some of these uses some post-collection processing (and potentially gate, or processing, fees) would be required. A group of interested local organisations, the Poole Harbour Catchment Initiative (2016) has also assessed the potential for harvesting and concluded that pilot studies should be supported. Capuzzo and Forster (2014) explored this in detail and considered some of the advantages and disadvantages of using a boat to harvest sea-weed. Boat-based sea-lettuce harvesting, with the objective of avoiding hypoxia events in shellfish production areas, has been successfully trialled at Prince Edward Island, Canada (Crane and Ramsay 2011). A comparison of the advantages and disadvantages of boat and land-based harvesting from both reports is outlined in Table 20.1.

Poole Harbour experiences a somewhat unusual double high tide, which reduces the time available for land-based harvesting. Moreover, given the generally fine/soft substrate and large intertidal area in Poole Harbour, boat-based harvesting appears to be the most practical approach. Once seaweed has been harvested, it must be quickly disposed of or processed before it begins to rot, though partial decomposition may not be a major prob-lem if anaerobic digestion is the final destination. There is a significant cost associated with sending seaweed to landfill. Ideally, economic uses should be identified for the sea-weed. For this to happen, the seaweed must generally be pre-treated. Seaweed harvest-ing can also capture unwanted material such as plastics and other marine litter, though filtering the seaweed by passing it through a screen or net could potentially capture larger items of rubbish (Bruton *et al.* 2009). The seaweed can also be washed to remove salt and sediment, all of which can seriously add to processing costs.

The sheet-like thalli of green algal species such as *Ulva lactuca* (now known to be the species *U. fenestrata* in the North Atlantic; Hughey *et al.* 2019) create a large surface area for the rapid uptake of nutrients such as nitrogen and phosphorous from the surrounding seawater (Msuya and Neori 2008). The green algal mats in Poole Harbour are known to support five species of *Ulva* sea lettuce: *U. lactuca* (Jones and Pinn 2006), *U. rigida, U. com-pressa, U. intestinalis* and *U. clathrata* (Thornton 2016). These species are notoriously dif-ficult to identify. Any possible economic use for the algal mats would have to work with all species as it would be impractical to separate them. Furthermore, the protein, lipid and sugar contents of *Ulva* show high seasonal variability (Mohy El-Din 2019).

Table 20.1 Advantages and disadvantages of boat-based and land-based seaweed harvesting

	Advantages	Disadvantages
Boat	Can cover large area May not disturb potentially fragile inter-tidal communities Reduced capture of unwanted sand/mud May collect both free and attached thalli	Requires specialised equipment Re-circulated sediment could reduce visibility Could re-circulate pollutants in sediment Restricted to intertidal Training required to operate By-catch (seagrass as particular concern) possible Large costs associated with boat storage, mooring and transportation
Tractor/ Excavator	Minimal specialist equipment required Effective on beaches Can operate in a wider range of weather conditions Minimal training required to operate Does not re-circulate sediment Reduced by-catch	Can capture sand/mud Can sink in soft sediments Can trample/scour sediment Can only harvest free algae Confined to the intertidal area Limited to low tide

(adapted from Capuzzo and Forster 2014, and Crane and Ramsay 2011.

It has been suggested that seaweed harvesting could be used as a bioremediation tool to combat eutrophication (Fernand *et al.* 2017). According to the local Poole Harbour management plans (e.g. citation in Capuzzo and Forster 2014), nitrogen (N) input into Poole Harbour needs to be reduced from 2,100 tonnes per annum to 1,730 tonnes per annum. Boat-based harvesting (see previous assumptions) could be expected to remove 15 to 30 tonnes of N from the system each year, though there are substantial uncertainties (e.g. Watson *et al.* 2020). While this level of removal is not sufficient on its own, repeated removal over several years could reduce the numbers of overwintering thalli fragments and thereby decrease *Ulva* coverage in the long term (Capuzzo and Forster 2014). Investigating the potential for harvesting green algal mats has seen significant investment in recent years, including in EU-funded projects such as BEACON, which works in partnership with a UK company (GreenSeas resources) to harvest and process green algae into biofuel and also RaNTrans. RaNTrans is a collaboration between the University of Portsmouth, Bournemouth University, the French Algae Technology and Innovation Centre (Centre d'Etude et de Valorisation des Algues – CEVA), Université de Bretagne Occidentale and the University of Caen (Normandy), among others. RaNTrans will investigate the harvesting of green algal mats in the Channel area, including Poole Harbour. Holes Bay has been identified as a suitable site for algal harvesting due to the density of green macroalgal mats and its proximity to the centre of Poole. Holes Bay exhibits good growth conditions for green macroalgae due to its shallow waters and low flushing rates. However, it has a legacy of receiving discharges from a sewage treatment plant and industry, including a former chemical plant. The low flushing rates also mean that a variety of pollutants, including metals, are of concern (Langston *et al.* 2003), though metals do not in general appear to be worse in Holes Bay compared with the rest of Poole Harbour. Only arsenic and mercury were found to exceed the UK environmental quality standards for estuarine and coastal waters in a study by Aly *et al.* (2013), which was similar to the rest of the harbour. However, considerable contamination of the sediment by cadmium, copper, mercury, lead and zinc exists, particularly in the northern area of the bay (Hübner 2009). Indeed, the concentrations of these metals exceeded the probable effect level values set in the Canadian Sediment Quality Guidelines (CCME 2002), indicating that adverse effects on aquatic biota may occur frequently. *Ulva* is known to bioaccumulate heavy metals from contaminated water and sediments (Kamala-Kannan *et al.* 2008), including zinc, lead, copper and cadmium (Haritonidis and Malea 1999). This may limit the use of *Ulva* in foodstuffs if the concentration of metals/metalloids exceeds the statutory limits for human food (European Union 2006) or animal consumption (European Union 2002). From first principles, however, green macroalgal mats are thought safe for consumption (Li *et al.* 2018). In some parts of the world *Ulva* is actively consumed by people, though consumption in Europe is very low (Capuzzo and Forster 2014).

Metal levels may also limit the use of algal mats as compost/organic fertiliser due to the potential transfer of metals to the crop (Greger *et al.* 2016) and effects on soil microbiology (Green *et al.* 2020). The legal controls on the application of organic materials containing metals to soils are weak (Green *et al.* 2020) despite agricultural soils receiving metal inputs from multiple sources (Nicholson *et al.* 2006). Care is therefore needed when applying materials to soil that contain elevated levels of metals. A small study undertaken in 2018 found that heavy metal concentrations in *Ulva* from Holes Bay (chromium, cadmium, copper, nickel, lead and zinc) were higher than in other parts of Poole Harbour (Browne 2019). However, these concentrations were found to be below the acceptable threshold concentrations specified for compost use in agriculture (BSI 2018), suggesting that application to soil would be safe. Moreover, increased levels of essential micronutrients, especially zinc, may make algal mat material a particularly useful source of these elements in areas of calcareous soils naturally containing low levels of available micronutrients.

Potential uses of green macroalgal mats

Bioenergy

As our society moves away from fossil fuels, biomass has been cited as a promising renewable source of energy (Reid *et al.* 2020). Bioenergy is harnessed from sources including wood, agricultural crops and sewage. Bioenergy from seaweed may be a particularly attractive option in the UK due to its extensive coastline (Roberts and Upham 2012). Land and water use are key concerns associated with the expansion of the bioenergy industry. Crops that are grown specifically for fuel can reduce the available resources for food crops and conservation management (Holmatov *et al.* 2019). This provokes ethical concerns around food security and biodiversity (Gamborg *et al.* 2012). Fuel made from 'waste' seaweed would not require dedicated land or water resources to grow. Additionally, green algal mats do not require the application of artificial fertilisers, unlike most terrestrial biofuel crops. Algae do not invest resources into forming lignin, roots or flowers. Energy resources in algae are dedicated primarily to growth, repair and reproduction. This allows algae to proliferate rapidly and makes them effective capturers of carbon (Patel *et al.* 2012). The combustion of algal biomass releases the sequestered carbon back into the atmosphere, theoretically resulting in a circular, net carbon neutral system, although this does not account for the CO_2 released during harvesting and processing (Subhadra and Edwards 2010). A systematic review found that biofuel is the most feasible use for algal biomass (Joniver *et al.* 2021).

Unlike other renewable energy sources such as solar, hydro or wind, biomass can be converted to liquid fuel. This biofuel can easily be stored and transported, potentially making it an attractive option for the transportation sector (Adams *et al.* 2009). In particular, the aviation industry has been identified as a promising market for biofuel, as electric batteries are not currently advanced enough to power large aircraft (Jiang *et al.* 2016). Bioethanol is the most promising potential liquid biofuel from green algae due to the high levels of carbohydrates and low levels of triglycerides (fats) in its cells (Chia *et al.* 2018), and the use of bioethanol within the EU was obliged by the renewable energy directive in 2009. Seaweeds may be more readily converted to refined fuel than land plants due to the absence of lignin and low cellulose content in their tissues (Kraan 2013). Sugars derived from *Ulva lactuca* can be hydrolysed and fermented (Potts *et al.* 2011). As part of this process, microbes such as *Clostridium* bacteria (van der Wal *et al.* 2013) or the yeast *Saccharomyces cerevisiae* (Jung *et al.* 2013) digest the seaweed and produce ethanol as waste (Chen *et al.* 2015). This ethanol can then be transported, stored and later combusted as biofuel. Most conventional cars can run on fuel mixtures of up to 5% biofuel (James 2010) and now 10% is mandated. As part of the EU-funded project MacroFuels, a modified Volkswagen car was driven with a fuel mixture of 10% ethanol from seaweeds; it was found to be compliant with European car emission standards (Frandson and Kolev 2020). *Ulva lactuca* can yield 15–20 kg of ethanol per tonne of biomass (Kostas *et al.* 2020). While this level of efficiency is insufficient for a large-scale seaweed cultivation operation, it could be commercially viable for waste seaweed or as part of a multi-scale biorefinery operation (Balina *et al.* 2017). Gegg and Wells (2017) note the importance of the role of the UK government in investing in, and possibly subsidising, the creation of a macroalgal biofuel industry and highlight the potential gains for the green economy and local job creation.

While most research on green algal bioenergy has focused on biofuel, there has been some exploratory research that has highlighted the potential for *Ulva lactuca* as a biodiesel, with oil yields of 10% (Suganya and Renganathan 2012). Biodiesel has similar physico-chemical properties to petroleum diesel and can be used in existing diesel engines (Daroch *et al*. 2013). However, algal biodiesel may not be industrially competitive, as commercial biodiesel crops, such as rapeseed, can yield upward of 40% oil (Matthaus *et al*. 2016). Biogas is another possibility as methane can be generated from the fermentation of seaweed. *Ulva lactuca* that has been washed and macerated can be processed to produce 16.5 mL of methane for every gram of algae; this can be increased by a further 48% when co-digested with animal manure (Nielsen and Heiske 2011). A 2013 economic assessment noted that while macroalgal biogas was not yet financially viable in the UK, it could be made feasible with green energy subsidisation and low seaweed cultivation costs (Dave *et al*. 2013). In Japan, *Ulva* (and *Laminaria*) has been used in a seaweed-only demonstration biogas plants (see Capuzzo and Forster 2014), and Nikolaisen *et al*. (2011) also highlighted the potential use of *Ulva* tissue as a co-digestate in manure-based anaerobic digestors. The seaweed from Poole Harbour would be a waste product and would not require cultivation, so conversion to biogas could be a workable option.

Bioenergy from algae is not a panacea to climate change. As a popular illustration, even if algae (cultivated in intensive 'algae-ponds') were grown with maximum efficiency, an average UK driver would require 420 m² of space dedicated to algal production to drive just 50 km (see assumptions in MacKay 2008). However, the use of 'waste' algae could help to expand the pool of available energy sources and contribute to the UK's energy independence while also helping to mitigate the environmental issues arising from algal accumulations.

Animal feed

Ulva can be a source of nutrition in animal agriculture and aquaculture. Broiler chickens fed on a diet of 3% *Ulva lactuca* were found to have increased muscle yield compared to a control group (Abudabos *et al*. 2013). *Ulva* can be a medium-level quality feed for ruminants (Ventura and Castañón 1998; Rjiba-Ktita *et al*. 2017). Maia *et al*. (2019) suggest that *Ulva* could comprise up to 25% of ruminant diets and could help to make livestock rearing a more sustainable industry. *U. clathrata* has been shown to be a promising source of protein and iron in animal feed (Peña-Rodríguez *et al*. 2011).

Green algae can also contribute to the diet of farmed fish and marine invertebrates. The animal aquaculture sector is heavily dependent on fish meal and fish oil from wild-caught fish, which limits the growth and sustainability of the industry (Tacon and Metian 2008). Protein from fishmeal is typically the most expensive part of fish feed (Shpigel *et al*. 2017), so finding a cheaper and more sustainable alternative is key to the growth of the sector. *U. lactuca* and *U. intestinalis* have both been found to contain high protein levels and the required compliment of amino acids required for growth in salmonids (Mæhre *et al*. 2014). Farmed salmon can be fed a diet of up to 5% *Ulva* without any impacts on their growth (Norambuena *et al*. 2015). Shrimp feed can include up to 50% *Ulva* without affecting growth (Pallaoro *et al*. 2016). *U. lactuca* has shown promise as a cost-saving feed for farmed seabream (Shpigel *et al*. 2017). Sea urchins, which are farmed for their roe, can benefit from a diet containing *U. rigida* as it is highly palatable to them and stimulates feeding and hence protein intake (Cyrus *et al*. 2015).

Technological advances in bioprocessing can increase the value of *Ulva* as a feed resource. Fermentation and enzymatic hydrolysis of *U. rigida* can increase protein content by up to 76% (Fernandes *et al*. 2019). Animal feed production can be combined with biofuel generation in a cascading biorefinery system (Bikker *et al*. 2016). This system takes sugars from the *Ulva* and refines them into biofuel, while the remaining substances form a high-protein animal feed that is of better quality than unprocessed *Ulva*.

Fertiliser

Seaweed has a long history of use as a fertiliser. In Jersey, seaweed has been collected and used in agriculture since the twelfth century (Blench 1966). *Ulva*-based fertilisers could offer several advantages (Ghaderiardakani 2019) as, in addition to nitrogen and phosphorous, seaweed-based fertiliser can contain plant growth regulators including auxins, cytokinins and gibberellins. These hormones are responsible for controlling growth in plants, and it is thought that they may be used to stimulate growth of particular plant tissues (Arioli *et al.* 2015). Seaweed must be processed before it can be used successfully as a large-scale fertiliser. Algal biomass from green tides in Brittany, France, has been successfully composted (Mazé *et al.* 1993). This process turns the seaweed into a basic fertilising agent. It can also be stored and transported more successfully than unprocessed seaweed as it significantly reduces unpleasant odours and fluid discharge. Nowadays, much of the research in this field explores the potential to produce liquid fertilisers. One such method, which has shown promise in *Ulva* in the south of Britain, is hydrothermal liquefaction (Raikova *et al.* 2017). Liquid biofertilisers can be easily stored, transported and marketed towards the agricultural and horticultural sectors.

Liquid biofertilisers can have advantages over synthetic or manure-based fertilisers as seaweeds contain biomolecules that can boost factors such as frost and drought resistance, increased germination rates and defence against pests and pathogens (Ghaderiardakani 2019). Liquid fertilisers from *Ulva* have been trialled successfully in numerous crops. Root and shoot growth were found to be superior in mung beans treated with liquid *Ulva* fertiliser, compared to traditional manure fertiliser (Akila *et al.* 2019). Similarly, peanuts were found to have greater yield and nutritive quality, and chilli peppers were found to have higher growth, yield and sugar levels (Sridhar and Rengasamy 2012). Closer to home, unprocessed seaweed is harvested from beaches in Jersey and used to promote growth and moisture retention in potato crops (Jersey Royal Company 2015).

Research into the potential of liquid *Ulva* fertilisers on farmlands in the south of Britain is lacking. Ghaderiardakani *et al.* (2019) note the importance of using the correct concentration of *Ulva* fertilisers, as overuse can inhibit crop development. A small-scale field trial of these fertilisers would be advantageous, particularly in the Frome or Piddle river basins. The mixture of livestock, arable and horticultural activities in the region creates a number of possible opportunities for biofertilisers. The green tides in Poole Harbour are caused, mostly, by run-off from agricultural fertilisers. Recycling some of these nutrients back into fertilisers would create a loop and ideally reduce the amount of external nutrients entering the catchment. One solution would be to create a cooperative, in which biofertilisers from Poole Harbour are supplied to farmers at a reduced rate. Biofertilisers have the added advantage of being organic. Organic produce can fetch a greater price from consumers, which would benefit the famers.

Bioprospecting

Bioprospecting is the search within living organisms for chemicals or genes that may have commercial or societal value. The bioprospecting of seaweeds has been growing rapidly since 1990 (Mazarrasa *et al.* 2014). Seaweeds have evolved chemical defences to combat herbivory, biofouling, sun damage and competition from other seaweed species (Pereira *et al.* 2017; Sudatti *et al.* 2020). These molecules can be extracted and refined, and can lead to the discovery of novel pharmaceuticals. Bioprospecting of green seaweeds has

Table 20.2 Proposed uses of *Ulva* biomolecules

Species	Target treatment	References
Ulva lactuca	Low mood	Allaert *et al.* 2018
	Breast cancer	Abd-Ellatef *et al.* 2017
	Liver cancer	Hussein *et al.* 2015
	Antimicrobial	Anjali *et al.* 2019
Ulva rigida	Oxidative stress from chemotherapy	Ksouri *et al.* 2020
	Antimicrobial	Ismail *et al.* 2018
	Alzheimer's disease	Olasehinde *et al.* 2019
Ulva compressa	Cancer	Maryati *et al.* 2020
Ulva intestinalis	Measles virus	Morán-Santibañez *et al.* 2016
Ulva clathrata	Antiviral (poultry)	Aguilar-Briseño *et al.* 2015

increased significantly in recent years; however, most of the trials have not yet reached the clinical stage. The polysaccharide ulvan, which is found in the cell walls of *Ulva*, is of particular interest to the pharmaceutical industry (Kidgell *et al.* 2019). The *Ulva* species found in Poole Harbour is no exception to this, and some of its potential pharmaceutical applications are outlined in Table 20.2. Additionally, ulvan has been targeted as a potential cutting-edge wound dressing and as a drug delivery system (Kuznetsova *et al.* 2020). While this field is still in its infancy, it offers some potentially lucrative uses for the waste algae in Poole Harbour.

Future outlook and the blue economy

The blue economy is an emerging concept, which links marine economic activities with responsible stewardship of our oceans (The Commonwealth 2020). As part of this, coastal communities will be encouraged to consider environmental issues such as carbon sequestration, coastal resilience and pollution. 'Blue carbon' is a term that refers to carbon that has been captured by marine vegetation, including green algal mats (Macreadie *et al.* 2019). As we continue to increase atmospheric CO_2, the potential for marine algae to lock carbon in their tissues becomes increasingly relevant. To date, most of the research into blue carbon sequestration has been focused around angiosperms, such as seagrasses, mangroves and saltmarshes (Krause-Jensen *et al.* 2018). Seaweeds had previously been less studied as it was thought that they fragment and erode too quickly, thereby quickly releasing carbon (Raven 2018), and their lack of root systems was thought to limit their ability to sequester carbon in situ (Queirós *et al.* 2019). In recent years, however, a greater focus has been placed on macroalgae. Wild macroalgae can sequester carbon at a rate of around 150 t CO_2 km^{-2} year^{-1}, but this can be significantly increased by harvesting and removing the algae (Chen and Xu 2020). *Ulva* species can sequester carbon at a rate of between 32 to 420 mg of CO_2 per gram (fresh weight) per day (Chung *et al.* 2011). Trevathan-Tackett *et al.* (2015) note the rapid growth rates of seaweeds as well as their ability to be easily transported to be important factors in their potential as carbon sequestration tools. Recently, macroalgal mat-specific parameters have been calculated (Watson *et al.* 2020).

Clearly, there could be benefits to removing nuisance green algal mats. The challenge going forward will be to do this with harvesting methods which are not in themselves harmful and in a way that generates economic benefit for local businesses and residents. The harvesting of green algal mats in Poole Harbour would seem to be a prime opportunity to embody the principles of the blue economy for the benefit of all of its stakeholders.

References

Abd-Ellatef, G.E.F., Ahmed, O.M., Reheim, E.S., and Abdel-Hamid, A.H.Z. 2017. *Ulva lactuca* polysaccharides prevent Wistar rat breast carcinogenesis through the augmentation of apoptosis, enhancement of antioxidant defence system, and suppression of inflammation. *Breast Cancer: Targets and Therapy* 9: 67–83. https://doi.org/10.2147/BCTT.S125165

Abudabos, A.M., Okab, A.B., Alijumaah, R.S., Samara, E.M., Abdoun, K.A., and Al-Haidary, A.A. 2013. Nutritional value of green seaweed (*Ulva lactuca*) for broiler chickens. *Italian Journal of Animal Science* 12(2): 177–81. https://doi.org/10.4081/ijas.2013.e28

Adams, J.M., Gallagher, J.A., and Donnison, I.S. 2009. Fermentation study on *Saccharina latissima* for bioethanol production considering variable pre-treatments. *Journal of Applied Phycology* 21: 569. https://doi.org/10.1007/s10811-008-9384-7f

Aguilar-Briseño, J.A., Cruz-Suárez, L.E., Sassi, J.F., Ricque-Marie, D., Zapata-Benavides, P., Mendoza-Gamboa, E., Rodríguez-Padilla, C., and Trejo-Avila, L.M. 2015. Sulphated polysaccharides from *Ulva clathrata* and *Cladosiphon okamuranus* seaweeds both inhibit viral attachment-entry and cell-cell fusion in NDV infection. *Marine Drugs* 13(2): 697–712. https://doi.org/10.3390/md13020697

Akila, V., Manikandan, A., Sukeetha, D.S., Balakrishnan, S., Ayyasamy, P.M., and Rajakumar, S. 2019. Biogas and biofertilizer production of marine macroalgae: An effective anaerobic digestion of *Ulva* sp. *Biocatalysis and Agricultural Biotechnology* 18: 101035. https://doi.org/10.1016/j.bcab.2019.101035

Allaert, F.A., Demais, H., and Collén, P.N. 2018. A randomized controlled double-blind trial comparing versus placebo the effect of an edible algal extract (*Ulva lactuca*) on the component of depression in healthy volunteers with anhedonia. *BMC Psychiatry* 18: 215. https://doi.org/10.1186/s12888-018-1784-x

Aly, W., Williams, I.D., and Hudson, M.D. 2013. Metal contamination in water, sediment and biota from a semi-enclosed coastal area. *Environmental Monitoring and Assessment* 185(5): 3879–95. https://doi.org/10.1007/s10661-012-2837-0

Anjali, K.P., Sangeetha, B.M., Devi, G., Raghunathan, R., and Dutta, S. 2019. Bioprospecting of seaweeds (*Ulva lactuca* and *Stoechospermum marginatum*): The compound characterization and functional applications in medicine – A comparative study. *Journal of Photochemistry and Photobiology, B: Biology* 200: 111622. https://doi.org/10.1016/j.jphotobiol.2019.111622

Arioli, T., Mattner, S.W., and Winberg, P. 2015. Applications of seaweed extracts in Australian agriculture. *Journal of Applied Phycology* 27: 2007–15. https://doi.org/10.1007/s10811-015-0574-9

Balina, K., Romangnoli, F., and Blumberga, D. 2017. Seaweed biorefinery concept for sustainable use of marine resources. *Energy Procedia* 128: 504–11. https://doi.org/10.1016/j.egypro.2017.09.067

Bikker, P., van Krimpen, M.M., van Wikselaar, P., Houweling-Tan, B., Scaccia, N., van Hal, J.W., Huijgen, W.J.J., Cone, J.W., and López-Contreras, A.M. 2016. Biorefinery of the green seaweed *Ulva lactuca* to produce animal feed, chemical and biofuels. *Journal of Applied Phycology* 28: 3511–25. https://doi.org/10.1007/s10811-016-0842-3

Blench, B.J.R. 1966. Seaweed and its use in Jersey agriculture. *The Agricultural History Review* 14(2): 122–8.

Browne, N. 2019. Metal content of seaweed tissue in Poole Harbour. Undergraduate dissertation. Bournemouth University, UK.

Bruton, T., Lyons, H., Lerat, Y., Stanley, M., and Rasmussen, M.B. 2009. A review of the potential of marine algae as a source of biofuel in Ireland. Technical report, Sustainable Energy Ireland, Dublin, Ireland.

BSI. 2018. *Publicly Available Specification: Specification for Composted Materials* (PAS 100:2018). British Standards Institution, London, England.

Capuzzo, E. and Forster, R.M. 2014. Removal and subsequent uses of opportunistic green macroalgae (*Ulva lactuca*) in Poole Harbour. CEFAS contract report C6285.

CCME. 2002. Canadian sediment quality guidelines for the protection of aquatic life. Summary Tables. Update 2002, in Canadian environmental quality guidelines, 1999, Canadian Council of Ministers of the Environment.

Chen, H., Zhou, D., Luo, G., Zhang, S., and Chen, J. 2015. Macroalgae for biofuels production: Progress and perspectives. *Renewable and Sustainable Energy Reviews* 47: 427–37. https://doi.org/10.1016/j.rser.2015.03.086

Chen, Y. and Xu, C. 2020. Exploring new blue carbon plants for sustainable ecosystems. *Trends in Plant Science* 25(11): 1067–70. https://doi.org/10.1016/j.tplants.2020.07.016

Chia, S.R., Ong, H.C., Chew, K.W., Show, P.L., Phang, S.L., Ling, T.C., Nagarajan, D., Lee, D.J. and Chang, J.S. 2018. Sustainable approaches for algae utilisation in bioenergy production. *Renewable Energy* 129(B): 838–52. https://doi.org/10.1016/j.renene.2017.04.001

Chung, I.K., Beardall, J., Mehta, S., Sahoo, D., and Stojkovic, S. 2011. Using marine macroalgae for carbon sequestration: A critical appraisal. *Journal of Applied Phycology* 23: 877–86. https://doi.org/10.1007/s10811-010-9604-9

Crane, C. and Ramsay, A. 2011. Sea lettuce harvest project report. PEI department of environment, labour and justice and PEI Department of Fisheries, Aquaculture and Rural Development. p. 84.

Cyrus, M.D., Bolton, J.J., Scholtz, R., and Macey, B.M. 2015. The advantages of *Ulva* (Chlorophyta) as an additive in sea urchin formulated feeds: Effects on palatability, consumption and digestibility. *Aquaculture Nutrition* 21(5): 578–91. https://doi.org/10.1111/anu.12182

Daroch, M., Geng, S., and Wang, G. 2013. Recent advances in liquid biofuel production from algal feedstocks. *Applied Energy* 102: 1371–81. https://doi.org/10.1016/j.apenergy.2012.07.031

Dave, A., Huang, Y., Rezvani, S., McIlveen-Wright,D., Novaes, M., and Kewitt, N. 2013. Techno-economic assessment of biofuel development by anaerobic digestion of European marine cold-water seaweeds. *Biosource Technology* 135: 120–7. https://doi.org/10.1016/j.biortech.2013.01.005

European Union. 2002. Commission directive 2002/32/EC of the European parliament and of the council of 7 May 2002 on undesirable substances in animal feed, OJ 2002 L 140/10.

European Union. 2006. Commission regulation (EC) No. 1881/2006 of 19 December 2006 setting maximum levels for certain contaminants in foodstuffs, OJ 2006 L 364/5.

Fernand, F., Isreal, A., Skjermo, J., Wichard, T., Timmermans, K.R., and Goldberg, A. 2017. Offshore macroalgae biomass for bioenergy production: Environmental aspects, technological achievements and challenges. *Renewable and Sustainable Energy Reviews* 75: 35–45. https://doi.org/10.1016/j.rser.2016.10.046

Fernandes, H., Salgado, J.M., Martins, N., Peres, H., Oliva-Teles, A., and Belo, I. 2019. Sequential bioprocessing of *Ulva rigida* to produce lignocellulolytic enzymes and to improve its nutritional value as aquaculture feed. *Bioresource Technology* 281: 277–85. https://doi.org/10.1016/j.biortech.2019.02.068

Frandson, S. and Kolev, D.Y. 2020. *Seaweed Derived Biofuels and Blend Derived from Fermentation.* H2020-LCE-11-2015 D5.1 MacroFuels Deliverable Report, Danish Technological Institute [Online]. Available from: https://www.macrofuels.eu/deliverables-1 (accessed 15 July 2020).

Gamborg, C., Millar, K., Shortall, O., and Sandøe, P. 2012. Bioenergy and land use: Framing the ethical debate. *Journal of Agricultural and Environmental Ethics* 25(6): 909–25. https://doi.org/10.1007/s10806-011-9351-1

Gegg, P. and Wells, V. 2017. UK macro-algae biofuels: A strategic management review and future research agenda. *Journal of Marine Science and Engineering* 5(3): 32. https://doi.org/10.3390/jmse5030032

Ghaderiardakani, F. 2019. Ulva growth, development and applications. PhD Thesis. University of Birmingham, UK.

Ghaderiardakani, F., Collas, E., Damiano, D.K., Tagg, K., Graham, N.S., and Coates, J.C. 2019.

Effects of green seaweed extract on *Arabidopsis* early development suggest roles for hormone signalling in plant responses to algal fertilisers. *Scientific Reports* 9: 1–13. https://doi.org/10.1038/s41598-018-38093-2

Green, I., Ginige, T., Demir, M., and Van Calster, P. 2020. Accumulation of potentially toxic elements in agricultural soils. In Pozzo, B. and Jacometti, V. (eds) *Environmental Loss and Damage in a Comparative Law Perspective: Attribution, Liability, Compensation and Restoration.* Intersentia, 87–107. https://doi.org/10.1017/9781839701191.007

Greger, M., Malmb, T., and Kautsky, L. 2016. Heavy metal transfer from composted macroalgae to crops. *European Journal of Agronomy* 26(3): 257–65. https://doi.org/10.1016/j.eja.2006.10.003

Haritonidis, S. and Malea, P. 1999. Bioaccumulation of metals by the green alga *Ulva rigida* from Thermaikos Gulf, Greece. *Environmental Pollution* 104: 365–72. https://doi.org/10.1016/S0269-7491(98)00192-4

Holmatov, B., Hoekstra, A.Y., and Krol, M.S. 2019. Land, water and carbon footprints of circular bioenergy production systems. *Renewable and Sustainable Energy Reviews* 111: 224–35. https://doi.org/10.1016/j.rser.2019.04.085

Hübner, R. 2009. Sediment geochemistry – A sediment approach. PhD thesis. Bournemouth University, UK.

Hughey, J.R., Maggs, C.A., Mineur, F., Jarvis, C., Miller, K.A., Shabaka, S.H., and Gabrielson, P.W. 2019. Genetic analysis of the Linnaean *Ulva lactuca* (Ulvales, Chlorophyta) holotype and related type specimens reveals name misapplications, unexpected origins, and new synonymies. *Journal of Phycology* 55: 503–8. https://doi.org/10.1111/jpy.12860

Hussein, U.K., Mahmoud, H.M., Farrag, A.G., and Bishayee, A. 2015. Chemoprevention of diethylnitrosamine-initiated and phenobarbital-promoted hepatocarcinogenesis in rats by sulphated polysaccharides and aqueous extract of *Ulva lactuca*. *Integrative Cancer Therapies* 14(6): 525–45. https://doi.org/10.1177/1534735415590157

Ismail, A., Ktari, L., Ben Redjem Romdhane, Y., Aoun, B., Sadok, S., Boudabous, A., and El Bour, M. 2018. Antimicrobial fatty acids from green alga *Ulva rigida* (Chlorophyta). *BioMed Research International* 2018: 1–12. https://doi.org/10.1155/2018/3069595

James, M.A. 2010. A review of initiatives and related R&D being undertaken in the UK and internationally regarding the use of macroalgae as a basis for biofuel production and other non-food uses relevant to Scotland. *Report Commissioned by Marine Scotland.*

Jersey Royal Company. 2015. Collecting and harvesting seaweed [Online]. Available from: https://www.jerseyroyal.co.uk/news/latest-news/2015/collecting-and-spreading-seaweed/ (accessed 2 July 2020).

Jiang, R., Ingle, K.N., and Goldberg, A. 2016. Macroalgae (seaweed) for liquid transportation biofuel production: What is next? *Algal Research* 14: 48–57. https://doi.org/10.1016/j.algal.2016.01.001

Jones, M. and Pinn, E. 2006. The impact of a macroalgal mat on benthic biodiversity in Poole Harbour. *Marine Pollution Bulletin* 53(1–4): 63–71. https://doi.org/10.1016/j.marpolbul.2005.09.018

Joniver, C.F.H. Photiades, A., Moore, P.J., Winters, A.L., Woolmer, A., and Adams, J.M.M. 2021. The global problem of nuisance macroalgal blooms and pathways to its use in the circular economy.. *Algal research* 58: 102407.

Jung, K.A., Lim, S.R., Kim, Y., and Park, J.M. 2013. Potentials of macroalgae as feedstocks for biorefinery. *Biosource Technology* 15: 182–90. https://doi.org/10.1016/j.biortech.2012.10.025

Kamala-Kannan, S., Batvari, B.P.D., Lee, K.J., Kannan, N., Krishnamoorthy, R., Shanthi, K., and Jayaprakash, M. 2008. Assessment of heavy metals (Cd, Cr and Pb) in water, sediment and seaweed (*Ulva lactuca*) in the Pulicat Lake, South East India. *Chemosphere* 71(7): 1233–40. https://doi.org/10.1016/j.chemosphere.2007.12.004

Kidgell, J.T., Magnusson, M., de Nys, R., and Glasson, C.R.K. 2019. Ulvan: A systematic review of extraction, composition and function. *Algal Research* 39: 101422. https://doi.org/10.1016/j.algal.2019.101422

Kostas, E.T., White, D.A., and Cook, D.J. 2020. Bioethanol production from UK seaweeds: Investigating variable pre-treatment and enzyme hydrolysis parameters. *BioEnergy Research* 13: 271–85. https://doi.org/10.1007/s12155-019-10054-1

Kraan, S. 2013. Mass-cultivation of carbohydrate rich macroalgae, a possible solution for sustainable biofuel production. *Mitigation and Adaptation Strategies for Global Climate Change* 18: 27–46. https://doi.org/10.1007/s11027-010-9275-5

Krause-Jensen, D., Lavery, P., Serrano, O., Marba, N., Masque, P., and Duarte, C.M. 2018. Sequestration of macroalgal carbon: The elephant in the blue carbon room. *Biology Letters* 14(6): 20180236. https://doi.org/10.1098/rsbl.2018.0236

Ksouri, R., Rabah, S., Mezghani, S., and Hamlaoui, S. 2020. The protective effect of grape seed and skin extract and *Ulva rigida* against oxidative stress. *Sakarya Journal of Science* 24(2): 312–23. https://doi.org/10.16984/saufenbilder.638725

Kuznetsova, T.A., Andryukov, B.G., Besednova, N.N., Zaporozhets, T.S., and Kalinin, A.V. 2020. Marine algae polysaccharides as a basis for wound dressings, drug delivery, and tissue engineering: A review. *Journal of Marine Science and Engineering* 8(7): 481. https://doi.org/10.3390/jmse8070481

Langston, W.J., Chesman, B.S., Burt, G.R., Hawkins, S.J., Readman, J., and Worsford, P. 2003. Characterisation of European marine sites. Poole Harbour Special Protection Area. Occasional Publication. Marine Biological Association of the United Kingdom, 12, 111.

Li, J-Y., Yang, F., Jin, L., Wang, W., Yin, J., He, P., and Chen, Y. 2018. Safety and quality of the green tide algal species *Ulva prolifera* for option of human consumption: A nutrition and contamination study. *Chemosphere* 210: 1021–28. https://doi.org/10.1016/j.chemosphere.2018.07.076

MacKay, D.J.C. 2008. *Sustainable Energy – Without the Hot Air*. UIT Cambridge, Cambridge.

Macreadie, P.I., Anton, A., Raven, J.A., Beaumont, N., Connolly, R.M., Friess, D.A., Kelleway, J.J., Kennedy, H., Kuwae, T., Lavery, P.S., Lovelock, C.E., Smale, D.A., Apostolaki, E.T., Atwood, T.B., Baldock, J., Bianchi, T.S., Chmura, G.L., Eyre, B.D., Fourqurean, J.W., Hall-Spencer, J.M., Huxham, M., Hendriks, I.E., Krause-Jensen, D., Laffoley, D., Luisetti, T., Marba, N., Masque, P., McGlathery, K.J., Megonigal, J.P., Murdiyarso, D., Russell, B.D., Santos, R., Serrano, O., Watanabe, K., and Duarte, C.M. 2019. The future of blue carbon science. *Nature Communications* 10: 1–13. https://doi.org/10.1038/s41467-019-11693-w

Mæhre, H.K., Malde, M.K., Eilertsen, K.E., and Elvevoll, E.O. 2014. Characterization of protein, lipid and mineral contents in common Norwegian seaweeds and evaluation of their potential as food and feed. *Journal of the Science of Food and Agriculture* 94(15): 3281–90. https://doi.org/10.1002/jsfa.6681

Maia, M.R.G., Fonseca, A.J.M., Cortez, P.P., and Cabrita, A.R.J. 2019. *In vitro* evaluation of macroalgae as unconventional ingredients in ruminant animal feeds. *Algal Research* 40: 101481. https://doi.org/10.1016/j.algal.2019.101481

Maryati, M., Saifundin, A., Wahyuni, S., Rahmawati, J., Arrum, A., Priyunita, O., Aulia, A., Putra, F.P.H., As'hari, Y., Rasyidah, U.M., Fadhilah, A., Muflihah, C.H., and Da'I, M. 2020. Cytotoxic effect of *Spirulina platensis* extract and *Ulva compressa* Linn. on cancer cell lines. *Food Research* 4(4): 1018–23. https://doi.org/10.26656/fr.2017.4(4).389

Matthaus, B., Özcan, M.M., and Al Juhaimi, F.A. 2016. Some rape/canola seed oils: Fatty acid composition and tocopherols. *Zeitschrift für Naturforschung C* 71(3–4): 73–7. https://doi.org/10.1515/znc-2016-0003

Mazarrasa, I., Olsen, Y.S., Mayol, E., Marbá, N., and Duarte, C.M. 2014. Global unbalance in seaweed production, research effort and biotechnology markets. *Biotechnology Advances* 32: 1028–36. https://doi.org/10.1016/j.biotechadv.2014.05.002

Mazé, J., Morand, P., and Potoky, P. 1993. Stabilisation of 'green tides' *Ulva* by a method of composting with a view to pollution limitation. *Journal of Applied Phycology* 5: 183–90. https://doi.org/10.1007/BF00004015

Ménesguen, A., Cugier, P., and Leblond, I. 2006. A new numerical technique for tracking chemical species in a multi-source, coastal ecosystem, applied to nitrogen causing *Ulva* blooms in the bay of Brest (France). *Limnology and Oceanography* 51(1.2): 591–601. https://doi.org/10.4319/lo.2006 .51.1_part_2.0591

Mohy El-Din, S.M. 2019. Temporal variation in chemical composition of *Ulva lactuca* and *Corallina mediterranea*. *International Journal of Environmental Science and Technology* 16: 5783–96. https://doi.org/10.1007/s13762-018-2128-6

Morán-Santibañez, K., Cruz-Suárez, L.E., Rique-Marie, D., Robledo, D., Freile-Pelegrín, Y., Peña-Hernández, M.A., Rodríguez-Padilla, C., and Trejo-Avila, L.M. 2016. Synergistic effects of sulphated polysaccharides from Mexican seaweeds against measles virus. *BioMed Research International* 2016: 1–11. https://doi.org/10.1155 /2016/8502123

Msuya, F.E. and Neori, A. 2008. Effect of water aeration and nutrient load level on biomass yield, N uptake and protein content of the seaweed *Ulva lactuca* cultured in seawater tanks. *Journal of Applied Phycology* 20: 1021–31. https://doi.org/10 .1007/s10811-007-9300-6

Nicholson, F., Smith, S., Alloway, B.J., Carlton-Smith, C., and Chambers, B. 2006. Quantifying heavy metal inputs to agricultural soils in England and Wales. *Water and Environment Journal* 20: 87–95. https://doi.org/10.1111/j.1747-6593.2006.00026.x

Nielsen, H.B. and Heiske, S. 2011. Anaerobic digestion of macroalgae: Methane potentials, pre-treatment, inhibition and co-digestion. *Water Science and Technology* 64(8): 1723–9. https://doi .org/10.2166/wst.2011.654

Nikolaisen, L., Daugbjerg, P., Svane, B.K., Dahl, J., Busk, J., Brødsgaard, T., Rasmussen, M.B., Bruhn, A., Bjerre, A.B., Bangsø N.H., Albert, K.R., Ambus, P., Kadar, Z., Heiske, S., Sandar, B., Schmidt, E.R.. 2011. Energy production from Marine Biomass (*Ulva lactuca*) PSO project no. 2008-1-0050.

Norambuena, F., Hermon, K., Skrzypczyk, V., Emery, J.A., Sharon, Y., Beard, A., and Turchini, G.M. 2015. Algae in fish feed: Performances and fatty acid metabolism in juvenile salmon. *PLoS ONE* 10(4): e0124042. https://doi.org/10.1371/ journal.pone.0124042

Olasehinde, T.A., Olaniran, A.O., and Okoh, A.I. 2019. Phenolic composition, antioxidant activity, anticholinesterase potential and modulatory effects of aqueous extracts of some seaweeds on β-amyloid aggregation and disaggregation. *Pharmaceutical Biology* 57(1): 460–9. https://doi .org/10.1080/13880209.2019.1634741

Pallaoro, M.F., Vieira, F.D.M., and Hayashi, L. 2016. *Ulva lactuca* (Chlorophyta Ulvales) as co-feed for Pacific white shrimp. *Journal of Applied Phycology* 28: 3659–65. https://doi.org/10.1007/s10811-016 -0843-2

Patel, B., Tamburic, B., Zemichael, F.W., Dechaiwongse, P., and Hellgardt, K. 2012. Algal

biofuels: A credible prospective? *International Scholarly Research Notes: Renewable Energy* 2012: 63157. https://doi.org/10.5402/2012/631574

Peña-Rodríguez, A., Mawhinney, T.P., Ricque-Marie, D., and Cruz-Suárez, L.E. 2011. Chemical composition of cultivated seaweed *Ulva clathrata* (Roth) C. Agardh. *Food Chemistry* 129(2): 491–8. https://doi.org/10.1016/j.foodchem.2011.04.104

Pereira, R.C., da Silva Costa, E., Sudatti, D.B., and da Gama, B.A.P. 2017. Inducible defences against herbivory and fouling in seaweeds. *Journal of Sea Research* 122: 25–33. https://doi.org/10.1016/j .seares.2017.03.002

Perrot, T., Rossi, N., Ménesguen, A., and Dumas, F. 2014. Modelling green macroalgal blooms on the coasts of Brittany, France to enhance water quality management. *Journal of Marine Systems* 132: 38–53. https://doi.org/10.1016/j.jmarsys.2013 .12.010

Poole Harbour Catchment Initiative. 2016. Algal harvesting workshop report.

Potts, T., Du, J., Paul, M., May, P., Beitle, R., and Hestekin, J. 2011. The production of butanol from Jamaica Bay macro algae. *Environmental Progress and Sustainable Energy* 31: 29–36. https:// doi.org/10.1002/ep.10606

Queirós, A.M., Stephens, N., Widdicombe, S., Tait, K., McCoy, S.J., Ingles, J., Ruhl, S., Airs, R., Beesley, A., Cazenave, P., Dashfield, S., Hua, E., Jones, M., Lindeque, P., McNeill, C.L., Nunes, J., Parry, H., Pascoe, C., Widdicombe, C., Smyth, T., Atkinson, A., Krause-Jensen, D., and Somerfield, P.J. 2019. Connected macroalgal-sediment systems: Blue carbon and food webs in the deep coastal ocaean. *Ecological Monographs* 89(3): e01366. https://doi .org/10.1002/ecm.1366

Raikova, S., Le, C.D., Beacham, T.A., Jenkins, R.W., Allen, M.J., and Chuck, C.J. 2017. Towards a marine biorefinery through the hydrothermal liquefaction of macroalgae native to the UK. *Biomass and Bioenergy* 107: 244–53. https://doi.org /10.1016/j.biombioe.2017.10.010

Raven, J. 2018. Blue carbon: Past, present and future, with emphasis on macroalgae. *Biology Letters* 14(10): 20180336. https://doi.org/10.1098/rsbl .2018.0336

Reid, W.V., Ali, M.K., and Field, C.B. 2020. The future of bioenergy. *Global Change Biology* 26: 274–86. https://doi.org/10.1111/gcb.14883

Rjiba-Ktita, S., Chermiti, A., Bodas, E., France, J., and López, S. 2017. Aquatic plants and macroalgae as potential feed ingredients in ruminant diets. *Journal of Applied Phycology* 29: 449–58. https://doi .org/10.1007/s10811-016-0936-y

Roberts, T. and Upham, P. 2012. Prospects for the use of macro-algae for fuel in Ireland and the UK: An overview of marine management issues. *Marine Policy* 36: 1047–53. https://doi.org/10.1016 /j.marpol.2012.03.001

Shpigel, M., Guttman, L., Shauli, L., Odinstov, V., Ben-Ezra, D., and Harpaz, S. 2017. *Ulva lactuca* from an integrated multi-trophic aquaculture (IMTA) biofilter system as a protein supplement in gilthead seabream (*Sparus aurea*) diet. *Aquaculture* 481: 112–18. https://doi.org/10.1016/j.aquacek.2017.08.006

Smetacek, V. and Zingone, A. 2013. Green and golden seaweed tides on the rise. *Nature* 504: 84–8. https://doi.org/10.1038/nature12860

Sridhar, S. and Rengasamy, R. 2012. The effects of seaweed liquid fertiliser of *Ulva lactuca* on *Capsicum annum*. *Algological Studies* 138(1): 75–88. https://doi.org/10.1127/1864-1318/2012/0012

Subhadra, B. and Edwards, M. 2010. An integrated renewable energy park approach for algal biofuel production in United states. *Energy Policy* 38: 4897–902. https://doi.org/10.1016/j.enpol.2010.04.036

Sudatti, D.B., Duarte, H.M., Soares, A.R., Salgado, L.T., and Pereira, R.C. 2020. New ecological role of seaweed secondary metabolites as autotoxic and allelopathic. *Frontiers in Plant Science* 11: 347. https://doi.org/10.3389/fpls.2020.00347

Suganya, T. and Renganathan, S. 2012. Optimization and kinetic studies on algal oil extraction from marine macroalgae *Ulva lactuca*. *Bioresource Technology* 107: 319–26. https://doi.org/10.1016/j.biortech.2011.12.045

Tacon, G.J.A. and Metian, M. 2008. Global overview on the use of fish meal and fish oil in industrially compounded aquafeeds: Trends and future prospects. *Aquaculture* 285: 146–58. https://doi.org/10.1016/j.aquaculture.2008.08.015

The Commonwealth. 2020. Blue economy [Online]. Available from: https://thecommonwealth.org/blue-economy (accessed 9 August 2020).

Thornton, A. 2016. The impact of green macroalgal mats on benthic invertebrates and overwintering wading birds. PhD Thesis. Bournemouth University, UK.

Thornton, A., Herbert, R.J.H., Stillman, R.A., and Franklin, D.J. 2020. Macroalgal mats in a eutrophic estuarine marine protected area: Implications for benthic invertebrates and wading birds. In: Humphreys, J. and Clark, R.W.E. (eds) *Marine Protected Areas*. Elsevier, 703–27. https://doi.org/10.1016/B978-0-08-102698-4.00036-8

Trevathan-Tackett, S.M., Kelleway, J., Macreadie, P.I., Beardall, J., Ralph, P., and Bellgrove, A. 2015. Comparison of marine macrophytes for their contributions to blue carbon sequestration. *Ecology* 96(11): 3043–57. https://doi.org/10.1890/15-0149.1

Van der Wal, H., Sperber, B.L.H.M., Houweling-Tan, B., Bakker, R.R.C., Brandenburg, W., and López-Contreras, A.M. 2013. Production of acetone, butanol, and ethanol from biomass of the green seaweed *Ulva lactuca*. *Bioresource Technology* 128: 431–7. https://doi.org/10.1016/j.biortech.2012.10.094

Ventura, M.R. and Castañón, J.I.R. 1998. The nutritive value of seaweed (*Ulva lactuca*) for goats. *Small Ruminant Research* 29: 325–7. https://doi.org/10.1016/S0921-4488(97)00134-X

Watson, G.J., Murray, J.M., Schaefer, M., and Bonner, A. 2017. Bait worms: A valuable and important fishery with implications for fisheries and conservation management. *Fish and Fisheries* 18(2): 374–88. https://doi.org/10.1111/faf.12178

Watson, S.C.L., Grandfield, F.G.C., Herbert, R.J.H., and Newton, A.C. 2018. Detecting ecological thresholds and tipping points in the natural capital assets of a protected coastal ecosystem. *Estuarine, Coastal and Shelf Science* 215: 112–23. https://doi.org/10.1016/j.ecss.2018.10.006

Watson, S.C.L., Preston, J., Beaumont, N.J., and Watson, G.J. 2020. Assessing the natural capital value of water quality and climate regulation in temperate marine systems using a EUNIS biotope classification approach. *Science of the Total Environment* 744: 140688. https://doi.org/10.1016/j.scitotenv.2020.140688

PART V

Conclusion

CHAPTER 21

Conservation and Regulation in an Industrialised Estuary

JOHN HUMPHREYS and ALICE E. HALL

Abstract

The chapters that make up this book demonstrate that while Poole Harbour contains many valuable species and habitats, it also offers many social and economic benefits. However, it also suffers – in particular, from serious chemical and biological contamination. Here we examine the statutory context behind this predicament and describe a regulatory system for Poole Harbour, which is complex in both legislative and structural terms. We then summarise the case of water quality as a barrier to good environmental status. While recognising significant efforts to reduce pollution from the agricultural and water industries, we focus on biological contamination from wastewater sources. We contrast the government's approach to the regulation of water quality with its efforts to conserve salmon stocks: a species of fish known to have declined due in part to water pollution. In this case we suggest that aspects of regulatory response appear to relate to the extent to which an industry is amenable to regulation, as much as it is to environmental risk or the tenets of sustainable development. We further suggest that amenability to regulation is a function of economic power and influence, operating within a fragmented regulatory structure. In these circumstances, systemic weaknesses of regulation can lead to an emphasis on initiatives of uncertain utility while leaving relatively undisturbed the major underlying practices standing in the way of environmental protection and remediation.

Keywords: Harbour conservation; estuary conservation; sustainable development; marine pollution; marine regulation

Correspondence: jhc@jhc.co

Introduction

The chapters of this book demonstrate that while Poole Harbour contains many valuable species and habitats, it also offers many social and economic benefits. However, these chapters also clearly show that the harbour suffers from serious chemical and biological contamination. In this final chapter we examine the regulatory system within which this has occurred. To this end we examine the questions posited in Chapter 1: What does sustainable development mean in relation to an estuarine environment, To what extent is it achieved and What are the factors affecting its achievement? Our intention is to see what the case of Poole Harbour may tell us more generally about the conservation of estuaries and coastal Marine Protected Areas.

John Humphreys and Alice E. Hall, 'Conservation and Regulation in an Industrialised Estuary' in: *Harbour Ecology*. Pelagic Publishing (2022). © John Humphreys and Alice E. Hall. DOI: 10.53061/UPWW8690

In a 2011 draft National Policy Framework the British government stated that a presumption in favour of sustainable development should be seen as a 'golden thread' informing all plans and decisions. Scrutiny of the draft by a parliamentary committee pointed out that such a statement would require an agreed definition, suggesting that without one the framework could in effect become simply a presumption in favour of development (CLGC 2011). In response the committee drew attention to the principles in the 2005 Sustainable Development Strategy (Defra 2005), in which the government set out five guiding principles of sustainable development: Living Within Environmental Limits, Ensuring a Strong, Healthy and Just Society, Achieving a Sustainable Economy, Promoting Good Governance and Using Sound Science Responsibly.

Phrases such as 'Living Within Environmental Limits' and the general idea of sustainable development are reflected in other government instruments. For example, in the UK Marine Strategy, 'Good Environmental Status' (GES) is defined in terms of 'ecologically diverse and dynamic ocean and seas which are clean, healthy and productive, within their intrinsic conditions, and the use of the marine environment is at a level that is sustainable, thus safeguarding the potential for uses and activities by current and future generations' (Defra 2019).

The concepts of Sustainable Development and Good Environmental Status allow for a level of exploitation for human benefit and therefore a degree of anthropogenically induced change but only within 'environmental limits' or 'intrinsic conditions'. In a sense they imply a sort of 'carrying capacity' for humans somewhat analogous to that attributed to an estuary's ability to support protected bird populations.

Despite its general applicability in government policy, in practice sustainable development is most often applied as a planning principle requiring proposed development to be scrutinised in terms of 'sustainability'. For historical and existing environmental problems, retrospective remediation tends to be governed by frameworks in which avoiding further deterioration and restoration to good environmental status are objectives.

Environmental regulation in Poole Harbour and its catchment

It is the responsibility of government, through statutory regulators and associated organisations, to achieve its various environmental policy commitments. Front-line regulatory organisations typically operate through statutory powers that enable them to enforce primary legislation, establish secondary legislation (byelaws), licence activities and, if necessary, prosecute offenders. In the river catchment of Poole Harbour, the Water Environment Regulations (WER), transposed from the EU Water Framework Directive, establish a legal framework for protecting inland waters with the goal of achieving GES. The WER is an example of an 'ecosystem approach' to conservation, which, as such, requires interested groups to cooperate in an integrated response. Chapter 17 provides a good example of this approach in practice through the Poole Harbour Catchment Initiative.

In fact, the scope of the WER includes coastal waters and therefore estuaries such as Poole Harbour, but here other frameworks such as the UK Marine Strategy, which extends to the mean high-water level of spring tides, also apply. Although similar to the WER in respect to an ecosystem approach and the ostensibly integrated goal of GES, this overlap exemplifies a more complex position for estuaries which is further complicated where an estuary is a designated Marine Protected Area as in the case of Poole Harbour, where each designation (SSSI, SAC, etc.) protects specified features, such as habitats and species selected for particular conservation attention.

Corresponding to these various conservation frameworks are a complex of statutory organisations tasked collectively with the management and regulation of estuaries but each with a different role and different priorities. Table 21.1 shows the organisations whose regulatory

Table 21.1 The management and regulation of Poole Harbour

Organisation	Role	Relevant powers	Parent organisation
Central			
Department of Agriculture, Environment and Rural Affairs (Defra)	Government ministry Broad remit for safeguarding the environment, supporting fisheries, etc.	National legislation, policy, strategic consents, etc.	
Department of Business, Energy and Industrial Strategy	Government ministry Broad remit for business and climate change	Licences for oil and gas exploitation in the English inshore and offshore areas	
Crown Estate Commissioners	Effectively owns much of the foreshore and seabed out to 12 nm	Leases the seabed for ports, harbours, infrastructure, marinas pipelines, outfalls and aquaculture, etc.	
Marine Management Organisation (MMO)	Regulation and planning of marine activities in England for sustainable development	Licences of fishing vessels; coastal and marine developments; wind, wave and tidal power; removal and disposal of dredged material	Defra
Environment Agency (EA)	Protecting the environment through ensuring sustainable development ('growth duty'), including the seas out to 3 miles	Regulatory and licensing authority issuing various consents and licenses. Law enforcement and sanctions including prosecutions	Defra
The Water Services Regulation Authority (Ofwat)	Regulates aspects of the operation of privatised water companies including capital investment in sewerage treatment facilities	Sets limits for charges to consumers for water and sewerage services, taking into account proposed capital investment schemes	Defra
Natural England (NE)	Statutory advisor for the natural environment in England	Adviser to government and non-departmental public bodies on marine biodiversity, national nature reserves, etc.	Defra
Joint Nature Conservation Committee (JNCC)	Advisor for the natural environment to UK government and devolved administrations	UK wide advice to central and devolved governments	Defra
Centre for Environment, Fisheries and Aquaculture Science (CEFAS)	Applied scientific research on the marine environment and fisheries	Inspection, monitoring and advice (major fisheries laboratory in Weymouth)	Defra
Regional			
Southern Inshore Fisheries and Conservation Authority (Southern IFCA)	Monitoring (stock assessments. etc.), regulation and management of inshore fisheries including MPAs	Legislation, enforcement, sanctions (including prosecutions)	Coastal local governments
Local			
Port authority: Poole Harbour Commissioners	Managing the business of the Port of Poole and Regulation of Harbour use by vessels	Various, including byelaw legislation and enforcement for navigation, safe and clean harbours	Trust port
Bournemouth, Christchurch and Poole Council (BCP)	Administrative control and jurisdiction normally to the 'low water mark'. Represented on, and part funds, Southern IFCA. Involved in Harbour Management Plans	Public health (water quality sampling), etc. Commercial vessel licences within the harbour	

decisions are of conservation significance for Poole Harbour. With respect to industries reliant on ecological services (as we have defined them in Chapter 1), the regulators include the 'Defra family' Environment Agency (EA), for example, on water quality and freshwater fisheries, and the joint local government Southern Inshore Fisheries and Conservation Authority (Southern IFCA) on marine fisheries. The case of species that migrate between freshwater and marine habitats, such as Atlantic Salmon *Salmo salar* and Sea Trout *Salmo trutta* (salmonids), is instructive as both organisations have an interest: the EA with a specific duty to protect salmonids and the Southern IFCA through its duty to protect salmonids as part of the marine environment (Defra 2014). The matter of salmonid regulation will be considered further in order to elucidate some of the characteristics of the regulatory system as a whole.

Water quality in Poole Harbour

Of the 19 rivers and other surface waterbodies feeding Poole Harbour, all 19 are categorised as 'fail' in terms of chemical status. From 12 possible named industrial sectors two are identified as the causes of this failure: agriculture/rural land management and the water industry (EA 2020).

Chapters 15–20 examine the particular problem of an excess of nitrates, originating mostly but not exclusively from agricultural fertilisers in the catchment. A second major problem is microbiological contamination including coliform bacteria (from sewage effluent and farm livestock/slurry).

In effect, Poole Harbour functions as a recipient for polluting waste from the agriculture and water industries. As such, in so far as the estuary can naturally treat these problems it provides 'ecological services', but in practice the harbour's 'environmental limits' are overwhelmed by the volume of such effluents, with negative consequences for both the ecology and the economy of the harbour.

Excess nitrate, particularly in combination with other elements, results in over-fertile (eutrophic) water with consequences including opportunistic macroalgal growth, which in summer forms extensive green 'algal mats' over sediment habitats (Chapters 18 and 19). These algal mats impede bird feeding on the harbour mud flats while also choking commercial fishing gear (Chapter 11). Chapter 20 describes a possible approach to remediating this problem by algal 'harvesting'. The harbour's eutrophic waters also suffer from phytoplankton blooms which in addition to increasing turbidity can also lead to oxygen depletion which is thought to be associated with oyster mortality events (Chapter 5).

Microbiological contamination, aside from causing general hygiene issues, also impacts the shellfisheries for which the harbour has long been famous (Chapter 14). Government 'Classified Shellfish Beds' are microbiologically monitored by BCP Council and the Food Standards Agency. As a precaution, in common with many other estuaries in England, certain shellfish are purified in sterilised seawater systems before sale for consumption. Sometimes acute episodes of *Norovirus* contamination from human effluent stop shellfisheries altogether until the problem has dissipated.

So, to our second question: Is the policy commitment of sustainable development being achieved in Poole Harbour? For reasons of chemical and biological contamination, the simple answer is no. Neither is the harbour achieving the GES goals of the water and marine strategy frameworks. More illuminating, however, is what Poole Harbour may tell us about the factors which determine this negative outcome. To elucidate this, we will focus further on water quality and in particular how government regulation responds to the challenges it presents.

On the positive side of the equation, Chapters 16 and 17 describe significant efforts to alleviate the extent of pollution from catchment farms and sewage treatment works.

Furthermore, the utility of both industries should be recognised. They both provide essential services which no community would expect to do without. Both industries also provide economic benefits such as livelihoods. The problem is that they also both demonstrate historical 'unsustainable development', so to speak, which continues in the present day. In delivering the economic side of the equation they are arguably overwhelming the estuary's 'environmental limits' and therefore failing on the environmental side.

However, simply blaming these industries would represent a superficial analysis. In a market economy, a purpose of regulation is to ensure that such for-profit industries operate within environmental legal frameworks. Also, we accept that neither landowners nor the agricultural industry is monolithic in terms of either practices or attitudes to environmental conservation. Nevertheless, to the extent that the exploitation of Poole Harbour as a recipient of effluents currently takes it beyond its environmental limits, it represents a failure of regulation. Below we contrast aspects of water quality and salmonid regulation in order to reach conclusions on some of the characteristics of environmental regulation which contribute to this situation.

The case of sewage overflows

Antiquated foul water infrastructure combined with inadequate treatment facilities result in rainfall events in which the capacity of sewage treatment works to process and/or hold effluent is overwhelmed. As a result, untreated 'raw' sewage is released directly into estuaries and their catchments. Investigative reporting by the *Guardian* newspaper has revealed that in 2019, water companies discharged raw sewage from 'storm overflows' for a total of 1.5 million hours over 204,000 release episodes. In 2020 the reported number of raw sewage discharges was over 420,000.

In Poole Harbour for example a total of 713 such overflows from seven outfalls occurred between 19 December 2019 and 17 February 2020. Although some were of short duration many were not. On some days a single outflow was releasing into the harbour for over 22 hours. These overflows are legal in so far as they remain within the limits expressed in permits from the Environment Agency. Although the EA do monitor water quality directly (Chapter 15), raw sewage overflow episodes have a 'lighter touch' regulatory regime involving 'operator self-monitoring'.

Accepting that 'water infrastructure has not kept pace with population growth' Defra (2021), under growing pressure from civil society organisations, announced 'measures to reduce harm from storm overflows', in the form of duties on government to publish a plan and on water companies to publish data. It has not escaped our notice that these are not in themselves measures that reduce harm from storm overflows.

The case of salmonids

Another ecological and economic impact of water pollution relates to Salmon and Sea Trout (salmonid) populations, which migrate through the estuary from and to their upstream spawning grounds. The Atlantic Salmon *Salmo salar* has been in decline since the industrialisation of the eighteenth century with severe impacts caused by human activities: notably water pollution (WWF 2001). Salmon thrive in well oxygenated water with low levels of organic and inorganic pollution. In the UK 86% of English rivers are not of 'good ecological status' (EA 2020).

Noting that salmon stocks in many rivers across England had failed to meet their minimum safe levels, the EA proposed its 'Five point approach high level commitments 2016-2021'. These included 24 actions across five categories: 'Improve marine survival; Further reduce exploitation by nets and rods; Remove barriers to migration and enhance habitat; Safeguard sufficient flows; Maximise spawning success by improving water quality.' Commitment 1 is to 'Improve marine survival' with a subclause *1.2. Optimize survival of adult salmon and smolts in estuarine and coastal environments.* These five points are effectively a list of potential causal factors in salmonid decline but they are not analysed or ranked in terms of relative detriment or material risk.

In seeking to fulfil Commitment 1.2, EA has turned its attention to Poole Harbour, in which netting for marine fish is long established. These fisheries do not target salmonids. For the EA the issue is by-catch. Their concern is that nets set to target marine species may intercept salmonids migrating through estuaries up to their freshwater spawning grounds. The EA accepts that quantification of this impact is difficult but nevertheless asserts that there is 'scope for the illegal catch to take a significant proportion of the salmon population' (EA 2017). On the basis that netting in Poole Harbour has the 'potential to seriously compromise and reduce the potential of a migratory salmonid fishery', the EA, with minor exceptions, "are of the opinion that closing all sea fish netting . . . represents the only logical and pragmatic means to manage this issue in the longer term' (EA 2017).

Some insight into the sentiment underlying this assertion is provided by an EA opinion relating to nearby Southampton Water in which, after observing returning adult salmon to have been below that estimated to be sufficient to sustain the population, the EA have indicated that the loss of a single adult female to be sufficient to conclude (using the technical language of conservation legislation) 'that no adverse effect on site integrity could not be demonstrated'. This 'one female salmon' approach has been raised in relation to netting in Poole Harbour.

So, the commitment to 'optimize' the survival of adult salmon in estuaries implies the cessation of all legal marine fish netting, in order that even a single salmon should not be accidentally caught. In contrast, the improvement of water quality requires actions of only indirect relevance, including '5.5 Assess priorities for action in addressing water pollution impacts on salmon', another ostensibly good intention but many years overdue and a long way short of action.

Discussion: power, politics and pollution

In contrasting regulatory responses to water pollution and salmonids, we are not asserting that estuarine water quality is currently the highest risk causal factor in the downward trend of the North Atlantic stock. There are reasons to believe that 'marine survival' may be significant for the current sustained decline, due to oceanic water temperature trends with consequences for food distributions (e.g. Nicola *et al.* 2018). However, water quality is a major general environmental problem with well understood detriment across many habitats and species, including salmonids. In contrast, already regulated marine netting fisheries are not causing ecosystem wide impacts and the extent to which marine fish netting by-catch would have any material effect on salmonid stocks is a contentious point, in terms of both the quantification of by-catch and the materiality of by-catch on a salmon population already compromised by many other challenges, including chronic water pollution.

In essence, it is doubtful whether cessation of netting for other species will have any measurable beneficial effect in terms of salmon stock recovery. The response to this situation is to cite the 'precautionary principle', which, in the context of EA's expressed desire for a total ban on nearshore netting, will collaterally damage existing artisanal livelihoods in a coastal community of the sort that the government is committed to 'levelling up'.

In this context the precautionary principle justifies attention to netting, effectively to compensate for either an inability (marine survival) or apparent disinclination (water quality) to act on more evidenced causes of decline. It appears that regulatory attention is determined as much by amenability to regulation as it is to environmental risk or the tenets of sustainable development. In the case of Poole Harbour, we suggest that water quality is not so amenable to regulation but artisanal netting for other species is. We further suggest that amenability to regulation is a function of economic power and influence, operating within a fragmented regulatory structure.

Cui bono, who benefits? In terms of avoiding the substantial costs of reducing water pollution, the polluters do: agricultural landowners and water companies. Whereas agricultural landowners and water companies are relatively organised, economically powerful and politically well represented, in contrast, inshore fishers are politically weak and economically significant only in the local context of coastal communities. There appears to be an inverse correspondence: the strong are weakly regulated while the weak are strongly regulated.

However, it would be superficial to attribute the persistent water quality problems in Poole Harbour (and many other estuaries) to a simple failure of regulation by the Environment Agency. Above the EA is a government manifestly reluctant to push too hard against economic interests to which they may for various reasons be politically committed and below it are commercial water companies and agricultural landowners for whom profit is a legitimate goal.

Moreover, solutions to raw sewage releases will require large injections of capital funding for infrastructure projects and the regulatory structure is such that privatised water companies wishing to protect their profits will effectively expect compensation in the form of increases in charges to consumers, something another government regulator, the Water Services Regulation Authority (Ofwat, which sets limits on charges to consumers) is reluctant to facilitate. In this context the EA's response to chronic pollution is weak.

It would be naive to suppose that power, politics and cost would not be factors in the determination of environmental outcomes. But the case of salmonids sheds light on the current state of statutory conservation. The citing of a single fish as the basis for control or cessation of an activity is revealing, not in itself but in so far as it is not applied to the larger risk of water pollution. Neither does the 'one female salmon' argument appear to be applied to the recreational salmon fisheries, for which the EA is responsible and which, even in the context of a catch-and-return policy, must result in a less than 'optimal' level of fish mortality.

Despite the stated 'logic' of the Environment Agency, it is unlikely that inshore netting will cease on the south coast. Others, less close to central government, seem more attentive to the balance between cause and effect, precaution and proportionality.

In any event the government's admission that infrastructure has lagged behind population growth is a tacit acceptance that the processes of environmental conservation and planning have become disconnected from the principle of sustainable development. This certainly does not represent a planned scheme to benefit larger firms while closing relatively sustainable artisan inshore fisheries. Rather, it suggests a systemic problem of environmental regulation, in which intelligent, committed professionals seek to apply often good science and always good intentions in a fundamentally flawed statutory system, characterised by structural fragmentation, piecemeal priorities and political compromise.

Despite its many natural assets, Poole Harbour elucidates a system of environmental conservation without effective strategy, in which amenability to regulation appears to have greater influence than environmental risk and where attention to lesser risks leaves relatively undisturbed the major underlying economic practices standing in the way of environmental protection and remediation.

Disclaimer. The views expressed in this chapter do not represent those of other chapter authors, our peer reviewers, any of the organisations with which we are or have been affiliated or the Poole Harbour Study Group.

References

CLGC. 2011. The National Planning Policy Framework – Communities and Local Government Committee. 4. The definition of sustainable development. Available from: https://publications.parliament.uk/pa/cm201012/cmselect/cmcomloc/1526/152607.htm

Defra. 2005. UK sustainable development strategy, securing the future.. http://www.defra.gov.uk/sustainable/government/publications/ukstrategy/ (accessed 13 October 2009).

Defra. 2014. Protection of freshwater and migratory species. Letter from DEFRA to IFCA Chief Executive Officers. 13 May 2014.

Defra. 2019. Marine strategy part one: UK updated assessment and good environmental status. Defra October 2019.

Defra. 2021. Measures to reduce harm from storm overflows to be made law. 29 March. Available from: www.gov.uk/government/news (accessed 13 October 2021).

EA. 2017. Environment agency evidence for nearshore netting. Letter from EA to Southern IFCA 10 February.

EA. 2020. Environment agency catchment data explorer: South West; Dorset; Poole Harbour rivers. Updated 17 September 2020.

Nicola, G.G., Elvira, B., Jonsson, B., Ayllon, D., and Almodovar, A. 2018. Local and global climatic drivers of Atlantic salmon decline in Southern Europe. *Fisheries Research* 198: 78–85. https://doi.org/10.1016/j.fishres.2017.10.012

WWF. 2001. The status of wild Atlantic Salmon: A river by river assessment. WWF May 2001.

Index

Printed in the USA
CPSIA information can be obtained
at www.ICGtesting.com
LVHW082147201023
761596LV00009B/1383

9 781784 274030